A TEXTBOOK OF OSCILLATIONS, WAVES AND ACOUSTICS

For B.Sc. Students

Dr. M. GHOSH
D.Sc.
Griffith Memorial Prize Man, Calcutta University; Associate Fellow of Institute of Physics, London; Principal and Head of the Physics Department (Retd.), City College, Calcutta; Formerly, Head of the Physics Department, Burdwan Raj College; Lecturer (BES), Presidency College, Calcutta, Lecturer; Post-Graduate Physics Department, Calcutta University, Formerly, Professor (part time), Ramakrishna Mission Residential College, Narendrapur

Revised by

Dr. D. BHATTACHARYA
Ph.D.
Head of the Department
Reader of Physics (Retd.), Barrackpore
Rastraguru Surendranath College,
Formerly Lecturer, Ramsaday College
Amta, Howrah

S. CHAND
PUBLISHING

S Chand And Company Limited
(ISO 9001 Certified Company)

S Chand And Company Limited

(ISO 9001 Certified Company)

Head Office: Block B-1, House No. D-1, Ground Floor, Mohan Co-operative Industrial Estate, New Delhi – 110 044 | Phone: 011-66672000

Registered Office: A-27, 2nd Floor, Mohan Co-operative Industrial Estate, New Delhi – 110 044

Phone: 011-49731800

www.**schandpublishing.com**; e-mail: **info@schandpublishing.com**

Branches

Ahmedabad	:	Ph: 27542369, 27541965; ahmedabad@schandpublishing.com
Bengaluru	:	Ph: 22354008, 22268048; bangalore@schandpublishing.com
Bhopal	:	Ph: 4274723, 4209587; bhopal@schandpublishing.com
Bhubaneshwar	:	Ph: 2951580; bhubaneshwar@schandpublishing.com
Chennai	:	Ph: 23632120; chennai@schandpublishing.com
Guwahati	:	Ph: 2738811, 2735640; guwahati@schandpublishing.com
Hyderabad	:	Ph: 40186018; hyderabad@schandpublishing.com
Jaipur	:	Ph: 2291317, 2291318; jaipur@schandpublishing.com
Jalandhar	:	Ph: 4645630; jalandhar@schandpublishing.com
Kochi	:	Ph: 2576207, 2576208; cochin@schandpublishing.com
Kolkata	:	Ph: 23357458, 23353914; kolkata@schandpublishing.com
Lucknow	:	Ph: 4003633; lucknow@schandpublishing.com
Mumbai	:	Ph: 25000297; mumbai@schandpublishing.com
Nagpur	:	Ph: 2250230; nagpur@schandpublishing.com
Patna	:	Ph: 2260011; patna@schandpublishing.com
Ranchi	:	Ph: 2361178; ranchi@schandpublishing.com
Sahibabad	:	Ph: 2771238; info@schandpublishing.com

First Edition 1976
Subsequent Editions and Reprints 1980, 1987, 2006, 2007, 2010
Fourth Revised & Enlarged Edition 2012; Reprint 2014
Fifth Revised Edition 2016; Reprint 2018
Reprint 2020

ISBN: 978-93-856-7615-4 **Product Code:** H6WOS68PHYS10ENAE16O

PRINTED IN INDIA

By Vikas Publishing House Private Limited, Plot 20/4, Site-IV, Industrial Area Sahibabad, Ghaziabad – 201 010 and Published by S Chand And Company Limited, A-27, 2nd Floor, Mohan Co-operative Industrial Estate, New Delhi – 110 044

PREFACE TO THE FIFTH REVISED EDITION

This book is an introduction to acoustics. It is mainly written for undergraduate students of science and engineering. UGC curriculum is based on the Acoustics requires a good knowledge of mathematics. This book is based on the U.G.C. curriculum. This edition contends an additional chapter 26. Some new articles are also included. A few university problems are solved and some questions are also included in this edition.

I am thankful to the management and the editorial team of S. Chand & Company Pvt. Ltd., New Delhi for help, support and guidance in publication of this book.

Dr. D. Bhattacharya

PREFACE TO THE FIRST EDITION

The author while teaching the Honours course in sound of the students preparing for degree examination felt the difficulties of these students for want of proper text books. They are to consult a large number of voluminous books for their short course, or they are required to keep themselves satisfied with books dealing with selected topics on the subject.

The book is written in order to satisfy the needs of the Honours students of different Indain Universities. The subject matter is developed gradually so that the students will not find any difficulty in going through different chapters of the book and they will not be required to consult the elementary text books of pass course to supplement their course of study.

The book is developed from the lecture lessons delivered to honours students and in guiding research work of students preparing for Doctorate degrees of different universities. In that connection, the author consulted freely the available standard books on sound and some journals relating to particular problems, such as Helmholtz and Sensation of tone translated Ellies, Rayleigh – Sound Vol. I, II, K.C. Kar— Theoretical Physics Vol. II, and also of E.H. Barton, H. Lamb, A. B. Wood, Alex Wood, E.G. Richardson, R.B. Lindsay, Olson and Massa, C.V. Raman Bull, 15 (I.A.C.S.) and Admiralty Hand Book of Wireless, and others. I gratefully acknowledge my indebtedness to all these authors.

The author finally thanks Prof. J.C. Sarkar (Mukerjee) of Calcutta University for going through the part of the manuscript and Dr. S.K. Ghosh of Jadavpur University and Dr. B.B. Banerjee of Sambalpur University (Burla) for suggesting me improvements and Dr. P.K. Sen of City College Calcutta for encouraging me in writing the book.

Any suggestions for improvement of any article will be cordially received.

The students are advised to read the mathematics given in appendix before they actually take up the subject matter of the book.

The author's labour in writing this book will be duly rewarded if the students find the book useful.

5 A Gopinath Saha Street
Serampore (E.R.)
West Bengal
1976

CONTENTS

1

CHAPTER

SOUND AND ITS CHARACTER

INTRODUCTION

1.1. SOUND AND ITS CHARACTER

Sound is a form of energy which stimulates that acts on the sense of hearing. It is the sensation that is received by our ear.

The science of sound generally known as Acoustics, is that branch of physics which deals with production and propagation of sound.

Sonorous body is found always in the state of vibration. It may be very rapid with small to and fro displacement. This vibratory motion of a material body owing to its rapidity in motion, may not be seen by our naked eyes, but can be very easily felt if it is touched. Any vibration of a body may not be sufficient to give sensation of sound. The ear can recognise the vibration of a body, provided the number of vibration per second lies between 20 Hz to 20000 Hz called the limit of audibility. The source giving out audible sound is called sonic source. The limit of course varies for different persons.

When a tuning fork is struck sound is produced. If we now touch lightly the fork the vibration can be felt. The experiment shows that the bodies emitting sound are in a state of vibration.

In order to affect the sense of hearing it is not enough to have vibrating body simply, but there must be an uninterrupted communication between the body and the ear. This communication is essential between the body and ear which is ordinarily, by surrounding air. Sound is not propagated through vacuum. An electric bell is suspended by connecting wire in a rubber stoppered bell jar which can be evacuated by an air pump. Sound given out by electric bell is distinctly heard from a distance, when the inside of the bell jar is in communication with that of the outside. As the air inside the bell jar is gradually pumped out, the intensity of sound gradually falls off. When the air inside is completely exhausted no sound will be perceived. As air is gradually introduced, the intensity of sound is gradually restored. Thus a medium of material nature is necessary for the propagation of sound.

But the medium is not bodily carried from sonorous body to our ear. This is observed from the following : A long wide tube open at one end and other end made conical with a nozzle. The tube is placed horizontally on a table. A steady candle flame is placed just in front of the nozzle. The space inside the other end of the tube is filled with dense fume (a watch glass containing strong hydrochloric acid and another containing ammonium hydroxide are placed to produce dense fume). On producing a report by clapping two blocks of wood near the open end, the flame is found to be struck after a very short time, though no fume comes out through the nozzle. This leads us to infer that, when sound moves from one point to another in a material medium, the medium is not bodily carried but it remains at rest.

The passage of sound from a source to our ear is not instantaneous. The report of a gun fired at a distance is heard after the flash is seen, though they occur simultaneously. Light travels with a velocity of 3×10^8 m/sec. So the time interval between the seeing of the flash and hearing of sound report is practically the time taken by the sound to reach from the source to the ear. Thus sound travels with a finite velocity (velocity of sound at 0°C is 332 m/sec).

The medium through which the sound is transmitted must be continuous and elastic. It may be solid, liquid or gas. Feeble sound of watch placed in contact at one end of a table is distinctly heard by placing ear in contact with other end of the table, though the ticks of the watch are inaudible through air from the same distance. The sound of an approaching train can be heard, applying ear on the rail, long before the sound heard in air. This gives evidence of propagation of sound through solid.

A diver inside water distinctly hears, sound produced in water or in air. This gives an evidence of propagation of sound through water or liquid. In bell jar experiment it is found that sound is transmitted through air which is a gas.

The propagation of sound through a medium takes place by transfer of disturbance from particle to particle by their to and fro motion. There are numbers of particles linked together elastically by imaginary elastic chain, any one of them cannot be displaced without disturbing others. The motion imparted to one is communicated to the adjacent particles surrounding it, which will thus execute similar motion and in doing so will in their turn transfer their energy to the surrounding particles and so on. The motion imparted to a single particle will move uniformly in all direction from particle to particle with a definite velocity. The propagation of disturbance of this kind due to transfer of energy from particle to particle is called wave motion.

REVIEW QUESTION

1. What are the characteristics of sound? [Barkatullah University 2011]

CHAPTER 2

Simple Harmonic Motion

2.1. INTRODUCTION

A particle is said to execute simple harmonic motion when the motion is a periodic motion and acceleration of the moving particle at any time is proportional to the displacement and is directed to mean position of rest. The characteristics of such a motion are :

(*a*) The motion is oscillatory and traces and retraces the same path over and over again.

(*b*) The particle moves in a straight line.

(*c*) The restoring force is always directed towards the mean position of rest.

(*d*) The restoring force is proportional to the displacement of the particle from the mean position of rest.

The particle which satisfies the condition (*a*) only is said to execute periodic motion and which satisfies conditions (*a*) and (*b*) is said to possess vibratory motion. The particles executing Simple Harmonic Motion (S.H.M.) must satisfy all the four conditions.

The motion of the earth round the sun is periodic motion.

The motion of a pendulum of a clock, the motion of the prong of a tuning fork etc., are the examples of vibratory periodic motion.

A pendulum whose length is made large and allowed to oscillate, such that the arc of oscillation is made so small that it approximates to straight line is said to be a simple pendulum and it executes S.H.M.

2.2. MOTION OF A SIMPLE PENDULUM

A heavy bob of mass m is suspended by a long and a very fine thread from an unyielding support C and when it oscillates the bob comes to A, traces an arc OA of a circle, C being the centre and l the length of the pendulum from the point of suspension C to the centre of gravity of the bob, being the radius of the circle. At A, mg the weight of the bob acts downwards. This is resolved into two components, one along the string, $mg\cos\theta$, and the other is $mg\sin\theta$ along the tangent to the arc OA of the circle, θ is the angle through which it is displaced and is of very small value, $mg\cos\theta = T$, the tension of the string. The force $mg\sin\theta$ acts towards O which tries to restore the bob back to its mean position.

$$\text{Restoring force} = -mg\sin\theta$$

$$= -mg\theta$$

$$\sin\theta = \theta \text{ as } \theta \text{ is a very small value}$$

$$\text{Acceleration } f = \frac{\text{Restoring force}}{\text{Mass}}$$

$$= -g\theta = -g\frac{\text{Arc } OA}{l}$$

$$= -\frac{g}{l} \times \text{displacement}$$

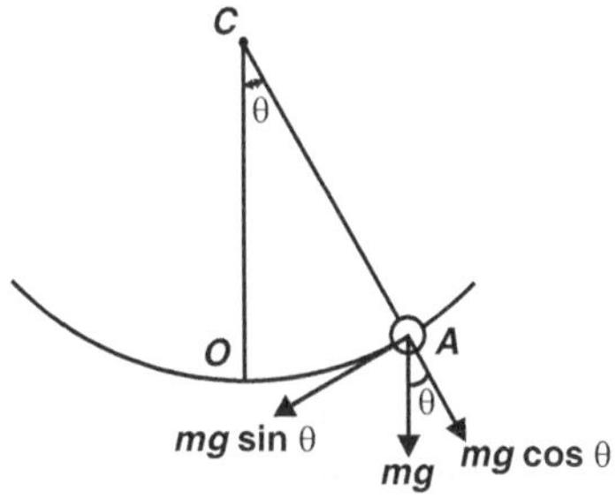

Fig. 1.

OA approximated to a straight line and is equal to y

$$f = -\omega^2 y$$

where $\omega^2 = \frac{g}{l}$. Time period $T = \frac{2\pi}{w} = 2\pi\sqrt{\frac{l}{g}}$

$\therefore$ Acceleration is proportional to the displacement as $\frac{g}{l}$ is a constant quantity at a given place for a pendulum and acceleration is always directed towards O, the mean position of rest.

2.3. PROJECTION OF UNIFORM CIRCULAR MOTION OF A PARTICLE GIVES A SIMPLE HARMONIC MOTION

Let us consider any particle P move with uniform angular velocity $\omega = \frac{v}{a}$ (v velocity of the particle) along the circumference of the circle $XYX'Y'$ of radius a. If θ be the angle swept out in time t, T the time for one revolution then $\omega = \frac{\theta}{t} = \frac{2\pi}{T}$.

Let YOY' be any diameter and N be the foot of the perpendicular drawn from any position of the moving particle P to the diameter YOY'. As P moves with uniform speed round the circle, N the foot of the perpendicular on the diameter YOY' moves along OY to O again Y' to O and so on. As the particle P goes round the circle the foot N traces and retraces the path YOY' over and over again. Any other point on the circle, starting from a different point on the circle, will have its foot of the perpendicular execute S.H.M.

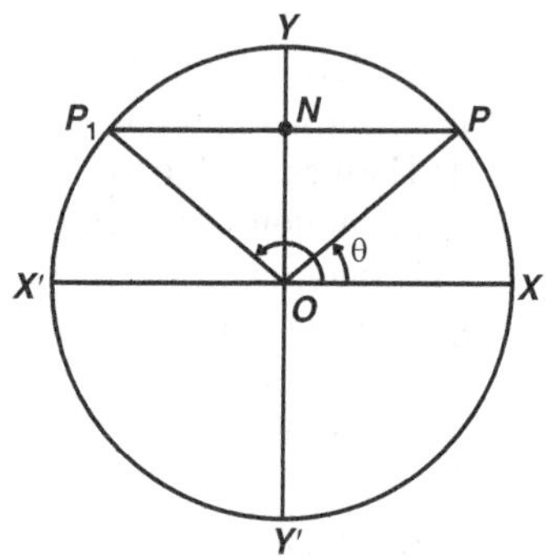

Fig. 2.

The motion of N satisfies first two conditions as it traces and retraces the path YOY' and direction is one time $Y \rightarrow Y'$ and other time $Y' \rightarrow Y$ at regular interval of time.

Let the particle start motion from X at any instant it occupies the position P. The foot of the perpendicular starts from O and comes to N in the same time t, OP sweeps out angle θ.

The displacement ON is given by

$$ON = y = a \sin\theta$$

or

$$y = a \sin \omega t \quad \text{...(1)}$$

as $\omega = \frac{\theta}{t} = \frac{2\pi}{T} = \frac{v}{a}$, T being the time that the particle takes to go once round the circle which is also equal to time taken by N to make a complete oscillation $O \rightarrow Y \rightarrow Y' \rightarrow O$.

So the velocity of the particle $\frac{dy}{dt} = a\,\omega \cos \omega t$.

and acceleration of the particle $\frac{d^2y}{dt^2} = -\omega^2 a \sin \omega t$

or

$$\frac{d^2y}{dt^2} = -\omega^2 y \quad \text{...(2)}$$

Thus it is found that the conditions (c) and (d) are satisfied.

Some Definitions:

Time period (T) of oscillation is the time taken by the particle in executing S.H.M to make a complete oscillation. Time taken by N to move from O to Y to Y' to O again.

Time period $T = \frac{2\pi}{\omega}$ where ω is the angular velocity.

Frequency (n) of a particle executing S.H.M. is the number of oscillations per second. Hence

$$n = \frac{I}{T}$$

Also $$\omega = \frac{2\pi}{T} = 2m$$

Amplitude of a particle executing S.H.M. is the maximum displacement on any side about the mean position of rest. Amplitude is $OY = OY' = a$, the radius of the circle known as circle of reference.

Phase of a vibrating particle executing S.H.M. at any instant denotes the state of motion that represents the position and direction of the particle at that instant. It is measured by angle equivalent to θ swept out by the particle, known as generating particle moving in the circle of reference. Any quantity which can give both the position and velocity of a particle in S.H.M. may serve as a measure of the phase.

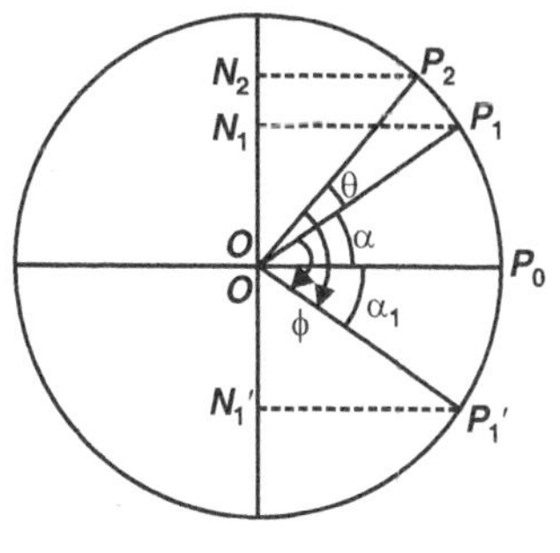

Fig. 3.

In equation (1) for S.H.M. the angle ωt is capable of so doing at the instant t. It is called the phase angle. Epoch is the initial phase of the particle executing S.H.M. at the commencement of the motion.

Sometimes the position of the particle executing S.H.M. is not its motion at the moment from which time is recorded. Let P_1 be the position of the particle at $t = 0$ on the circle of the reference whose centre represents the mean position of rest for N executing S.H.M. This gives N is at N_1 at $t = 0$. Let the angle already swept out is $\angle P_0OP_1 = \alpha$ say. The particle P comes from P_1 to P_2 in time t *i.e.*, N comes from N_1 to N_2 in time t, the angle θ between $\angle P_1OP_2 = \omega t$, so the angle $P_0OP_2 = \theta + \alpha\ \omega t + \alpha$. So the equation (1) becomes $y = a\sin(\omega t + \alpha)$. If however the particle starts motion from P_1', and comes to P_1 in time t, that is N starts from N_1' and comes to N_1 in time t the angle $\angle P_1'OP_1 = \phi = \omega t$ and $\angle P_1'OP_0 = \alpha_1$ gives $\angle P_0OP_1 = \phi - \alpha_1 = \omega t - \alpha_1$. The equation (1) can be written as $y = a\sin(\omega t - \alpha_1)$. Here α and $-\alpha_1$ are the initial phases *i.e.*, epoch of the particle executing S.H.M.

2.4. DIFFERENTIAL EQUATION OF MOTION EXECUTING S.H.M.

(a) Let y be the displacement from the mean position of rest at any instant t. The force is $\frac{d^2y}{dt^2}$. In the case of a particle executing S.H.M. force is proportional to the displacement and is always oppositely directed to it.

That is $$m\frac{d^2y}{dt^2}\alpha - y$$

or $$m\frac{d^2y}{dt^2} = -\mu y$$

where μ is the proportionality constant is referred to as the spring constant or the stiffness factor.

or $$\frac{d^2y}{dt^2} = \frac{\mu}{m}y$$

or $$\frac{d^2y}{dt^2} + \omega^2 y = 0 \qquad ...(3)$$

where $$\omega^2 = \frac{\mu}{m}$$

(*b*) *From energy consideration*

In any conservative system, the sum of the kinetic and potential energies is constant. The particle is displaced through a distance y in time t during S.H.M.

$$\text{Kinetic energy} = \frac{1}{2}m\left(\frac{dy}{dt}\right)^2$$

Potential energy is the total work the particle can do in changing the position. It is the force multiplied by the distance along the force. If the distance traversed is dy

Potential energy = $\mu y dy$

when it is displaced through y

$$\text{Potential energy} = \int_0^y \mu y dy = \frac{\mu y^2}{2}$$

$$\text{Total energy} = \frac{1}{2}m\left(\frac{dy}{dt}\right)^2 + \frac{\mu y^2}{2} = \text{Constant}$$

Differentiating both sides with respect to t

$$\frac{1}{2}m \cdot 2\frac{dy}{dt}\frac{d^2y}{dt^2} + \frac{\mu}{2}2y\frac{dy}{dt} = 0$$

or $$\frac{d^2y}{dt^2} + \omega^2 y = 0 \qquad ... (3)$$

where $$\omega^2 = \frac{\mu}{m}$$

This is the required equation of motion.

2.5. SOLUTION OF THE DIFFERENTIAL EQUATION OF MOTION

The equation of motion is

$$\frac{d^2y}{dt^2} + \omega^2 y = 0 \qquad ...(3)$$

Let the solution of equation (3) be assumed to be

$$y = ae^{mt}$$

$$\therefore \quad \frac{d^2y}{dt^2} = m^2ae^{mt} = m^2y$$

So the equation (3) becomes

$$(m^2+\omega^2)y = 0$$

Since $y \neq 0$ $$m^2 + \omega^2 = 0$$

or $m = \pm i\omega$

So $y = a_1 e^{i\omega t} + b_1 e^{-i\omega t}$

or $y = (a_1 + b_1)\cos\omega t + i(a_1 - b_1)\sin\omega t$

as $e^{\pm i\omega t} = (\cos\omega t \pm i\sin\omega t)$

where $i = \sqrt{-1}$

Put $(a_1 + b_1) = A$ and $i(a_1 - b_1) = B$

$\therefore$ $y = A\cos\omega t + B\sin\omega t$ ... (4)

Where a_1, b_1, A and B are arbitrary constants.

If we put $A = P\sin\alpha$ $B = P\cos\alpha$ δ_{sm} the equation (4) becomes

$$y = P\sin(\omega t + \alpha) \quad \text{... (5)}$$

where $P = \sqrt{A^2 + B^2}$ and $\alpha = \tan^{-1}\left(\frac{A}{B}\right)$

2.6. EVALUATION OF THE CONSTANTS A AND B

It is found on solving the equation (3)

$$y = \text{A}\cos\omega t + \text{B}\sin\omega t \quad \text{...(4)}$$

The particle may start motion either by (*i*) initial displacement or by (*ii*) initial velocity or by (*iii*) both.

At $t = 0$ $y = y_0$ initial displacement

$$\frac{dy}{dt} = v \text{ initial velocity} \quad \text{... (5)}$$

From equation (4) with the help of equation (5)

$$y_0 = A.1 + B.\,0 \text{ gives } A = y_0$$

$$v = A\omega.0 + B\omega.1 \text{ gives } B = \frac{v}{w}$$

$\therefore$ $$y = y_0\cos\omega t + \frac{v}{w}\sin\omega t \quad \text{... (6)}$$

If the motion is started by initial velocity without initial displacement

$$y = \frac{v}{\omega}\sin\omega t$$

and when there is initial displacement y_0 only

$$y = y_0 \text{ cost } \omega t$$

2.7. ENERGY OF A PARTICLE EXECUTING S.H.M.

A particle oscillates, total energy of a particle at any stage is the sum of the kinetic and the potential energy of the particle at that stage.

$$\text{Kinetic energy} = \frac{1}{2}m\left(\frac{dy}{dt}\right)^2$$

$$\text{Potential energy} = \text{mass} \times \text{acceleration} \times \text{distance}$$

$$= \int_0^y m\omega^2\, y dy = \frac{1}{2}m\omega^2 y^2$$

$$\therefore \qquad \text{Total energy} = \frac{1}{2}m\left(\frac{dy}{dt}\right)^2 + \frac{1}{2}m\omega^2 y^2 \qquad \text{... (7)}$$

For S.H.M.

$$y = \alpha \sin \omega t$$

$$\frac{dy}{dt} = a\omega \cos \omega t$$

Equation (7) becomes

$$\text{Total energy} = \frac{1}{2}m\omega^2 a^2 \cos^2 \omega t + \frac{1}{2}m\omega^2 a^2 \sin^2 \omega t$$

$$= \frac{1}{2}m\omega^2 a^2 = = 2\pi^2 ma^2 n^2$$

as

$$\omega = 2\pi n$$

This can also be calculated as follows. The total energy of a conservative system is constant so when kinetic energy is maximum potential energy is zero. So total energy is the maximum kinetic energy but maximum value of $\frac{1}{2}ma^2\omega^2\cos^2\omega t$ is $\frac{1}{2}ma^2\omega 2$.

2.8. COMPOSITION OF TWO SIMPLE HARMONIC MOTIONS ALONG THE SAME DIRECTION OF THE SAME FREQUENCY.

A particle is acted upon by two Simple Harmonic Motions in the same direction, they have same frequency of vibration but they differ in phase.

Let y_1 and y_2 be the displacement produced by two S.H.M. They are given by at any instant t

$$y_1 = a \sin \omega t \qquad \text{... (8)}$$

$$y_2 = b \sin(\omega t + \alpha) \qquad \text{... (9)}$$

At any instant ωt is the phase of the first S.H.M. and $(\omega t + \alpha)$ is that of the second, so that the phase difference is α.

The resultant displacement at any instant t is the algebraic sum of individual displacement y_1 and y_2.

From equations (8) and (9)

$$\begin{aligned} y &= y_1 + y_2 \\ &= \alpha \sin \omega t + b\sin(\omega t + \alpha) \\ &= a \sin \omega t + b \sin \omega t \cos\alpha + b \cos \omega t \sin\alpha \\ &= (a + b \cos \alpha) \sin \omega t + (b \sin \alpha) \cos \omega t \qquad \text{... (10)} \end{aligned}$$

Putting $a + b \cos \alpha = A \cos \in$ and $b \sin \alpha = A \sin \in$ the equation (10) becomes

$$y = A \sin(\omega t + \in) \qquad \text{... (11)}$$

where

$$A^2 = a^2 + b^2 + 2ab \cos \alpha$$

and

$$\in = \tan^{-1} \frac{b \sin \alpha}{a + b \cos \alpha}$$

The resultant motion (equation (11)) is Simple Harmonic Motion of the same frequency as the components.

Special case :

(i) If $\alpha = 0, 2\pi, 4\pi$ or even multiple of π

$A = a + b, \in = 0$ and $y = (a + b) \sin \omega t$

(ii) If $\alpha = \pi, 3\pi, 5\pi$ or odd multiple of π

$A = a - b, \in = 0$ and $y = (a - b) \sin \omega t$

If $a = b$, the particle in the case will not execute S.H.M. but $y = 0$ for all values of t.

(iii) The displacement ~ time curve are two motions drawn in the cases (i) the periods are same and the phase difference is 0, 2π, 4π etc. that is the phases are same and resultant curve shown by dotted line (Fig. 4) is obtained by simply adding the ordinates of the two curves at the same instant whether above or below the time axis. The nature of the motion will be same but the amplitude will be the sum of the amplitudes of the component motions.

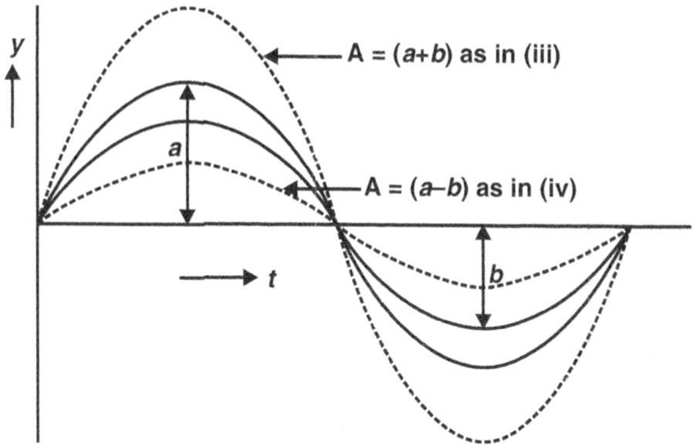

Fig. 4.

(iv) Since the phase difference $\alpha = \pi, 3\pi, 5\pi$ etc. is any odd multiple of π, we as before draw the time displacement curve and draw the resultant curve by taking the algebraic sum of the ordinates at a given time. The dotted curve represents the resultant motion. Here the amplitude is given by $(a - b)$ and when $a = b$ represent amplitude will be zero, there is no motion.

2.9. TWO SIMPLE HARMONIC MOTIONS ACT UPON A PARTICLE SIMULTANEOUSLY IN THE SAME DIRECTION HAVING NO PHASE DIFFERENCE BUT THEY DIFFER IN FREQUENCY BY VERY SMALL AMOUNT

The two S.H.M. acting upon a particle is given by

$$y_1 = a \sin \omega t = a \sin 2\pi nt \quad \text{... (12)}$$

$$y_2 = b \sin \omega_1 t = b \sin 2\pi(n + v)t \quad \text{... (13)}$$

where $\omega = 2\pi n$ and $\omega_1 = 2\pi(n + v)$, v is the difference of frequencies of two motions, y_1 and y_2 are the displacement at any instant and a and b are the amplitudes. The second one is of higher frequency by v. The resultant motion is given by

$$y = y_1 + y_2$$

$$= \alpha \sin 2\pi nt + b \sin 2\pi(n + v)t$$

$$= a \sin 2\pi nt + b \sin 2\pi nt \cos 2\pi vt + b \cos 2\pi nt \sin 2\pi vt$$

$$= (a + b\cos 2\pi vt) \sin 2\pi nt + (b\sin 2\pi vt) \cos 2\pi nt = A\sin (2\pi nt + t)$$

$$= A \sin (2\pi nt + \in) \quad \text{... (14)}$$

where $a + b \cos 2\pi vt = A \cos \in$ and $b \sin 2\pi vt = A \sin \in$

and $$A^2 = a^2 + b^2 + 2ab \cos 2\pi vt \quad \text{... (15)}$$

$$\in = \tan^{-1} \frac{b \sin 2\pi vt}{a + b \cos 2\pi vt} \quad \text{... (16)}$$

where A is the amplitude of the resultant motion which in this case [vide equation (15)] depends upon time, so also the phase difference $\in$ (vide equation 16). They change with time.

Putting different values of time in equation (15) it is found

Time	Amplitude A	Nature of A	Time Difference Between the Two
0	$a + b$	maximum	
$\frac{1}{2v}$	$a - b$	minimum	$\frac{1}{v}$
$\frac{1}{v}$	$a + b$	maximum	$\frac{1}{v}$
$\frac{3}{2v}$	$a - b$	minimum	
$\frac{2}{v}$	$a + b$	maximum	$\frac{1}{v}$

Thus the maximum amplitudes of magnitude $(a + b)$ or $2a$ when $a = b$ of the resultant motion is occurring at $t = 0, \frac{1}{v}, \frac{2}{v}$ *etc.*, that is at regular interval of time $\frac{1}{v}$. The minimum amplitudes of $(a - b)$ or 0 when $a = b$ are occurring at time $\frac{1}{2v}, \frac{3}{2v}$ *etc.*, that is, at regular interval of time $\frac{1}{v}$.

In order to notice these changes the difference between the frequencies v of the two component motions should be very small.

The resultant time displacement curve of two component displacement curves of two component motions differing in frequency by very small amount is obtained as before by algebraic sum of displacements at a given time. Fig. 5.

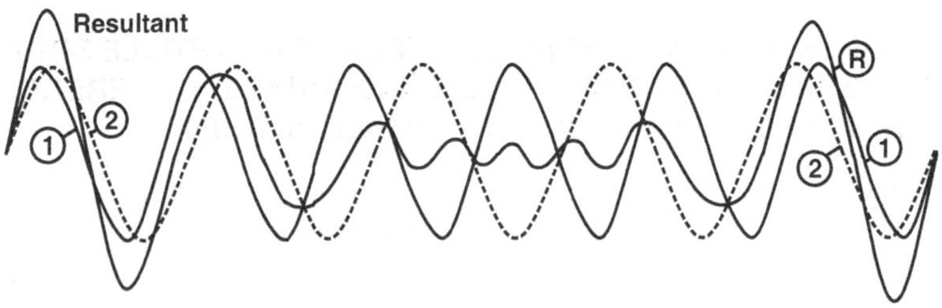

Fig. 5.

The amplitude of the resultant vibration is some time maximum and other time minimum, the time interval between the two maxima or the time interval of two minima is equal to $\frac{1}{v}$, hence frequency of occurrence is v, the difference of the frequency of two vibrations.

2.10. COMPOSITION OF TWO SIMPLE HARMONIC MOTIONS ACTING UPON A PARTICLE SIMULTANEOUSLY AT RIGHT ANGLES TO EACH OTHER, SAME TIME PERIOD BUT DIFFERENT IN PHASE.

A particle is acted upon by two S.H.M. at right angles to each other be represented by

$$x = a \sin(\omega t + \alpha) \quad \text{... (17)}$$

$$y = b \sin \omega t \quad \text{... (18)}$$

where $\omega = \frac{2\pi}{T}$, T being the time period, α the phase difference. They also differ in amplitudes which are a and b respectively.

From equations (17) and (18)

$$\frac{y}{b} = \sin \omega t \qquad \text{... (19)}$$

and
$$\frac{x}{a} = \sin \omega t \cos \alpha + \cos \omega t \sin \alpha$$

or
$$\frac{x}{a} = \frac{y}{b}\cos\alpha + \sqrt{1-\frac{y^2}{b^2}}\sin\alpha \qquad \text{... (20)}$$

or
$$\frac{x}{a} - \frac{y}{b}\cos\alpha = \sqrt{1-\frac{y^2}{b^2}}\sin\alpha \qquad \text{... (21)}$$

Squaring both sides and simplifying we get

$$\frac{x^2}{a^2} + \frac{y^2}{b^2} - \frac{2xy}{ab}\cos\alpha = \sin^2\alpha \qquad \text{... (22)}$$

This is the equation of an ellipse in general confined within a rectangle of sides $2a$ and $2b$ (Fig. 6(*a*))

Special cases:

(*a*) When phase difference $\alpha = 0$ equation (22) becomes

$$\frac{x^2}{a^2} + \frac{y^2}{b^2} - \frac{2xy}{ab} = 0$$

or
$$\left(\frac{x}{a} - \frac{y}{b}\right)^2 = 0 \qquad \text{... (23)}$$

or
$$y = \frac{b}{a}x \qquad \text{... (24)}$$

This is an equation of a pair of coincident straight lines. It lies in the first and third quadrants. It is traced twice in opposite direction in the same period. The equation (24) represents straight line which passes through the origin and has an inclination $a = \tan^{-1}\left(\frac{b}{a}\right)$ to the X-axis [Fig. 6(*b*)].

(*a*) (*b*)

Fig. 6.

(*b*) When $\alpha = \pi$ the equation (22) becomes

$$\left(\frac{x}{a}+\frac{y}{b}\right)^2 = 0 \qquad \text{or} \qquad y = y = -\frac{b}{a}x \qquad \text{... (25)}$$

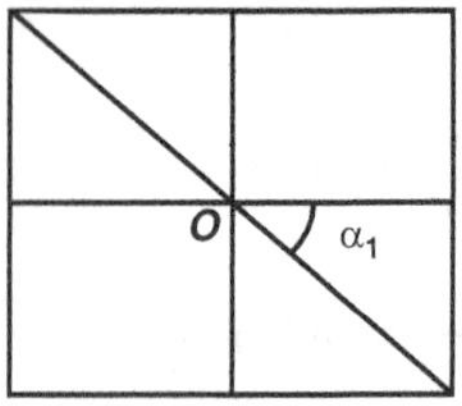

Fig. 6 (*c*)

This also represents a pair of its coincident straight lines passing through the origin O in the second and fourth quadrants. It is also traced twice in opposite directions in the same period. This straight line has an inclination $\alpha_1 = \tan^{-1}\left(-\frac{b}{a}\right)$ [Fig 6 (*c*)].

(*c*) When $\alpha = \frac{\pi}{2}$ the equation (22) becomes

$$\frac{x^2}{a^2}+\frac{y^2}{b^2} = 1 \qquad \text{... (26)}$$

It is a regular ellipse with semi-axes a and b along the coordinate axes [Fig 6 (*d*)]. If in addition $a = b$, the equation (22) reduces to an equation of a circle [Fig. 6(*e*)].

$$x^2 + y^2 = a^2$$

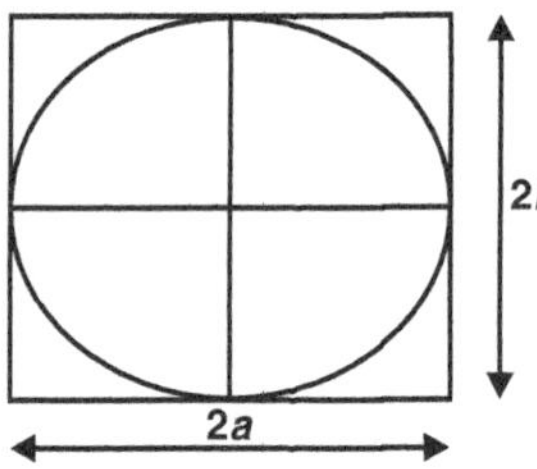

Fig. 6 (*d*)

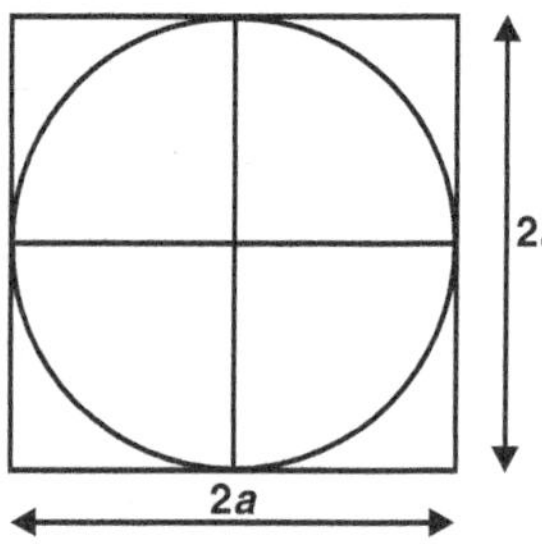

Fig. 6 (*e*)

(*d*) When $\alpha = \frac{\pi}{4}$ equation (22) becomes

$$\frac{x^2}{a^2}+\frac{y^2}{b^2}-\sqrt{2}\,\frac{xy}{ab} = \frac{1}{2} \qquad \text{... (27)}$$

This represents an oblique ellipse

2.10. (a) GRAPHICAL REPRESENTATION

Geometrical method: Let a particle be acted upon by two S.H.M. of amplitude a and b respectively occur in perpendicular directions XOX' and YOY' represented in a framework, O being the central position of two motions. Draw straight lines through $X\,X'$, and $Y\,Y'$ parallel to XOX' and YOY' as shown in Fig. 7. The projection Q describing the lower circle on EF and projection of P describing the right hand circle on RS will execute S.H.M. along EF parallel to XOX' and RS parallel to YOY'. The quadrants of the circles are divided into equal parts and draw straight lines through these points of divisions on the lower circles parallel to YOY' and similarly through the points of divisions on the right hand circle parallel to XOX'.

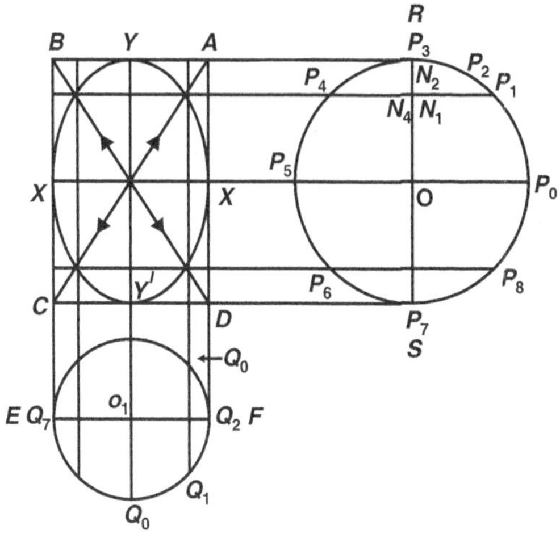

Fig. 7

Case I. $\alpha = 0$. Here foot to the projection from Q is at O, from point P moves along the positive direction OX, P is at O moves along the positive direction OY. The corresponding positions of P and Q are given by $P_0\ P_1\ P_2$..... and $Q_0\ Q_1\ Q_2$.... The point of intersection at the foot of perpendiculars drawn from P on YOY' and Q on XOX' will intersect at different points in the framework $ABCD$. These points constitute a straight line inclined at an angle equal to $\tan^{-1} b/a$ traced twice in opposite directions.

Case II. $\alpha = \pi$. Here Q is at O, moves along the negative direction OX' and P at O moves in the positive direction OY. Proceeding in the similar way a straight line represented by broken line is obtained in the framwork inclined at an angle equal to $-\tan^{-1} b/a$. This is also traced twice in opposite directions.

Case III. $\alpha = \pi/2$. Here Q is at X moves along the negative direction X to O_1 and P at O moves in the positive direction O to Y. In similar way as above the points of intersection in the framework are obtained and the curve obatined by joining these points is an ellipse.

2.11. LISSAJOUS FIGURE

Curves formed by the resultant of two simple harmonic motions at right angles to each other are known as Lissajous figure. The shape of the curve depends on the relative amplitude; frequencies and phase relation at the initial moment. It is found that when two S.H.M. having frequency ratio 1 : 1, 1 : 2 *etc.*, and vibrating at right angles to each other are compounded together the resultant path due to motion is represented by a curve which remains constant for a particular frequency ratio so long as the phase difference between the vibration is constant. The form of the curves for the resultant path due to motion will assume different shapes as phase difference changes from O through $\frac{\pi}{2}, \pi, \frac{3\pi}{2}$. The set of curves will be obtained for the same range of phase difference *i.e.*, $0 \rightarrow 2\pi$ the ratio of frequencies differ 1 : 1, 1 : 2, 1 : 3, 2 : 3 *etc.*

The curves are shown in Fig. 8 for various phases for frequency ratio 1 : 1, 1 : 2, 1 : 3.

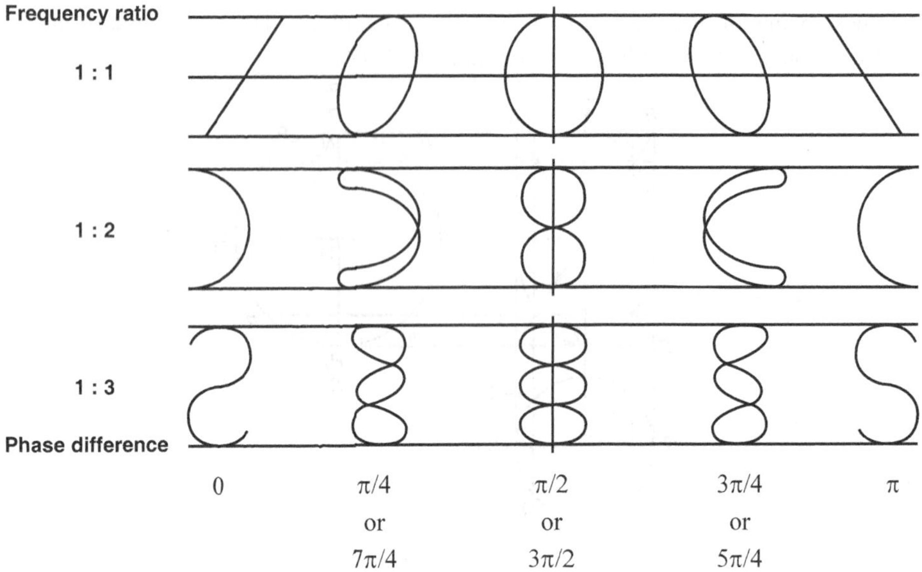

Fig. 8.

2.12. EXPERIMENTAL DETERMINATION OF LISSAJOUS FIGURES

(*a*) **Black Burns Pendulum** Fig. 9. It consists of a heavy metal ring with a glass funnel filled inside it. The metal ring and funnel are suspended from Y shaped suspension of thread fixed with a vertical wooden stand. The two upper ends of the string pass through the corners *A* and *B* of the framework and wound round the screw pegs. The metal ring with the funnel filled with dry fine sand is suspended from the lower prolonged portion of the string branches out into three portions which carries the ring with funnel containing sand; this acts as bob of the pendulum. The arrangement combines in itself two pendulums. If pulled in the plane of the framework and let go, it will execute SHM like a simple pendulum with length *CM*, in the plane of the framework. If it is pulled in the perpendicular direction to the framework and is released, the ring which behaves like a bob of the pendulum of length equivalent to prolongation of *MC* where it meets *AB*. As the pendulum has two motions, one in the plane of the framework, and other perpendicular to the framework, the sand from the funnel trickles down on the base and traces out a Lissajous' figure. As the time period is proportional to the square root of the length so the time period can be made in the ratio 1 : 2 by adjusting *CM* and above in the ratio 1 : 4.

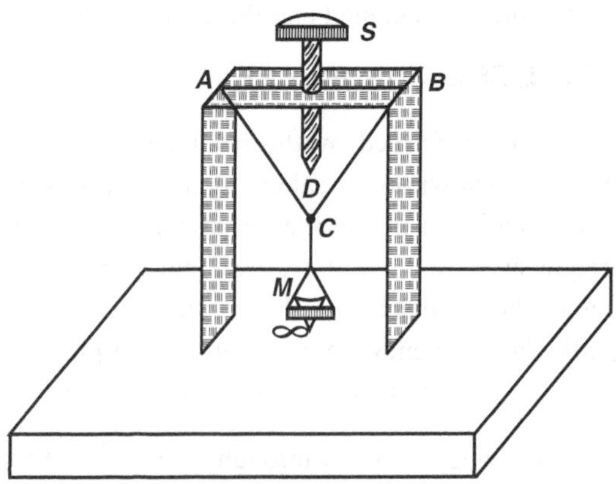

Fig. 9.

(b) Cathode Ray Oscilloscope (CRO)

For demonstration of Lissajous figures CRO is most suitable instrument. The essential requirement for a CRO is

(i) A glass envelope electron gun assembly.

(ii) Arrangement to focus the electron beam into a narrow pencil.

(iii) Deflection plate assembly.

(iv) A fluorescent screen.

Glass envelope is a conical highly evacuated glass bulb and supports the various electrodes. Electron gun produces a focussed beam of electron. It essentially consists of an indirectly heated cathode, a control grid, a focussing anode and accelerating anode. The focussing electrode consists of two metal cylinders with large aperture. Deflection plate assembly consists of one set of vertical deflection plate and the other set are the horizontal deflecting plates. The fluorescent screen consists of a glass coated with fluorescent material such as zinc oxide.

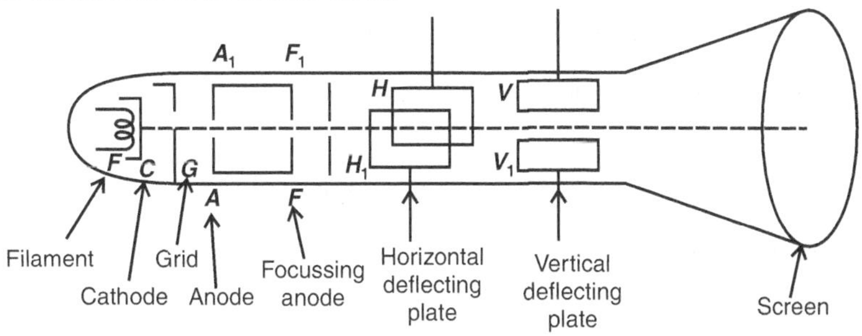

Fig. 10.

A beam of electron from cathode C passed through the metal anode AA_1, focussing device FF_1, HH_1 horizontal deflecting plates and VV_1 vertical deflecting plates and finally to a focus on the screen.

An alternating potential difference between the plates HH_1 will spread the beam horizontally and we get a horizontal line instead of a spot. The alternating voltage along VV_1 deflects the spot vertically. The combination of the two perpendicular fields at right angles to each other traces out Lissajous figures.

(c) Optical Method

A strong beam of light from an electric arc S is made convergent by a converging lens L and is made to fall upon a mirror M_1 attached to one of the prongs of the tuning fork T_1 in such a way that after reflection from the mirror M_1 it is again reflected from the mirror M_2 attached to a second fork T_2.

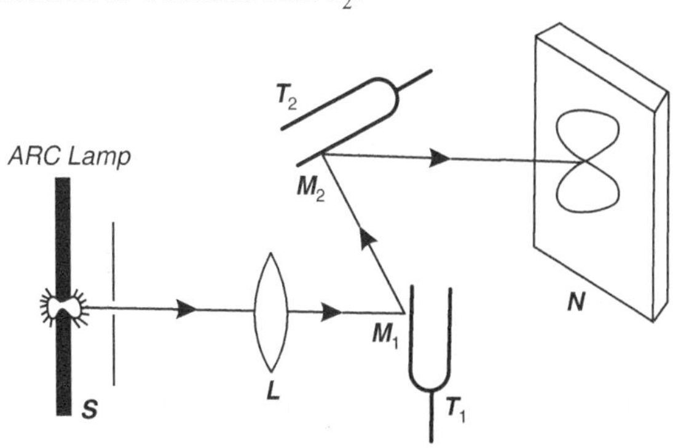

Fig. 11.

When both the forks vibrate in directions at right angles to one another, the resulting motion of the beam of light will trace out a figure on a white opaque screen N. The shape of the figure depends upon the frequencies and phase difference of the vibration of the tuning forks.

SOLVED PROBLEMS

1. A particle moving with S.H.M. has a period 0.001 sec amplitude 0.05 m. Show that the acceleration is 7.9×10^4 m/sec^2 when it is 0.002 m away from the mean position of rest and the maximum velocity it will attain is $\pi \times 10^2$ m/sec.

Given Amplitude $a = 0.05$ m

Time period $T = 0.001$ sec

$$\therefore \quad \text{Angular velocity } \omega = \frac{2\pi}{T} = \frac{2\pi}{0.001} \text{ rad/sec}$$

$$= 2\pi \times 10^3 \text{ rad/sec}$$

Displacement $y = 0.002$ m

We know that the acceleration of the particle $\frac{d^2y}{dt^2} = \omega^2 y$

$$= 4\pi^2 \times 10^6 \times 0.002 \text{ m/sec}^2$$
$$= 7.9 \times 10^4 \text{ m/sec}^2$$

Velocity of the particle $\frac{dy}{dt} = a\omega \cos \omega t$

$$= \omega\sqrt{a^2 - y^2}$$

$\frac{dy}{dt}$ will be maximum when $y = 0$

$$\therefore \quad \left(\frac{dy}{dt}\right)_{max} = \omega a$$

$$= 2\pi \times 10^3 \times 0.05 \text{ m/sec}$$
$$= \pi \times 10^2 \text{ m/sec}$$

1 (a). The acceleration of the particle executing simple harmonic motion is $\frac{\pi^2}{3}$ cm/second^2 when its displacement is 3 cm. Calculate its time period of motion.

[Ravisankar Shukla University 2010]

According to the problem acceleration $f = \frac{\pi^2}{3}$ cm/sec^2

but we know acceleration $f = \omega^2 x$

since $x = 3$ cm

$$\therefore \quad \frac{\pi^2}{3} = \omega^2 3$$

or $$\omega = \frac{\pi}{3}$$

But the time period $T = \frac{2\pi}{\omega} = \frac{2\pi}{\pi/3}$ sec $= 6$ sec.

1 (*b*). A body of mass 15×10^{-3} kg oscillate about a fixed point with SHM of amplitude 0.08 m. If the body is attracted towards the fixed point when at a distance of 0.04 m with a force equal to the weight of 0.01 kg.

Find

(*i*) the period of oscillation of the mass about the fixed point and

(*ii*) the maximum kinetic energy. Assume $g = 9.8\ ms^{-2}$

[Jain University Nagpur 2010]

From the problem we get

$$\text{Mass } m = 15 \times 10^{-3} \text{ kg}$$

$$\text{Amplitude } a = 0.08 \text{ m}$$

$$\text{Distance } x = 0.04 \text{ m}$$

$$\text{Weight} = 0.01 \text{ kg}$$

$$\therefore \quad \text{Force } mf = 0.01 \times 9.8 \text{ Newton}$$

$$15 \times 10^{-3} f = 0.01 \times 9.8$$

$$\therefore \quad f = \frac{.01 \times 9.8}{15 \times 10^{-3}} = 6.53 \text{m/sec}^2$$

But we know $\quad f = \omega^2 x$

$$\therefore \quad 6.53 = \omega^2 \times .04$$

or $\quad \omega^2 = \dfrac{653}{4}$

or $\quad \omega = \sqrt{\dfrac{653}{4}} = 12.78 \text{ sec}^{-1}$

Time period $\quad T = \dfrac{2\pi}{\omega} = \dfrac{2\omega}{12.78} = 0.49$

Maximum kinetic energy

$$= \frac{1}{2} m\omega^2 a^2$$

$$= \frac{1}{2} \times 15 \times 10^{-3} \times \frac{653}{4} \times (.08)^2$$

$$= 7.836 \times 10^{-3} \text{ Joules.}$$

1 (*c*). The displacement of the particle executing simple harmonic motion is given by $x = 10 \cos (4\pi t + \pi/3)$mt. Find out the frequency and displacement after time 1 second.

[Osmania University 2010]

Since $\quad x = 10 \cos (4\pi t + \pi/3)$

$$\therefore \quad \omega = 4\pi$$

$$\therefore \quad 2\pi f = 4\pi$$

$$\therefore \quad \text{Frequency } f = 2 \text{ Hz}$$

The displacement after 1sec is

$$x = 10 \cos (4\pi + \pi/3)\text{mt}$$

$$= 10 \cos \pi/3 \text{ mt}$$

$$= 10.\frac{1}{2} = 5 \text{ mt.}$$

2. A uniform U-tube contains a very little mobile liquid to a height l in each limb. If the liquid in one limb is depressed and then released, find the period of oscillation of the liquid in the tube.

Let the cross-section of the U-tube be A and density of the mobile liquid ρ.

$\therefore$ The mass of the mobile liquid $= A \cdot 2l\rho$

Let the depression of the liquid level be y at a time t, then the difference of liquid level in the two limbs $= 2y$

Thrust on the liquid = $A2y\rho g$

$$\therefore \qquad A2l\rho\frac{d^2y}{dt^2} = -A\,2y\rho g$$

$$\text{or} \qquad \frac{d^2y}{dt^2} + \frac{g}{l}y = 0$$

$$\text{or} \qquad \frac{d^2y}{dt^2} + \omega^2 y = 0$$

where $\omega^2 = \frac{g}{l}$, time period $T = \frac{2\pi}{\omega} = 2\pi\sqrt{\frac{l}{g}}$

3. A particle is subjected to two simple harmonic motions

$$y_1 = a\sin\omega t$$

and

$$y_2 = b\sin\left(\omega t + \frac{\pi}{3}\right)$$

Find (a) the displacement at $t = 0$ (b) the maximum speed of the particle and (c) the maximum acceleration of the particle.

(*a*) At $t = 0$ $\qquad y_1 = a\sin\omega.\,0 = 0$

$$y_2 = b\sin\left(\omega.0 + \frac{\pi}{3}\right) = b\sin\frac{\pi}{3} = \frac{\sqrt{2}}{2}b$$

The resultant displacement

$$y = y_1 + y_2 = \frac{\sqrt{3}}{2}b$$

(*b*) The amplitude of the resultant motion (Vide Art. 2.8)

$$A = \sqrt{a^2 + b^2 + 2ab\cos\frac{\pi}{3}}$$

$$= \sqrt{a^2 + b^2 + ab}$$

The maximum speed of the particle $V_{max} = \omega A$

$$= \omega\sqrt{a^2 + b^2 + ab}$$

(*c*) The maximum acceleration of the particle

$$\left(\frac{d^2y}{dt^2}\right)_{max} = \omega^2 A = \omega^2\sqrt{a^2 + b^2 + ab}$$

4. A particle is subjected simultaneously to two simple harmonic motions in the same direction each of same frequency but of different amplitudes and a phase difference $\left(\frac{\pi}{4}\right)$. Find the amplitude of the resultant displacement and its phase relation to one of the components.

[Burd .U. 2001, 2005]

Let y_1 and y_2 be the displacements produced by two S.H.M. and are given by

$$y_1 = a\sin\omega t$$

$$y_2 = b\sin\left(\omega t + \frac{\pi}{4}\right)$$

The resultant displacement

$$y = y_1 + y_2$$
$$= A\sin(\omega t + \epsilon) \qquad \text{[Vide Art. 2.8 eq. (3)]}$$

where $A^2 = a^2 + b^2 + 2ab\cos\frac{\pi}{4} = a^2 + b^2 + \sqrt{2}ab$

or $A = (a^2 + b^2 + \sqrt{2}ab)^{1/2}$

and the phase angle $\epsilon = \tan^{-1}\left(\dfrac{b\sin\frac{\pi}{4}}{a + b\cos\frac{\pi}{4}}\right) = \tan^{-1}\left(\dfrac{b}{\sqrt{2}\,a + b}\right)$

5. Two vibrations at right angles to each other, are described by equations $x = 3\cos 4\pi t$ and $y = 3\cos\left(8\pi t + \frac{\pi}{3}\right)$ where x and y are expressed in centimeters and t in seconds. Construct the Lissajous figure of the combined motion.

We have

$$x = 3\cos 4\pi t$$

or $\dfrac{x}{3} = \cos 4\pi t$

and $y = 3\cos\left(8\pi t + \frac{\pi}{3}\right)$

or $\dfrac{y}{3} = \cos\left(8\pi t + \frac{\pi}{3}\right) = \cos 8\pi t\cos\frac{\pi}{3} - \sin 8\pi t\sin\frac{\pi}{3}$

or $\dfrac{y}{3} = (2\cos^2 4\pi t - 1).\frac{1}{2} - 2\sin 4\pi t\cos 4\pi t.\frac{\sqrt{3}}{2}$

or $\dfrac{y}{3} = \dfrac{x^2}{9} - \dfrac{1}{2}\sqrt{1 - \dfrac{x^2}{9}}\,.\dfrac{x}{\sqrt{3}}$

or $\dfrac{x}{3}\sqrt{1 - \dfrac{x^2}{9}} = \dfrac{x^2}{9} - \dfrac{1}{2} - \dfrac{y}{3}$

or $x^2\left(1 - \dfrac{x^2}{9}\right) = 3\left(\dfrac{x^2}{9} - \dfrac{1}{2} - \dfrac{y}{3}\right)^2$

6. Obtain the equation of motion due to superposition at two mutually perpendicular simple harmonic motion of time period in ratio 1 : 2. [Ravisankar Shukla University 2010]

From the problem we have the ratio of time period $\dfrac{T_1}{T_2} = \dfrac{1}{2}$

$$\therefore \qquad \frac{\omega_1}{\omega_2} = \frac{\frac{2\pi}{T_1}}{\frac{2\pi}{T_2}} = \frac{T_2}{T_1} = \frac{2}{1}$$

or $$\omega_1 = 2\omega_2$$

A particle is acted upon by two simple harmonic motion at right angle to each other be represented by

$$x = a \sin (\omega_1 t + \alpha) \qquad ...(1)$$

and $$y = b \sin \omega_2 t \qquad ...(2)$$

$$\therefore \qquad \frac{y}{b} = \sin \omega_2 t \qquad ...(3)$$

and $$\frac{x}{a} = \sin (2\omega_2 t + \alpha) \qquad \text{as } \omega_1 = 2\omega_2$$

or $$\frac{x}{a} = \sin 2\omega_2 t \cos \alpha + \cos 2\omega_2 t \sin \alpha$$

or $$\frac{x}{a} = 2\sin \omega_2 t \cos \omega_2 t \cos \alpha + (1 - 2\sin^2 \omega_2 t) \sin\alpha$$

Now substituting the value of sin $\omega_2 t$ from equation (3)

We get

$$\frac{x}{a} = 2\frac{y}{b}.\sqrt{1-\frac{y^2}{b^2}}.\cos\alpha + \left(1-\frac{2y^2}{b^2}\right)\sin \alpha$$

or $$\frac{x}{a} - \left(1-\frac{2y^2}{b^2}\right)\sin \alpha = 2\frac{y}{b}\sqrt{1-\frac{y^2}{b^2}}\cos \alpha$$

Squaring both sides we have

$$\frac{x^2}{a^2} + \left(1-\frac{2y^2}{b^2}\right)\sin^2 \alpha - \frac{2x}{a}\left(1-\frac{2y^2}{b^2}\right)\sin\alpha = \frac{4y^2}{b^2}\left(1-\frac{y^2}{b^2}\right)\cos^2 \alpha$$

Adding $\left(1-\frac{2y^2}{b^2}\right)^2 \cos^2 \alpha$ to both sides we have

$$\frac{x^2}{a^2} + \left(1-\frac{2y^2}{b^2}\right) - \frac{2x}{a}\left(1-\frac{2y^2}{b^2}\right)\sin \alpha = \cos^2 \alpha$$

$\therefore$ The resultant motion is represented as

$$\frac{x^2}{a^2} + \left(1-\frac{2y^2}{b^2}\right) - \frac{2x}{a}\left(1-\frac{2y^2}{b^2}\right)\sin \alpha = \cos^2 \alpha \qquad ...(4)$$

We consider the following particular cases

Case I when $\alpha = 0$ then the equation (4) will be as

$$\frac{x^2}{a^2} + \left(1-\frac{2y^2}{b^2}\right)^2 = 1$$

or $$\frac{x^2}{a^2} - \frac{4y^2}{b^2} + \frac{4y^4}{b^4} = 0$$

Case II when $\alpha = \pi/2$ then equation (4) becomes

$$\frac{x^2}{a^2}+\left(1-\frac{2y^2}{b^2}\right)^2-2x/a\left(1-\frac{2y^2}{b^2}\right)=0$$

or $$\frac{x^2}{a^2}+1+\frac{4y^4}{b^4}-\frac{4y^2}{b^2}-\frac{2x}{a}+\frac{4xy^2}{ab^2}=0$$

or $$\left(\frac{x^2}{a^2}-\frac{2x}{a}+1\right)+\frac{4y^4}{b^4}+\frac{4y^2}{b^2}\left(\frac{x}{a}-1\right)=0$$

or $$\left(\frac{x}{a}-1\right)^2+2\left(\frac{x}{a}-1\right)\frac{2y^2}{b^2}+\left(\frac{2y^2}{b^2}\right)^2=0$$

or $$\left(\frac{x}{a}-1+\frac{2y^2}{b^2}\right)^2=0$$

The above equation represents two coincidence parabola.

7. A cylinder of mass M and cross-sectional area A floats vertically in a liquid of density ρ. If it is depressed slightly and then released, find the frequency of vertical oscillation. [C.U. 1980]

From the problem we have

$$\frac{Md^2x}{dt^2}=-A\rho gx$$

or $$\frac{d^2x}{dt^2}=-\frac{A\rho g_x}{M}$$

or $$\frac{d^2x}{dt^2}=-\omega^2x$$

where $$\omega^2=\frac{A\rho g}{M}$$

∴ $$\omega=\sqrt{\frac{A\rho g}{M}}$$

But we know $$\omega=\frac{2\pi}{T}=\sqrt{\frac{A\rho g}{M}}$$

or $$\frac{1}{T}=\frac{1}{2\pi}=\sqrt{\frac{A\rho g}{M}}$$

or $$\text{Frequency } f=\frac{1}{T}=\frac{1}{2\pi}=\sqrt{\frac{A\rho g}{M}}$$

8. Two simple harmonic motions act in the same direction simultaneously on a particle. The SHM have the same frequency 5 Hz; amplitudes of 0.005 m and 0.002 m respectively; and the second component lead the first by 45°. Determine the amplitude of the resultant displacement and its phase with respect to the first component. Give the expression for the resultant displacement as a function of time. [Burd. U. 1997]

Let y_1 and y_2 be the displacements produced by two S.H.M. and are given by

$$y_1 = a \sin \omega t = 0.005 \sin(10\pi t) \qquad [\omega = 2\pi f = 2\pi .5 = 10\pi]$$

$$y_2 = b \sin (\omega t + \alpha) = 0.002 \sin\left(10\pi t + \frac{\pi}{4}\right) \qquad \left[\alpha = 45° = \frac{\pi}{4}\right]$$

The resultant displacement

$$y = y_1 + y_2$$

$$= 0.005 \sin 10\pi t + 0.002 \sin\left(10\pi t + \frac{\pi}{4}\right)$$

$$= 0.005 \sin 10\pi t + 0.002 \left[\sin 10\pi t \cos\frac{\pi}{4} + \cos 10\pi t \sin\frac{\pi}{4}\right]$$

$$= \left(0.005 + 0.002 \cos\frac{\pi}{4}\right) \sin 10\pi t + \left(0.002 \sin\frac{\pi}{4}\right) \cos 10\pi t$$

Put $\quad 0.005+0.002 \cos\frac{\pi}{4} = A \cos \in$

and $\quad 0.002 \sin\frac{\pi}{4} = A \sin \in$

$\therefore \quad y = A \cos \in \sin 10\pi t + A \sin \in \cos 10\pi t = A \sin (10\pi t + \in)$

$\therefore$

$$A^2 = (0.005)^2 + (.002)^2 + 2 \times 0.005 \times 0.002 \cos\frac{\pi}{4}$$

$$= 25 \times 10^{-6} + 4 \times 10^{-6} + 14.1 \times 10^{-6}$$

$$= 43.1 \times 10^{-6}$$

$\therefore$

$$A = (43.1)^{1/2} \times 10^{-3}$$

$$= 6.565 \times 10^{-3}$$

$$= 0.006565 \text{ m}$$

and the phase angle

$$\in = \tan^{-1}\left[\frac{0.002 \sin\frac{\pi}{4}}{0.005 + 0.002 \cos\frac{\pi}{4}}\right]$$

$$= \tan^{-1} (2.21 \times 10^{-1})$$

$$= 12.462°$$

$\therefore$ The expression for the resultant displacement is

$$y = 0.006565 \sin(10\pi t + 12.462°) \text{ m}$$

REVIEW QUESTIONS

1. Define simple harmonic motion explaining the terms Period, Amplitude and Phase.
2. Prove that acceleration of a particle executing S.H.M. is proportional to displacement and is always directed to the mean position of rest. Define the terms frequency and epoch.
3. Show that a pendulum which executes S.H.M. is simple pendulum. Derive the expression for the time period of the same.
4. Derive the differential equation for S.H.M. from the energy principle.
5. Two simple harmonic motions having the same phase but slightly differ in frequency act in the same direction upon a particle simultaneously. What will be the subsequent motion of the particle?

6. What are Lissajous figures? How can you demonstrate them experimentally?
7. What is a simple harmonic oscillator? Establish the differential equation for it and solve it to the expression for velocity, displacement, time period.
[Barkatullah University 2011]
8. Derive differential equation of simple harmonic motion and find expression for velocity and acceleration. [N. H. Bahuguna Garhwal University 2010]
9. Deduce an expression for potential energy, kinetic energy and total energy of a particle executing simple harmonic motion. [Pandit Ravisankar University 2010]
10. Two simple harmonic motions of the same frequency and acting at right angles to each other are superimposed. Show the resulting changes with time.
11. Two collinear S.H.M. acting simultaneously on a particle are given by
$$y_1 = a \cos \omega t$$
$$y_2 = b \cos (\omega t + \theta)$$
Show that the resultant motion of the particle is simple harmonic and also obtain the expression for the amplitude and phase constant of the resultant motion in terms of *a, b,* and θ.

PROBLEMS

1. A body of mass 0.01 kg executes S.H.M. with amplitude 3×10^{-2} m and time period 2 sec. Calculate the energy of the particle. [**Ans.** 4.43×10^{-5}J]
2. Mercury of density 13.6×10^3 kg/m^3 is contained in a U-tube with its arms vertical. Neglecting damping find the time period of oscillation of the mercury if the total length of the tube occupied by the mercury is 0.3 m. If the area of cross-section of the tube be 2×10^{-4} m^2, find the energy of the motion when the amplitude is 0.05 m.
[**Ans.** 0.78 sec, 6.7×10^{-2}J]
3. Three S.H.M. of same frequency act on a particle simultaneously in the same direction. Their amplitudes are 0.01, 0.015 and 0.02 m respectively. The phase angle of the second with respect to first is 60° and that of the third with respect to the second is 30°. Obtain the resultant amplitude and phase angle relative to the first. [Burd. U. 1999]
[**Ans.** Amp = 0.0375 m and phase $\phi = 62°$]
4. Two collinear simple harmonic motions acting simultaneously on a particle are given by
$$y_1 = 0.003 \sin 2\pi t$$
$$y_2 = 0.002 \sin \left(2\pi t - \frac{\pi}{3}\right)$$
where y is expressed in metre and t in seconds. Write down the expression for the resultant displacement as a function of time. [**Ans.** $y = 0.004347 \sin (2\pi t + \alpha)$]
$\alpha = 23°22'$
5. Construct the Lissajous figures for
$$x = a \cos 2\omega t$$
$$y = a \cos (2\omega t - \pi/4)$$
[**Ans.** $x^2 - \sqrt{2}y + y^2 = \frac{a^2}{2}$]

CHAPTER 3

FREE, DAMPED AND FORCED VIBRATION

3.1. INTRODUCTION

A mechanical system may be excited in a body, the body will vibrate after removing exciting force. This vibration will be simple harmonic and is called free vibration or natural vibration. The time period of the vibrating body is called natural period.

A simple pendulum set in oscillation or tuning fork set in vibration and the system will not oscillate or vibrate for infinite time. The amplitude of vibration of a system gradually becomes smaller and smaller and finally motion subsides as in (Fig. 3.) This is due to some resisting force due to viscosity or internal friction acting upon the oscillating system. This resisting force is offered by air. Such a motion is called damped vibration.

A vibrating system loses vibration due to damping. To maintain vibration an external periodic force is applied on the body or a system. The body or the system will try to vibrate with natural frequency and natural vibration dies away due to damping force and ultimately vibrates with the time period of the external periodic force. Such a type of vibration is called the forced vibration.

3.2. MOTION DUE TO A CONSTANT FORCE

A constant force of magnitude F is acting on a particle executing simple harmonic motion. The equation of motion of the particle is given by

$$m\frac{d^2y}{dt^2}+\mu y = F \qquad \text{... (1)}$$

where F is a constant force acting on the particle.

The equation (1) can be written as

$$\frac{d^2y}{dt^2}+\omega^2 y = \frac{F}{m} \qquad \text{... (1.1)}$$

where $\omega^2 = \frac{\mu}{m}$

or $$\frac{d^2y}{dt^2}+\omega^2\left(y-\frac{F}{m\omega^2}\right) = 0 \qquad \text{... (2)}$$

Putting $y-\frac{F}{m\omega^2} = Z$ we have

$$\frac{d^2Z}{dt^2}+\omega^2 Z = 0 \qquad \text{... (3)}$$

The solution of the equation (3) is given by [vide Art. 2.5 eq.(4)]

$$Z = A\sin\omega t + B\cos\omega t \qquad \text{... (4)}$$

or $$y = \frac{F}{m\omega^2}+A\sin\omega t+B\cos\omega t \qquad \text{... (5)}$$

Applying initial conditions

$$t = 0, y = y_0 \text{ and } \frac{dy}{dt} = v$$

We get
$$A = \frac{v}{\omega}, B = y_0 - \frac{F}{m\omega^2} \qquad \text{... (6)}$$

Hence the equation (5) becomes

$$y = \frac{F}{m\omega^2} + \frac{v}{\omega}\sin\omega t + \left(y_0 - \frac{F}{m\omega^2}\right)\cos\omega t \qquad \text{... (7)}$$

The effect of the constant force is to diminish the initial displacement and it alters the mean position of rest.

If $F = 0$ the equation (7) becomes

$$y = \frac{v}{\omega}\sin\omega t + y_0 \cos\omega t \qquad \text{... (8)}$$

being the solution of simple harmonic motion [vide Art. 2.6 eq. (4)]

3.3. THE FORCE ACTS FOR SHORT TIME AND TO FIND ITS EFFECT

Let the force act for a short time dt' at the time t' and to find the effect at the time t ($t > t'$)

As before the equation of motion is

$$\frac{d^2y}{dt^2} + \omega^2 y = \frac{F}{m} \qquad \text{... (1.1)}$$

The effect of $\frac{F}{m}$ which is an explicit function of time is to generate in time dt', a velocity

$\frac{F}{m}\,dt'$ whose result at a time t will be $y = \frac{F}{\omega m}dt' \sin\omega(t - t')$ ... (9)

The total effect in time 't' is given by

$$y = \int_0^t \frac{F}{\omega m}\sin\omega(t - t')dt' \qquad \text{... (10)}$$

The lower limit is taken zero for convenience without taking arbitrary t'

The complete solution is given by

$$y = y_0 \cos\omega t + \frac{v}{\omega}\sin\omega t + \frac{1}{\omega m_0}\int_0^t F \sin\omega(t - t')dt' \qquad \text{... (11)}$$

3.4. A PARTICLE EXECUTING S.H.M. IS ACTED UPON BY A HARMONIC FORCE *F* sin *pt* OF FREQUENCY $\frac{p}{2\pi}$ OR *F* COS *pt* OF FREQUENCY $\frac{p}{2\pi}$

The equation of motion is given by

$$m\frac{d^2y}{dt^2} + \mu y = F \sin pt \text{ or } F \cos pt \qquad \text{... (12)}$$

If there is no force acting on the particle $F \sin pt$ or $F \cos pt$ will be equal to zero, the solution of the equation (12) in such a case is

$$y = A \sin \omega t + B \cos \omega t \qquad \text{... (4)}$$

[Vide Art. 2.5 eq. (4)] where $\omega^2 = \frac{\mu}{m}$, when the force is acting the equation (12) becomes

$$\frac{d^2 y}{dt^2} \omega^2 y = f \sin pt \qquad \text{... (13)}$$

where $f = \frac{F}{m}$

Writing $D = \frac{d}{dt}$, an operator and $\frac{d^2}{dt^2} = D^2$

The equation (13) becomes

$$(D^2 + \omega^2) = f \sin pt$$

or $$y = \frac{f}{D^2 + \omega^2} \sin pt$$

or $$y = \frac{f}{\omega^2 - p^2} \sin pt \qquad \text{... (14)}$$

$$\begin{aligned}[D(\sin pt) = p \cos pt,\ D^2(\sin pt) &= D.\ (D \sin pt) \\ &= D\ (p \cos pt) \\ &= -p^2 \sin pt \\ D^2 &= -p^2\ ]\end{aligned}$$

while the force be acting the system will forget its own natural frequency ω but will begin to vibrate with the frequency of the force $\frac{p}{2\pi}$.

If the frequency of the system and the frequency of the force are equal $\omega = p$ the amplitude $\frac{f}{\omega^2 - p^2}$ becomes infinite and the system will break down.

3.5. MOTION IN A RESISTING MEDIUM

If a simple pendulum sets in oscillation or a tuning fork set in vibration, the system will not oscillate or vibrate for infinite time. The amplitude of vibration of a system gradually becomes smaller and smaller and finally motion subsides. This is due to some resisting force acting upon the oscillating system. Such a motion is called damped vibration.

Let a particle of mass m move in a straight line under a force proportional to the distance towards the fixed point, the initial position of rest, under a small resistance to its motion. The force due to resistance is proportional to the velocity of the particle for the small value of the velocity.

The equation of motion of the particle of mass m is given by

$$m\frac{d^2 y}{dt^2} = \mu y - R\frac{dy}{dt} \qquad \text{... (15)}$$

or $$\frac{d^2 y}{dt^2} + K\frac{dy}{dt} + \omega^2 y = 0 \qquad \text{... (16)}$$

where μ is a constant and is called the spring factor or stiffness factor, R frictional force per unit velocity, K the coefficient of friction given by $\frac{R}{m}$ and $\omega^2 = \frac{\mu}{m}$.

As it is the equation representing the oscillation affected by damping the solution is assumed to be of the form

$$y = A_1 e^{pt} \text{ so } \frac{dy}{dt} = pA_1 e^{pt}, \frac{d^2y}{dt^2} = p^2 A_1, e^{pt} = p^2 y$$

so the equation (16) on substituting these values becomes

$$(p^2 + Kp + \omega^2)y = 0 \qquad \text{... (17)}$$

or
$$p^2 + Kp + \omega^2 = 0 \qquad \text{... (18)}$$

gives
$$p = -\frac{K}{2} \pm \sqrt{\frac{K^2}{4} - \omega^2}$$

or
$$y = e^{-\frac{Kt}{2}} \left(Ae^{\sqrt{\frac{K^4}{4} - \omega^2}.t} + Be^{\sqrt{\frac{K^2}{4} - \omega^2}.t} \right) \qquad \text{... (19)}$$

where A and B are arbitrary constants.

The value of p is given by Eqn. (19) depends upon the magnitude of the expression under radical sign in the right hand side.

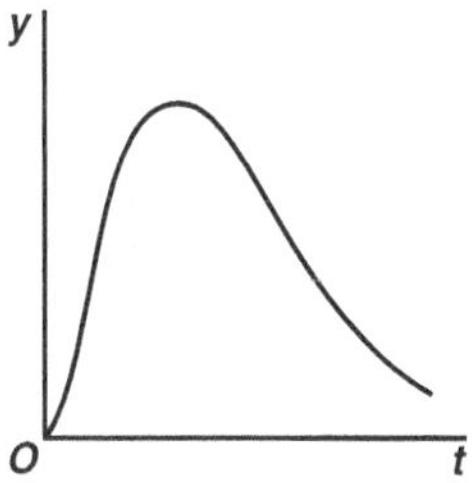

Fig. 1.

Case I. If $\frac{K^2}{4} > \omega^2$ *i.e.* $\sqrt{\frac{K^2}{4} - \omega^2}$ is real, the value of y consists of two terms both together falling exponentially to zero. The motion of a over damped *dead beat* galvanometer is one of the best examples in this case.

The variation of displacement y with time t is shown in Fig. 1.

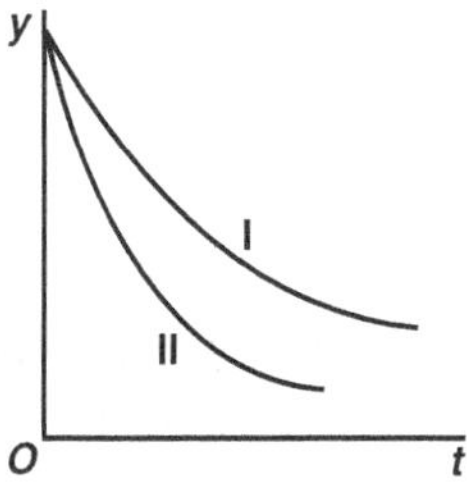

Fig. 2. Curve II : Critical damped curve

Case II. If $\frac{K}{2} = \omega$, which is known as *critical damping*, y becomes

$$y = (A + Bt)e^{\frac{-Kt}{2}}$$

Here decay is rapid. If ω is slightly greater than K the motion is oscillatory, if ω is slightly less than K, the motion is *dead beat.* The variation of displacement y with t is shown in Fig. 2.

Case III. If $\frac{K}{2} < \omega$, the expression under the radical sign will be imaginary, and the motion will be oscillatory. The expression under the radical sign

$$\sqrt{\frac{K^2}{4} - \omega^2} = i\sqrt{\omega^2 - \frac{K^2}{4}} = i\omega_1 \text{ where } \omega_1 = \sqrt{\omega^2 - \frac{K^2}{4}}$$

or Eqn. (19) becomes

$$y = e\frac{K}{2}t(Ae^{i\omega_1 t} + Be^{-i\omega_1 t})$$

$$= e^{\frac{K}{2}t}\{(A+B)\cos\omega_1 t + i(A-B)\sin\omega_1 t\}$$

$$= e^{\frac{K}{2}t}(A_1 \cos\omega_1 t + B_1 \sin\omega_1 t) \qquad \text{... (20)}$$

$$= e^{\frac{K}{2}t}\sqrt{A_1^2 + B_1^2}\sin(\omega_1 t + \tan^{-1} A_1 / B_1)$$

where A_1 and B_1 are constants and

$$\omega_1 = \sqrt{\omega^2 - \frac{K^2}{4}}$$

Here ω_1 is acting as natural frequency of the system. The effect of damping is to lower th frequency by exceeding small value. Since the amplitude $\sqrt{A_1^2 + B_1^2}e^{-\frac{K}{2}t}$ decays exponentially with time the motion is damped oscillatory. The measure of deeay of complitude is the damping coefficient K.

The variation of y with time t represents geometrically a simple harmonic motion as represented by the curve, Fig. 3. The amplitude

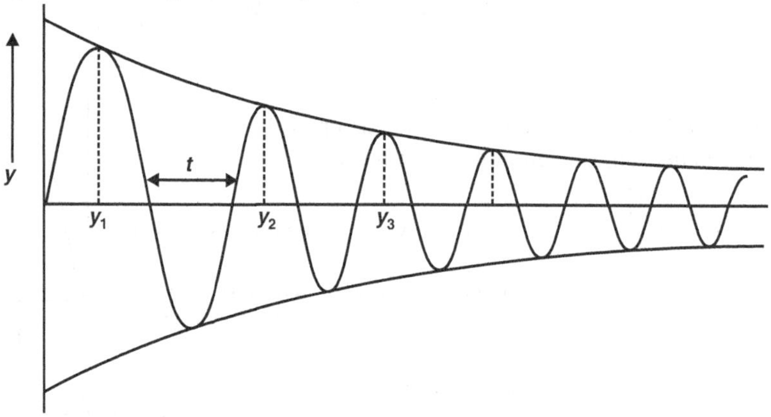

Fig. 3.

$\sqrt{A^2 + B^2}\, e^{-\frac{K}{2}t}$ diminishes exponentially to zero as time advances. The measure of decay of amplitude is the damping coefficient K.

Let, at any time t the first maximum displacement is given by

$$y_1 = ae - \frac{K}{2}t$$

where a stands for $\sqrt{A_1^2 + B_1^2}$ as

$\sin(\omega_1 t + \in) = \pm 1, \quad \in = \tan^{-1}\frac{A_1}{B_1}$

when the displacement is maximum. At the successive interval of time period T, the maximum displacement, on one direction are given by $y_{1,} y_2 ....$

Then
$$y_2 = ae^{-\frac{K}{2}(t+T)}$$
$$y_3 = ae^{-\frac{K}{2}(t-2T)}$$
$$y_4 = ae^{-\frac{K}{2}(t+3T)}$$
$$\dots \quad \dots \quad \dots \quad \dots \quad \dots$$

So
$$\frac{y_1}{y_2} = \frac{ae^{-\frac{K}{2}(t)}}{ae^{-\frac{K}{2}(t+T)}} = e^{\frac{K}{2}T}$$

$$\frac{y_2}{y_3} = e^{\frac{K}{2}T}$$

or $$\frac{y_3}{y_2} = e^{\frac{K}{2}T}$$

Hence $$\frac{y_1}{y_2} = \frac{y_2}{y_3} = \ldots\ldots\ldots\ldots\ldots D = e^{\frac{KT}{2}}$$

Taking logarithm on both sides

$$K = \frac{2}{T}\log_e D = \frac{2 \times 2.3}{T}\log_{10} D$$

This is log decrement of the damping oscillatory system. The damping coefficient is determined by adopting this method.

3.6. (a) Forced Vibration

A periodic force of frequency $\frac{p}{2\pi}$ and amplitude F is acting on a particle executing S.H.M. in a resisting medium, the resisting force being proportional to the velocity of the particle.

The equation of motion of the particle of mass m is given by

$$m\frac{d^2y}{dt^2} = F \sin pt - \mu y - R\frac{dy}{dt} \quad \ldots (21)$$

or $$\frac{d^2y}{dt^2} + K\frac{dy}{dt} + w^2 y = f \sin pt \quad \ldots (22)$$

where $K = \frac{R}{m}$ the damping coefficient and $\omega^2 = \frac{\mu}{m}$, μ the stiffness constant or spring constant, $\frac{\omega}{2\pi}$ the natural frequency of the system and $f = \frac{F}{m}$.

If it be a simple oscillatory system with damping without any external periodic force acting on the system, the solution of the equation (22) as before [Vide Eq. (20) Art 3.5)] is given by

$$y = e^{-\frac{K}{2}t}(A\cos\omega_1 t + B\sin\omega_1 t) \quad \ldots (20)$$

where $$\omega_1 = \sqrt{\omega^2 - \frac{K^2}{4}}$$

But the system will behave differently while external force is acting.

The equation (22) can be written in the form

$$(D^2 + KD + \omega^2)y = f \sin pt$$

where $$D = \frac{d}{dt} \text{ or } y = \frac{f \sin pt}{D^2 + KD + \omega^2} \quad \ldots (23)$$

Writing $-p^2$ for D^2

$$y = f\frac{\sin pt}{(\omega^2 - p^2) + KD}$$

or $$y = \frac{f[(\omega^2 - p^2) - KD]\sin pt}{[(w^2 - p^2)] - K^2D^2}$$

$$= \frac{f[(\omega^2 - p^2)\sin pt - Kp\cos pt]}{[(w^2 - p^2)^2 + K^2p^2]}$$

If $$\omega^2 - p^2 = A_1 \cos \in \quad ...(24)$$

$$Kp = A_1 \sin \in \quad ...(25)$$

$\therefore$ $$A_1 = [(\omega^2 - p^2)^2 + K^2 p^2]^{1/2} \quad ...(26)$$

and $$\in = \tan^{-1}\frac{Kp}{\omega^2 - p^2} \quad ...(27)$$

$\therefore$ $$y = \frac{f\sin(pt - \in)}{[(w^2 - p^2)^2 + K^2p^2]^{1/2}} \quad ...(28)$$

$$= \frac{F\sin(pt - \in)}{[(w^2 - p^2)^2 + K^2p^2]^{1/2}} \quad ...(29)$$

$$= \frac{F}{pZ}\sin(pt - \in) \quad ...(30)$$

where $$Z = \frac{m}{p}[(w^2 - p^2)^2 + K^2p^2]^{1/2} \quad ...(31)$$

$$= \left[m^2\left(\frac{\omega}{p} - \frac{p}{\omega}\right)w^2 + R^2\right]^{1/2} \quad ...(32)$$

as $Km = R$ or $$Z = \left[\left(\frac{m\omega^2}{p} - mp\right)^2 + R^2\right]^{1/2} \quad ...(33)$$

$$= \left[\left(\frac{\mu}{p} - mp\right)^2 + R2\right]^{1/2} \quad ...(34)$$

$$= \sqrt{X^2 + R^2} \quad ...(35)$$

where $$X = m\omega\left(\frac{\omega}{p} - \frac{p}{\omega}\right) = \left(\frac{\mu}{p} - mp\right) \quad ...(36)$$

Hence $$y = \frac{F}{p\sqrt{X^2 + R^2}}\sin(pt - \in) \quad ...(37)$$

$$= \frac{F}{pZ}\sin(pt - \in) \quad ...(37)$$

where Z is known as mechanical impedance given by $Z = \sqrt{X^2 + R^2}$, X is the Reactance and R is resistance.

(*i*) If μ is large Z reduces to $\frac{\mu}{p}$ system is said to be stiffness controlled

(*ii*) If R is large the system is resistance controlled when Z reduces to R

(*iii*) If m is large Z becomes equal to mp the system is said to be mass controlled.

(b) Alternative Method of Solution

A simple arrangement of four pendulums A, B, C and D is made (Fig.4). They are suspended from a light horizontal elastic support. The plane of oscillation is perpendicular to the plane of the paper. This will help us in assuming the solution of the above system. Pendulum B is heavy, A is of same length with light bob, C and D are of different lengths with light bob. A and B are of same time period of oscillation as are of same lengths. C is of shorter length so time period is less and D is of longer length so time period of oscillation is greater than those of A and B.

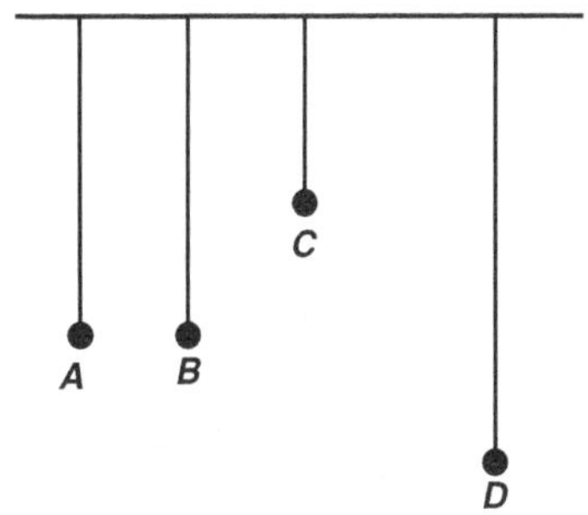

Fig. 4.

B is set in oscillation, so A having the same time period also sets in oscillation readily with longer amplitude, C and D set in oscillation with small swing that is with less amplitude but with same time period that of B which acts as a driver. Thus the driven system C and D oscillates with same time period or frequency of the driver.

Hence the solution may be assumed as

$$y = A \sin (pt - \in) \quad \text{... (38)}$$

where A and $\in$ are unknown amplitude and phase which are to be determined.

$$\therefore \quad \frac{dy}{dt} = Ap \cos (pt - \in) \quad \text{... (39)}$$

and

$$\frac{d^2y}{dt^2} = -Ap^2 \sin (pt - \in) \quad \text{... (40)}$$

substituting these values from equations (38), (39) and (40) in equation (22) we get

$$-Ap^2 \sin (pt - \in) + KAp \cos(pt - \in) + \omega^2 A \sin (pt - \in)$$
$$= f \sin pt$$
$$= f \sin (pt - \in + \in)$$

or

$$-Ap^2 \sin(pt - \in) + KAp \cos(pt - \in) + \omega^2 A\sin(pt - \in)$$
$$= f \sin (pt - \in) \cos \in + f \cos(pt - \in) \sin \in$$

Comparing the coefficient of sin $(pt - \in)$ and cos$(pt - \in)$on both sides

$$A(\omega^2 - p^2) = f \cos \in$$
$$AKp = f \sin \in$$

whenence on squaring and adding and solving

$$A = \frac{f}{[(\omega^2 - p^2)^2 + K^2p^2]^{1/2}}$$

and

$$\in = \tan^{-1} \frac{Kp}{(\omega^2 - p^2)}$$

Hence

$$y = \frac{f}{[(\omega^2 - p^2)^2 k^2 p^2]^{1/2}} \sin\left(pt - \tan^{-1} \frac{Kp}{\omega^2 - p^2}\right) \quad \text{... (41)}$$

which is identical as equation (28)

At the beginning when the force acting on the vibrating system, the natural vibration is superimposed on the forced vibration which of course dies away with time. The complete solution at the beginning becomes [From eqn. (20) in Art 3.5 and eqn. (41)]

$$y = e^{-\frac{Kt}{2}}[(A_1 \cos \omega_1 t + B_1 \sin \omega_1 t)] + \frac{f}{[(\omega^2 - p^2)^2 + k^2 p^2]^{1/2}} \sin\left(pt - \tan^{-1}\frac{Kp}{\omega^2 - p^2}\right) \quad ... (42)$$

The first term on the right hand side of equation (42) gives natural damped vibration of the particle and is transient and second term represents forced vibration.

At the beginning if ω_1 is close to p, during unsteady state of superposition, beats will be formed known as transient beat. After some time first expression will subside which depends on the value of K.

3.7. AMPLITUDE RESONANCE : MAXIMUM DISPLACEMENT OF THE SYSTEM

Displaced amplitude is maximum when p in equation (28) [Art. 3.6(a)] and equation (41) [Art. 3.6(b)] is minimum as regards the variation of forced frequency *i.e*, when $[(\omega^2 - p^2)^2 + K^2 p^2]^{1/2}$ is minimum

or $$\frac{d}{dp}[(\omega^2 - p^2)^2 + K^2 p^2]^{1/2} = 0$$

or $$\frac{1}{2}[\omega^2 - p^2)^2 + K^2 p^2]\left(-\frac{1}{2}\right)[-4p(\omega^2 - p^2) + 2K^2 p] = 0$$

or $$2K^2 p = 4p(\omega^2 - p^2)$$

or $$p = \sqrt{\omega^2 - \frac{K^2}{2}} \quad ... (43)$$

which is the condition for amplitude resonance.

Thus displacement amplitude is maximum when forcing frequency $\frac{p}{2\pi}$ is equal to $\frac{1}{2\pi}\sqrt{\omega^2 - \frac{K^2}{2}}$ that is when the forcing frequency is slightly lower than natural frequency $\frac{\omega}{2\pi}$ which is nearly equal to the frequency of the damped system $\frac{1}{2\pi}\sqrt{\omega^2 - \frac{K^2}{4}}$. As $\omega^2 \gg \frac{K^2}{2}$ then $p = \omega$ which gives

$$y_{max} = \frac{f}{K\omega} \quad ... (44)$$

Fig. 5 represents the curves showing the variation of amplitude with the change of frequency of the force with varying degree of damping. The amplitude is infinite which does not occur in the nature as K is never zero. For finite value of K the amplitude is maximum when $p = \omega$.

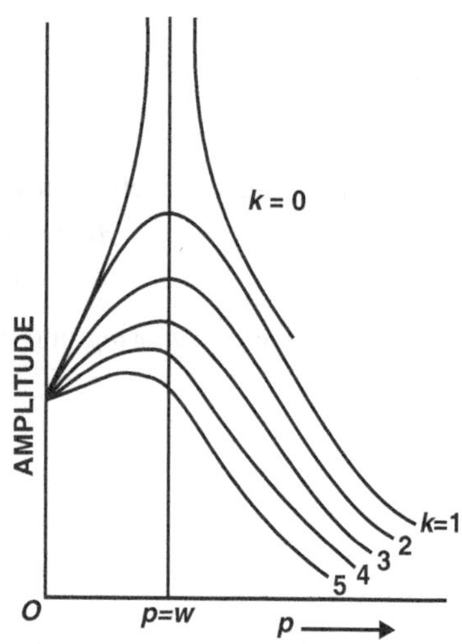

Fig. 5.

3.8. MAXIMUM ENERGY OF THE SYSTEM : VELOCITY RESONANCE

The energy of the driven system is maximum when the frequency of the driven system without damping is equal to the frequency of the driver or $\omega = p$. This is known as velocity resonance or energy resonance from Eqns. (31) to (37) and (42).

$$\frac{f}{\sqrt{\{(\omega^2 - p^2)^2 + K^2 p^2\}}} = \frac{F}{P_{sm}\sqrt{\left\{m^2\left(\frac{\omega}{p} - \frac{p}{\omega}\right)^2 \omega^2 + R^2\right\}}}$$

$$= \frac{f}{pZ} = \frac{F}{p\sqrt{X^2 + K^2}}$$

where X the mechanical reactance stands for $X = m\left(\frac{\omega}{p} - \frac{p}{\omega}\right)\omega$.

This is also defined as lack of coincidence or mistuning.

A set of curves showing the variation of energy E with X for different values of K is shown in Fig. 6.

Total energy is equal to the maximum kinetic energy

or
$$E = \frac{m}{2}\left(\frac{dy}{dt}\right)^2_{\max} = \frac{m}{2}\left\{\frac{Fp\cos(pt - \in)}{p\sqrt{X^2 + R^2}}\right\}^2_{\max}$$

$$= \frac{m}{2} \cdot \frac{F}{X^2 + R^2} \quad \text{... (45)}$$

as maximum value of $\cos(pt - \in) = \pm 1$ E is maximum when $X = 0$ *i.e.* $p = \omega$. The lack of coincidence of two frequencies produces diminution of energy E. Moreover the decrease of energy upon the lack of coincidence is same for a given ratio of frequencies whether the frequency p is too great or too small.

From Eqn. (45) when

$$X = 0, p = \omega$$

$$E_{\max} = \frac{m}{2}\frac{f^2}{K^2} = \frac{m}{2}\frac{F^2}{R^2} \quad \text{... (46)}$$

Hence
$$\frac{E}{E_{\max}} = \frac{R^2}{X^2 + R^2} \quad \text{... (47)}$$

when $p = \omega$, $X = 0$ and $\frac{E}{E_{\max}} = 1$ for all values of damping.

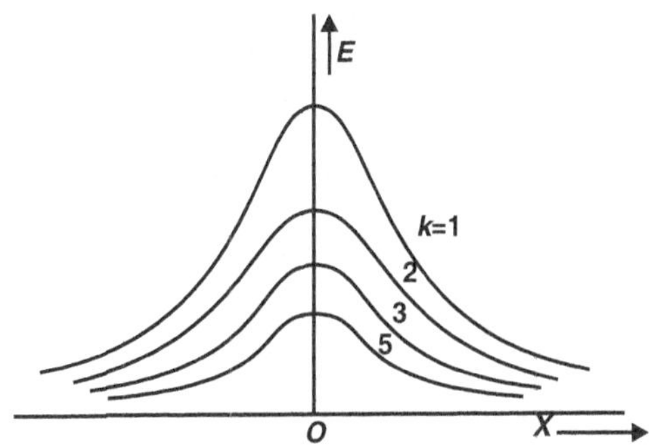

Fig. 6.

This is the condition for energy resonance. When X is very large $\frac{E}{E_{\max}} = 0$ for small values of R.

Sharpness of Resonance. The ratio of the fractional diminution of energy for a given value of the frequency of the force to the square of mistuning is known as the sharpness of resonance.

Sharpness
$$S = \frac{1}{X^2}\frac{E_{\max} - E}{E} = \frac{1}{X^2}\left(\frac{E_{\max}}{E} - 1\right)$$

$$= \frac{1}{X^2}\left(\frac{X^2+R^2}{R^2}-1\right)=\frac{1}{R^2} \quad \text{... (48)}$$

The curves in Fig. 7 represents the variation of E/E_{max} with X for different values of R, decreasing from upper to lower curve, here X measures the mistuning when $\omega > p$, X is positive and $\omega < p$, X is negative.

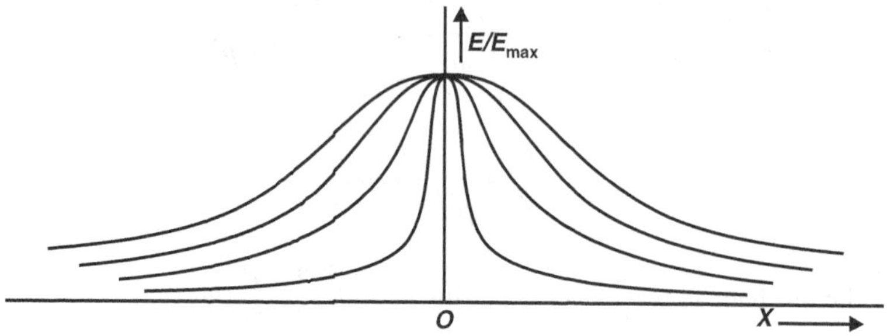

Fig. 7.

The resonance will be sharper, when greater is the dependence of energy of the forced vibration, on the exactitude of the frequencies ω and p. Thus the sharpness varies inversely as the square of the damping.

From the curves (Fig. 7) it is observed that for large values of damping, (vide upper curve) the variation of X on any direction of $X = 0$ produces slight change of E/E_{max} which is the case of *flat resonance*. For lowest value of damping (vide lowest curve) the slight change of X on either side of $X = 0$ produces large change of E/E_{max} where sharper is the resonance.

The effect of sharpness of resonance may be experimentally observed by placing a vibrating tuning fork on the sound board of a sonometer with a thin or a thick string. A paper rider will be thrown off from the thin string, only when there is resonance but from the thick string, it will be thrown off even when it is appreciably away from resonance, as damping in this case is large. In case of resonance column experiment, when the vibrating tuning fork is held over the open end of the tube, it begins to speak over a length of the tube without a particular length. This is due to the fact damping in air column is large.

3.9. PHASE OF THE FORCED VIBRATION

The phase of the steady motion is given by

$$\in = \tan^{-1}\frac{Kp}{\omega^2 - p^2} \text{ or } \frac{Kp}{\omega^2-p^2} = \tan\in$$

Three cases will arise

(*i*) $p < \omega, \in$ is positive, and lies between 0 and $\frac{\pi}{2}$ for all values of K.

(*ii*) $p > \omega$ $\in$ lies between $\frac{\pi}{2}$ and π and is negative for all values of K.

(*iii*) $p = \omega$ in case of resonance $\in = \frac{\pi}{2}$ for all values of K.

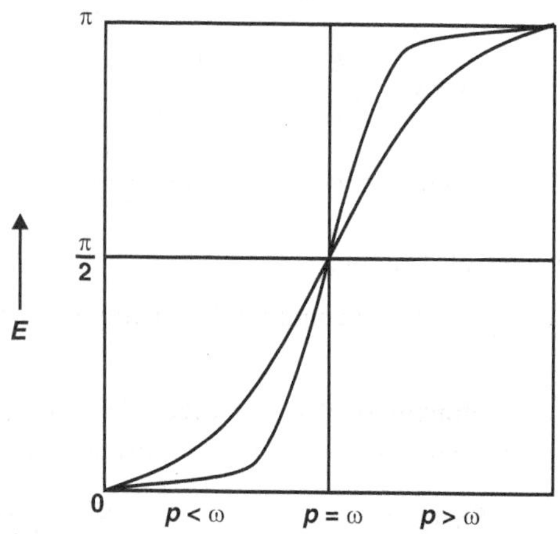

Fig. 8.

When damping is small and $p < \omega$, $\in$ will tend to be equal to zero and $p < \omega$ will tend to be equal to π. In the former case displacement will be in the same phase with the force and in the latter case they will be in the opposite phase *i.e.*, π. During resonance the phase difference $\frac{\pi}{2}$ *i.e.*, when $p = \omega$.

3.10. POWER SUPPLY AT STEADY STATE OF FORCED VIBRATION

At any time, the rate of supply of energy *i.e* power by forcing system to the forced system is

Power $$P = \text{Rate of work} = \frac{dW_1}{dt} = \text{Force} \times \text{Velocity}$$

$$= F \sin pt . \frac{dy}{dt}$$

Using equation (37.1) $$y = \frac{F}{pZ} \sin(pt - \in)$$

Power $$P = F \sin pt . \frac{Fp}{pZ} \cos(pt - \in) = \frac{F^2}{Z} \sin pt \cos(pt - \in)$$

On taking average over the whole time period

$$P_{av} = \frac{1}{T} \int_0^T P dt$$

$$= \frac{F^2}{Z} \frac{1}{T} \int_0^T \sin pt \cos(pt - \in) dt = \frac{1}{2} \frac{F^2}{Z} \sin \in$$

If $$\in = 0, P_{av} = 0$$

$$P_{av} = \frac{1}{2} \frac{F^2}{Z^2} R \qquad \text{... (49)}$$

as $$\sin \in = \frac{Kp}{\left[(\omega^2 - p^2)^2 + K^2 p^2 \right]^{1/2}} = \frac{Kmp}{pZ} = \frac{R}{Z}$$

Similarly, the rate of work done by dissipative force *i.e.*, average rate of energy loss due to work W_2 done against the dissipative force

$$P' = \frac{dW_2}{dt} = \text{Force} \times \text{velocity} = R \frac{dy}{dt} . \frac{dy}{dt} = R \left(\frac{dy}{dt} \right)^2$$

$$= \frac{RF^2}{Z^2} \cos^2 (pt - \in)$$

Average over the cycle $$P'_{av} = \frac{RF^2}{Z^2} \frac{1}{T} \int_0^T \cos^2 (pt - \in) dt = \frac{F^2 R}{2Z^2} \qquad \text{... (50)}$$

Thus the rate of supply of energy P given by equation (49) by the driving force is equal to the rate of energy dissipated P'_{av} given by equation (50) by the frictional force.

When resonance occurs *i.e.*, $p = \omega$ gives $X = 0$, hence $Z = R$ average power $\frac{1}{2} \frac{F^2 R}{Z^2}$ becomes $\frac{F^2}{2R}$. When the value of the average power becomes half the value of that of the value at resonance which is generally known as half power value, that is

$$\frac{F^2R}{2Z^2} = \frac{1}{2} \text{ the value of resonance} = \frac{1}{2}\frac{F^2}{2R}$$

which gives $Z^2 = 2R^2$

But $Z^2 = X^2+R^2 = 2R^2$

so $X^2 = R^2$

or $X = \pm R$

At this stage p has two values, one (p_1) is greater than ω and other (p_2) is less. Putting the value of X from equation (36)

$$\frac{\mu}{p_1} - mp_1 = R$$

or $$m\frac{\mu}{mp_1} - mp_1 = \text{R}$$

or $$\frac{\omega^2}{p_1} - p_1 = \frac{R}{m} \quad \text{... (51)}$$

and $$\frac{\omega^2}{p_2} - p_2 = = -\frac{R}{m} \quad \text{... (52)}$$

where $$\frac{\mu}{m} = \omega^2$$

We have by adding from equations (51) and (52)

$$\omega^2\left(\frac{1}{p_1}+\frac{1}{p_2}\right) - (p_1+p_2) = 0$$

or $$\frac{\omega^2}{p_1p_2} = 1 \quad \text{... (53)}$$

Eliminating ω^2 in equations (30) by equation (53)

$$\left(\frac{p_1p_2}{p_1} - p_1\right) = p_2 - p_1 = \frac{R}{m}$$

This is known as bandwidth or half power bandwidth. The same result can be obtained by putting $\frac{E}{E_{\max}} = \frac{1}{2}$ in equation (47).

SOLVED PROBLEMS

1. A mass of 10 gm is acted on by a restoring force 5 dyne/cm and a resistance 2 dyne-sec/cm. Find out whether the motion is oscillatory or aperiodic. Find the value of resistance for the motion to be critically damped.

[C.U. 1994, 1999]

Given mass m = 10 gm

Restoring force μ = 5 dyne/cm

Resistance R = 2 dyne. sec/cm

$\therefore \qquad \omega^2 = \frac{\mu}{m} = \frac{5}{10} = 0.5 \text{ rad}^2/\text{sec}^2$

and $\qquad K = \frac{R}{m} = \frac{2}{10} = 0.2$ neper/cm

(*i*) $\therefore \qquad \frac{K^2}{4} = \frac{0.04}{4} = 0.01$

so $\qquad \frac{K^2}{4} < \omega^2$, the motion will be oscillatory

(*ii*) For critically damping

$$\frac{K}{2} = \omega$$

or $\qquad K = 2\omega$

or $\qquad \frac{R}{m} = 2\sqrt{\frac{\mu}{m}}$

or $\qquad R^2 = 4\,\mu m$

or $\qquad R = 2\sqrt{\mu m}$

$$= 2\sqrt{5.10} = 14.14 \text{ dyne sec/cm}$$

2. Displacement of a particle of mass m executing underdamped oscillatory motion is given by $x_t = \rho e^{-bt}\cos(n't - \alpha)$ where $n' = \sqrt{n^2 - b^2}$ n is the natural frequency of the particle, b the damping constant. Calculate ρ and α when initial velocity $= V_0$ *[C.U. 2002]*

Given $\qquad x_t = \rho e^{-bt}\cos(n't - \alpha)$

$\therefore \qquad \frac{dx_t}{dt} = -b\,\rho e^{-bt}\cos(n't - \alpha) - n'\,\rho e^{-bt}\sin(n't - \alpha)$

$$\left.\frac{dx_t}{dt}\right|_{t=0} = V_0 = -b\rho\cos\alpha + n'\sin\alpha$$

or $\qquad V_0 = \rho(n'\sin\alpha - b\cos\alpha)$

Again $\qquad x_{ot} = \rho\cos\alpha$ or $\cos\alpha = x_{ot/p}$

and $\qquad \sin\alpha = \sqrt{1-\cos^2\alpha} = \frac{\sqrt{p^2 - x_{0t}^2}}{P_{sm}}$

$$\tan\alpha = \frac{\sqrt{\rho^2 - x_{0t}^2}}{x_{0t}}$$

$$\alpha = \tan^{-1}\frac{\sqrt{\rho^2 - x_{0t}^2}}{x_{0t}}$$

$\therefore \qquad V_0 = \rho\left[n'\frac{\sqrt{\rho^2 - x_{ot}^2}}{\rho} - b\frac{x_{ot}}{\rho}\right]$

or $\qquad V_0 = V_0 = \left[\left(n'\sqrt{\rho^2 - x_{ot}^2}\right) - bx_{ot}\right]$

or $$(V_0 + bx_{ot})^2 = n'^2 (\rho^2 - x^2_{ot})$$

$$n'^2 \rho^2 = V_0^2 + 2bV_0 x_{ot} + b^2x^2_{ot} + n'^2 x^2_{ot}$$

or $$\rho^2 = \frac{V_0^2 + 2bV_0x_{ot} + (b^2 + n'^2)x_{ot}^{2-}}{n'^2}$$

or $$\rho = \frac{1}{n'}\left[V_0^2 + 2bV_0x_{0t} + (b^2 + n'^2)x_{0t}^2\right]^{1/2}$$

3. The frequency of underdamped harmonic oscillator is adjusted to be equal to half the frequency experienced by the oscillator without damping. Calculate the logarithmic decrement of the system. *[Pb.U.95]*

From the given condition we get

$$\omega_1 = \frac{\omega}{2}$$

where ω is the natural frequency of the system and ω_1 is the frequency of the underdamped harmonic oscillator.

or $$\frac{\omega}{2} = \sqrt{\omega^2 - \frac{K^2}{4}}$$

or $$\frac{\omega^2}{4} = \omega^2 - \frac{K^2}{4}$$

or $$K = \sqrt{3}\omega$$

Now the logarithmic decrement $\lambda = \frac{KT}{2}$

$$T = \frac{2\pi}{\omega_1}$$

$\therefore$ $$\lambda = \frac{K}{2}\frac{2\pi}{\omega_1} = \frac{\sqrt{3}}{2}\omega\frac{2\pi}{\omega_1}$$

$$= \sqrt{3}\omega_1\frac{2\pi}{\omega_1} = 2\pi\sqrt{3}$$

4. If ω_1 and ω_2 are the half-power frequencies and ω_0 is the resonant frequency of a forced system, show that $\omega_0^2 = \omega_1\omega_2$.

From article 3.10 we get the average power at resonance is $\frac{F^2}{2R}$

and $\frac{F^2R}{2Z^2} = \frac{1}{2}$. The value of resonance $= \frac{1}{2}\frac{F^2}{2R}$

which gives $$Z^2 = 2R^2$$

But $$Z^2 = X^2 + R^2 = 2R^2$$

or $$X = \pm R$$

At this stage frequency has two values ω_1 and ω_2. $\omega_1 > \omega_0$ and $\omega_2 < \omega_0$.

$\therefore$ $$\frac{\mu}{\omega_1} - m\omega_1 = R$$

or $$m\frac{\mu}{m\omega_1} - m\omega_1 = R$$

or $$\frac{\omega_0^2}{\omega_1} - \omega_1 = \frac{R}{m} \quad \text{... (1)}$$

and $$\frac{\omega_0^2}{\omega_2} - \omega_2 = \frac{R}{m} \quad \text{... (2)}$$

From eq. (1) and (2) we get

$$\omega_0^2\left(\frac{1}{\omega_1}+\frac{1}{\omega_2}\right)-(\omega_1+\omega_2) = 0$$

or $$\frac{\omega_0^2}{\omega_1\omega_2} = 1$$

or $$\omega_2{}^0 = \omega_1\omega_2$$

5. A damped vibrating system, starting from rest, reaches a first amplitude of 500 mm which reduces to 50 mm in that direction after 100 oscillations each of period 2.3 sec. Find damping constant, relaxation time and correction for the first displacement for damping. (Given $\log_e 10 = 2.3$) (Indore U. 2001)

We know that the damping constant k is given by the relation

$$\log_e\left(\frac{y_1}{y_2}\right) = \log_e\left(\frac{y_2}{y_3}\right) = \text{........} = \frac{KT}{2}$$

where y_1, y_2, y_3,.... are the successive amplitude and T the time period. The amplitude n vibration after the first

$$\log_e\frac{y_1}{y_{n+1}} = \log_e\frac{y_1}{y_2}\times\log_e\frac{y_2}{y_3}\times\log_e\frac{y_3}{y_4}\times\text{.......}\times\log\frac{y_n}{y_{n+1}} = n\frac{KT}{2}$$

Now $y_1 = 500$ mm, $y_{101} = 50$ mm, $n = 100$, $T = 2.3$ sec

$$\therefore \quad \log_e\frac{500}{50} = 100\frac{K.2.3}{2}$$

$$\frac{K}{2} = \frac{1}{100} = 0.01$$

Relaxation time $$T_r = \frac{1}{K/2} = \frac{1}{0.01} = 100 \text{ sec}$$

log decrement $$\lambda = \frac{K}{2}T = 0.01\times 2.3 = 0.023 \text{ sec}$$

Amplitude y_0 when there is no damping is given by

$$y_0 = A_1\left(1+\frac{\lambda}{4}\right)$$

$$= 500\left(1+\frac{1\times 0.01\times 2.3}{4}\right) \text{ mm} = 502.875 \text{ mm}$$

correction for first displacement = (502.875 – 500) mm

= 2.875 mm

6. A mass of 20 gm is acted by a restoring force of 5 dynes/cm and a resisting force of 2 dyne-sec/cm.
   (*i*) Find whether the motion in aperiodic or oscillatory.
   (*ii*) What value of the resisting force will make the motion critically damped (with the restoring force remaining the same)? *[C.U. 2010]*

or $$V_0 = \rho(n' \sin\alpha - b\cos\alpha)$$

Given mass m = 20 gm

Restorning force μ = 5 dynes/cm

Resisting force R = 2 dyne-sec/cm

$$\therefore \quad \omega^2 = \frac{\mu}{m} = \frac{5}{20} = 0.25 \text{ rad}^2/\text{sec}^2$$

and $$K = \frac{R}{m} = \frac{2}{20} = 0.1 \text{ neper/cm}$$

(i) $\therefore$ $$\frac{K^2}{4} < \frac{(0.1)^2}{4} = 0.0025$$

so $$\frac{K^2}{4} < \omega^2$$

$\therefore$ The motion will be oscillatory.

(ii) For critical damping

$$\frac{K}{4} = \omega$$

$$\therefore \quad K = 2\omega$$

or $$\frac{R}{m} = 2\sqrt{\frac{\mu}{m}}$$

or $$R = 2\sqrt{\mu m}$$

$$= 2\sqrt{5.20}$$

$$= 20 \text{ dyne-sec/cm}$$

REVIEW QUESTIONS

1. Write the differential equation for the damped natural oscillation of a particle and solve it. Give the conditions for overdamped, critical damping and under damping motion.
2. Write down the equation of damped simple harmonic oscillatior. Find the expression for the displacement and discuss when we get oscillatory damped simple harmonic motion [Nagpur. U. 2001]
3. Set up the differential equation of motion of a simple harmonic oscillator subjected to a damping force and an external simple harmonic force. Obtain the expression for the amplitude and the phase angle of the displacement in the steady state. [C.U. 1996]
4. Obtain the expression for the velocity amplitude of the forced vibration. What is the value of frequency for velocity resonance. Show that the velocity of a particle is in phase with the driving force at velocity resonance.
5. Show that in the steady state the time averaged input power is equal to time averaged power dissipated through friction. [C.U. 1996]
6. Derive an expression for the average power supplied to a forced oscillator by an external driving force $F = F_0 \cos pt$.
7. Distinguish between amplitude resonance and velocity resonance. Show that at velocity resonance the maximum velocity is inversely proportional to damping and the velocity of the oscillator is in phase with driving force.

8. What is sharpness of resonance? Explain the effect of damping on sharpness of resonance. [Meerut U. 2003, Nagpur U. 2003, P.U. 1995]
9. Solve the differential equation $M\ddot{x} + R\dot{x} + Sx = F\cos\omega t$ for a forced vibration.
10. What is log decrement of a damped oscillatory motion. How can damping coefficient be determined of such a system?
11. Give the theory of forced vibration clearly distinguishing between steady and unsteady state. Show that at the steady state, the average rate of work done by the external force is equal to the rate of dissipation of energy.
12. Discuss the theory of forced vibration of a damped system. Explain from the physical consideration when the frequency of the forced vibration is the same as the external force. What is sharpness of resonance? Give examples of sharp resonance curve and flat resonance curve.
13. Explain the following terms relating to forced vibration (*a*) Transient beat (*b*) Mechanical impedance (*c*) Half Power value (*d*) Band width.
14. What is meant by a damped harmonic oscillator? Obtained an expression for average energy dissipation of damped harmonic oscillation. [Pandit Ravisankar University 2010]
15. What do you mean by band width of Resonance? Find an expression for it.

PROBLEMS

1. Show that for a forced vibration (from eqn. $M\ddot{x} + R\dot{x} + Sx = F\cos\omega t$ the total energy of the vibrating system is not constant. Prove that for such a case

$$\frac{\text{Average potential energy}}{\text{Average kinetic energy}} = \frac{\omega_0^2}{\omega^2}$$

$$\text{where } \omega_0 = \sqrt{\frac{S}{M}}$$

2. In one dimensional motion of a mass of 10 gm, it is acted on by a restoring force of 10 dyne per cm and a resisting force of 2 dyne sec/ cm
   (*i*) Find whether the motion is periodic or oscillatory.
   (*ii*) Find the value of resisting force which will make the motion critically damped.
   (*iii*) Find the value of mass for which given forces will make the motion critically damped.
   (*iv*) Suppose that the mass is given an initial impulse of 20 gm cm/sec while at rest. Obtain an expression for the displacement '*x*' at a time *t*. *[C.U. 1990]*
   [**Ans.** (*i*) Motion is oscillatory, (*ii*) Resisting force per unit velocity = 20 dyne sec/cm (*iii*) Mass = 0.1 gm and (*iv*) $x = \frac{2}{\sqrt{0.09}} e^{0.t} \sin\sqrt{0.99}t$ cm]
3. A particle of mass 0.002 kg is capable of executing a damped simple harmonic motion and is acted on by an impulsive force while at rest. Its successive maximum displacements on the opposite side immediately after it starts to move are 0.13 m and 0.1 m respectively. Find the logarithmic decrements and mechanical resistance if the time period is 10 sec. What would be the amplitude of oscillation when there is no damping?
   [**Ans.** 0.263, 0.0002 kg/sec., 0.1482 m]
4. Express amplitude, energy, logarithmic decrement and relaxation time in terms of Q the quality factor.
   [**Ans.** $A = A_0 e^{-\frac{\omega_1}{2Q}t}, E = E_0 e^{-\frac{\omega' t}{Q}}, \lambda = \frac{\pi}{Q}, t_r = \frac{2Q}{\omega'}$]
5. Find an expression for the power and quality factor of a damped harmonic oscillator.
   [N.H. Bahuguna Garhwal University 2010]

CHAPTER 4

COUPLED VIBRATION

4.1. COUPLED VIBRATION

It is seen that in the case of forced vibration the driving force is in no way affected by the driven system. The amplitude and frequency of the driving force remains unaltered.

But the role of the driver and driven will be changed, when two systems are comparable. Motion of each system will react on that of the other, when two systems are same in mass or inertia. The vibration is coupled vibration. In the coupled vibration two or more oscillators are linked together in such a way that an exchange of energy transfer takes place between them.

The phenomenon can be very easily demonstrated.

Two similar simple pendulums are suspended from an India rubber cord, *EF* attached between two rigid uprights. Now the pendulum *B* is set in motion by initial displacement. The oscillation of the pendulum *B* provides a periodic force to the elastic support *EF*. The vibration of the cord supplies periodic forces to the other pendulum *D* at the point of suspension. The pendulum *D* begins to oscillate. The amplitude of *D* gradually increases while the amplitude of *B* gradually diminishes and almost comes to rest. This goes on to a certain point when reverse process sets in, the amplitude of *D* slows down momentarily, comes to rest and that of *B* increases. This process will be repeated, *B* acts as driver and *D* is driven. Next *D* acts as a driver and *B* as driven.

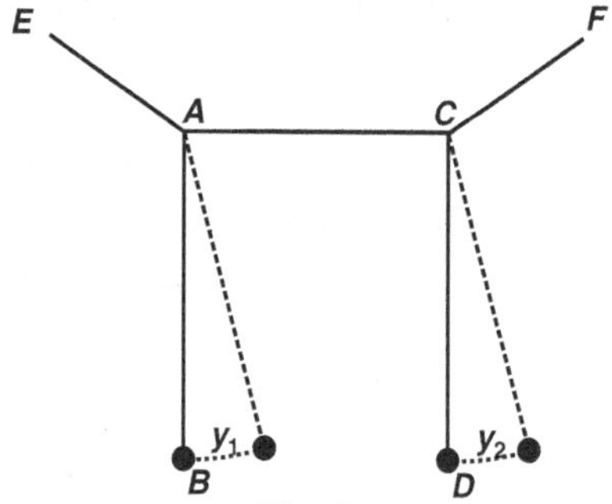

Fig. 1.

There is feedback of energy from driven to the driver. The feedback energy is lesser if the difference between the frequency or oscillations of the two is large and the coupling between the two is said to be 'loose'. If the feedback is large the coupling is said to be 'tight'.

The time displacement curves of these two pendulums are shown in Fig. 2(a) represents the time displacement curve of *B* started by initial displacement and Fig. 2(b) represents that of pendulum *D*.

Let us consider two systems coupled together, oscillating in a direction *y*, let their masses be m_1 and m_2 and, μ_1 and μ_2 be their forces of restitution for unit displacement μ_{12} is the force of restitution on the first or the second for the unit displacement of the second or the first, which may be called coupling coefficient.

The equation of motions of two systems are given by

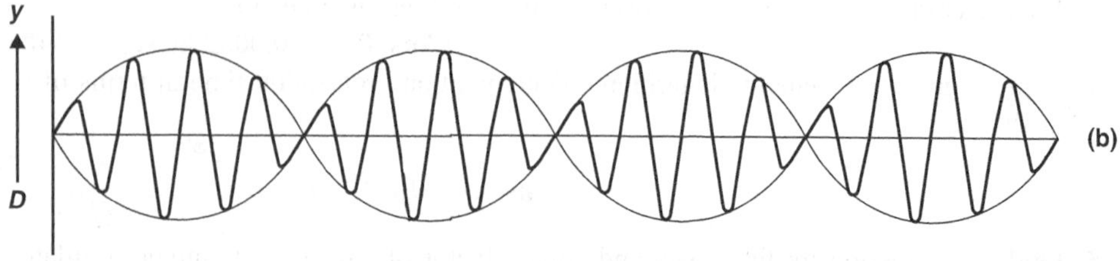

Fig. 2.

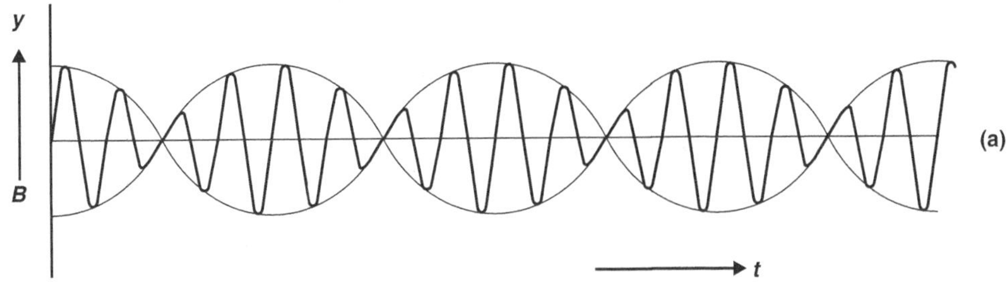

Fig. 2.

$$m_1 \frac{d^2 y_1}{dt^2} + \mu_1 y_1 + \mu_{12} y_2 = 0 \quad \text{... (1)}$$

$$m_2 \frac{d^2 y_2}{dt^2} + \mu_2 y_2 + \mu_{12} y_1 = 0 \quad \text{... (2)}$$

Dividing the equation (1) by $\sqrt{m_1}$ and (2) by $\sqrt{m_2}$ have

$$\sqrt{m_1} \frac{d^2 y_1}{dt^2} + \frac{\mu_1}{\sqrt{m_1}} y_1 + \frac{\mu_{12}}{\sqrt{m_1}} y_2 = 0 \quad \text{... (3)}$$

$$\sqrt{m_2} \frac{d^2 y_2}{dt^2} + \frac{\mu_2}{\sqrt{m_2}} y_2 + \frac{\mu_{12}}{\sqrt{m_2}} y_1 = 0 \quad \text{... (4)}$$

Put

$$Y_1 = \sqrt{m_1} y_1 \quad \text{... (5)}$$

$$Y_2 = \sqrt{m_2} y_2 \quad \text{... (6)}$$

So from the equations (3), (4), (5) and (6) we have

$$\frac{d^2 Y_1}{dt^2} + \omega_1^2 Y_1 + \omega^2 Y_2 = 0 \quad \text{... (7)}$$

$$\frac{d^2 Y_1}{dt^2} + \omega_2^2 Y_2 + \omega^2 Y_1 = 0 \quad \text{... (8)}$$

where $\omega_1^2 = \frac{\mu_1}{m_1}, \omega_2^2 = \frac{\mu_2}{m_2}$ and $\omega^2 = \frac{\mu_{12}}{\sqrt{m_1}, \sqrt{m_2}}$

where $\frac{\omega_1}{2\pi}$ and $\frac{\omega_2}{2\pi}$ are the natural frequency of the system respectively.

Both the systems are assumed to oscillate with the same frequency $\frac{p}{2\pi}$ when of course the oscillations are simple harmonic. We assume

$$Y_1 = A\, e^{ipt} \quad \text{...(9)}$$

$$Y_2 = B\, e^{ipt} \quad \text{...(10)}$$

The simple harmonic motion associated with each normal coordinate Y_1 or Y_2 is called a normal mode of the coupled system.

Substituting equations (9) and (10) in equations (7) and (8) we get

$$(\omega^2_1 - p^2)\, A = -\omega^2\, B \quad \text{... (11)}$$

$$(\omega_2^2 - p^2)\, B = -\omega^2 A \qquad \text{... (12)}$$

Multiplying equations (11) and (12) we get

$$(\omega_1^2 - p^2)\,(\omega_2^2 - p^2) = \omega^4$$

or $(\omega_1^2 + \omega_2^2)\,(\omega_1^2\,\omega_2^2 - \omega^4) = 0$ ... (13)

Solving equation (13)

$$p^2 = \frac{1}{2}\left(\omega_1^2 + \omega_2^2\right) \pm \frac{1}{2}\sqrt{\left(\omega_1^2 + \omega_2^2\right)^2 - 4\left(\omega_1^2\omega_2^2 - \omega^4\right)}$$

or

$$p^2 = \frac{1}{2}\left(\omega_1^2 + \omega_2^2\right) \pm \frac{1}{2}\sqrt{(\omega_1^2 - \omega_2^2)^2 + 4\omega^4} \qquad \text{... (14)}$$

So p has two values p_1 and p_2 and is given bu

$$p_1 = \left[\frac{1}{2}\left(\omega_1^2 + \omega_2^2\right) + \frac{1}{2}\sqrt{(\omega_1^2 - \omega_2^2)^2 + 4\omega^4}\right]^{1/2} \qquad \text{... (15)}$$

and

$$p_2 = \left[\frac{1}{2}\left(\omega_1^2 + \omega_2^2\right) - \frac{1}{2}\sqrt{(\omega_1^2 - \omega_2^2)^2 + 4\omega^4}\right]^{1/2} \qquad \text{... (16)}$$

p_1 is greater than and p_2 is less than either of ω_1 and ω_2. The effect of coupling is to spread apart the natural frequencies of the coupled system. Now when the natural frequencies of the two are equal i.e.,

$$\omega_1 = \omega_2$$

and

$$p^2 = \omega_1^2 \pm \omega^1$$

Hence

$$p_1 = \sqrt{\omega_1^2 + \omega^2} = \omega_1\left(1 + \frac{\omega^2}{\omega_1^2}\right)^{1/2}$$

$$= \omega_1 + \frac{\omega^2}{2\omega_1} \text{ approx.} \qquad \text{... (17)}$$

Similarly

$$p_2 = \sqrt{\omega_1^2 - \omega^2} = \omega_1 - \frac{\omega^2}{2\omega_1} \text{ approx} \qquad \text{.... (18)}$$

whence

$$p_1 + p_2 = 2\omega_1 \qquad \text{... (19)}$$

and

$$p_1 - p_2 = \frac{\omega^2}{\omega_1} \qquad \text{... (20)}$$

where $\frac{p_1}{2\pi}$ and $\frac{p_2}{2\pi}$ are two modified frequencies of the coupled system when coupled together. So the general solution can be written as

$$Y_1 = A_1 e^{ip_1 t} + A_2 e^{ip_2 t} \qquad \text{... (21)}$$

$$Y_2 = B_1 e^{ip_1 t} + B_2 e^{ip_2 t} \qquad \text{... (22)}$$

From equations (5) and (6) we get

$$y_1 = \frac{Y_1}{\sqrt{m_1}} = \frac{1}{\sqrt{m_1}}\left[A_1 e^{ip_1 t} + A_2 e^{ip_2 t}\right]$$

$$= \frac{1}{\sqrt{m_1}}\left[A_1\left(\cos p_1 t + i\sin p_1 t\right) + A_2\left(\cos p_2 t + i\sin p_2 t\right)\right]$$

or $$y_1 = a_1 \cos p_1 t + a_2 \sin p_1 t + a_3 \cos p_2 t + a_4 \sin p_2 t \quad \text{... (23)}$$

where $$a_1 = \frac{A_1}{\sqrt{m_1}}, a_2 = \frac{iA_1}{\sqrt{m_1}}, a_3 = \frac{A_2}{\sqrt{m_1}} \text{ and } a_4 = \frac{iA_2}{\sqrt{m_1}}$$

Similarly

$$y_2 = b_1 \cos p_1 t + b_2 \sin p_1 t + b_3 \cos p_2 t + b_4 \sin p_2 t \quad \text{... (24)}$$

and $$b_1 = \frac{B_1}{\sqrt{m_2}}, b_2 = \frac{iB_1}{\sqrt{m_2}}, b_3 = \frac{B_2}{\sqrt{m_2}} \text{ and } b_4 = \frac{iB_2}{\sqrt{m_2}}$$

The constants a_1, a_2, a_3 and a_4 and b_1, b_2, b_3 and b_4 are evaluated from initial conditions. If the particle m_1 starts from rest by initial displacement i.e., at

$$t = 0, y_1 = (y_1)_0, y_2 = 0, \frac{dy_1}{dt} = 0, \frac{dy_2}{dt} = 0$$

We have from equation (23)

$$a_1 + a_3 = (y_1)_0$$

At the beginning there is no coupling and $p_1 = p_2$ as a special case

$$a_1 = a_3 = \frac{(y_1)_0}{2}$$

$$a_2 = a_4 = 0$$

$$b_1 + b_3 = 0 \text{ gives}$$

or $$b_1 = -b_3 = b \text{ say}$$

$$b_2 = b_4 = 0$$

From equations (23) and (24)

$$y_1 = \frac{(y_1)_0}{2}(\cos p_1 t + \cos p_2 t)$$

$$y_2 = b(\cos p_1 t - \cos p_2 t)$$

or $$y_1 = (y_1)_0 \cos\frac{p_1 + p_2}{2}t \cos\frac{p_1 - p_2}{2}t \quad \text{... (25)}$$

and $$y_2 = 2b \sin\frac{p_1 + p_2}{2}t \sin\frac{p_1 - p_2}{2}t \quad \text{... (26)}$$

From equations (19), (20), (25) and (26)

$$y_1 = (y_1)_0 \cos\omega_1 t \cos\frac{\omega^2}{2\omega_1}t \quad \text{... (27)}$$

$$y_2 = 2b\sin\omega_1 t \sin\frac{\omega_2}{2\omega_1}t \quad \text{... (28)}$$

Substituting the values of p_1^2 and p_2^2 from equations (17) and (18) in equations (11) and (12) we get

$$\frac{A}{B} = +1 \quad \text{or} \quad A = B$$

In other words the amplitude of Y_1 is equal to the amplitude Y_2.

As $$Y_1 = \sqrt{m_1}y_1, \; Y_2 = \sqrt{m_2}y_2$$

So the amplitude of $\sqrt{m_1}y_1$ is equal to the amplitude of $\sqrt{m_2}y_2$. Hence the amplitude of $y_1 = \sqrt{\frac{m_2}{m_1}}$, amplitude of y_2. But the amplitude of $y_1 = \frac{(y_1)_0}{2}$ and $y_2 = b$

so $$\frac{(y_1)_0}{2} = \sqrt{\frac{m_2}{m_1}}b \text{ or } b = \frac{(y_1)_0}{2}\sqrt{\frac{m_1}{m_2}}$$

Hence $$y_1 = (y_1)_0 \cos\frac{\omega^2}{2\omega_1}t\cos\omega_1 t \quad ...(29)$$

$$y_2 = \sqrt{\frac{m_1}{m_2}}.(y_1)_0 \sin\frac{\omega^2}{2\omega_1}t.\sin\omega_1 t \quad ...(30)$$

These two equations (29) and (30) represent the oscillation of two systems where they are coupled together. The time displacement of the two systems are shown in Fig. 2.

4.2. COUPLED VIBRATION OF TWO LOADS ATTACHED TO A STRING UNDER TENSION.

Let two loads of masses m_1 and m_2 are attached at B and C of the string $ABCD$ fixed at A and D. The string is taken horizontal and weightless compared to the masses attached. The action due to gravity on the load is neglected. The length of the string is l stretched under tension T which assumed to remain constant during the motion of the load. Let $AB = a$, $BC = b$ and $CD = c$ such that $a + b + c = l$

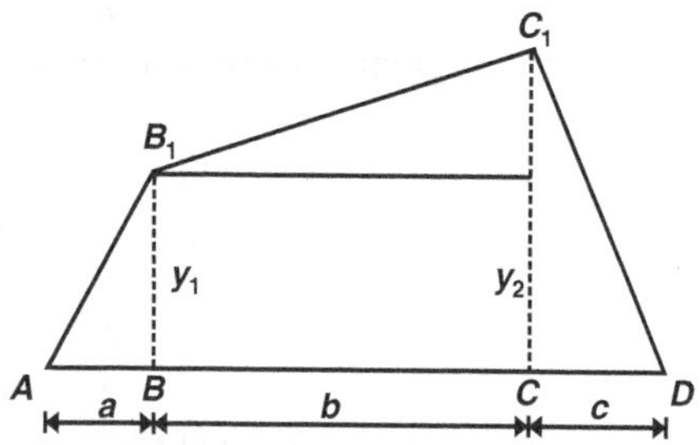

Fig. 3.

Let y_1 and y_2 be displacement of the loads at any time t which is small, B_1 and C_1 represent the displaced position of the load. Algebraic sum of the total vertical component of tension at B_1.

$$- T\sin\angle B_1AB + T\sin\angle C_1B_1N$$

$$= -T\frac{y_1}{a} + T\frac{y_2 - y_1}{b} \quad \text{as angles are small}$$

$$= T\frac{y_2}{b} - Ty_1\left(\frac{1}{a}+\frac{1}{b}\right) \quad ...(31)$$

Similarly, algebraic sum of the vertical components of tensions at C_1

$$-T\frac{y_2}{c} - T\frac{(y_2 - y_1)}{b}$$

$$= T\frac{y_1}{b} - Ty_2\left(\frac{1}{c}+\frac{1}{b}\right) \quad ...(32)$$

These are the restoring forces on the masses m_1 and m_2

so $$m_1\frac{d^2y_1}{dt^2} + T\left(\frac{1}{a}+\frac{1}{b}\right)y_1 - \frac{T}{b}y_2 = 0 \quad ...(33)$$

$$m_2\frac{d^2y_2}{dt^2} + T\left(\frac{1}{c}+\frac{1}{b}\right)y_2 - \frac{T}{b}y_1 = 0 \quad ...(34)$$

Putting $T\left(\frac{1}{a}+\frac{1}{b}\right)=\mu_1$, $T\left(\frac{1}{c}+\frac{1}{b}\right)=\mu_2$ and $-\frac{T}{b}=\mu_{12}$ the equations (33) and (34) become

$$m_1\frac{d^2y_1}{dt^2}+\mu_1y_1+\mu_{12}y_2 = 0 \quad ...(35)$$

$$m_2\frac{d^2y_2}{dt^2}+\mu_2y_2+\mu_{12}y_1=0 = 0 \quad ...(36)$$

The equations are identical to those of the equations (1) and (2) [vide Art. 4.1] where the solutions have been worked out and which can be easily followed in this case.

Special Case

The masses $m_1 = m_2 = m$ (say) and the segment of the string $a = b = c = \frac{l}{3}$. Putting these values in equations it can be easily obtained.

$$m\frac{d^2y_1}{dt^2}+\frac{6T}{l}y_1-\frac{3T}{l}y_2 = 0 \quad ...(37)$$

and $$m\frac{d^2y_2}{dt^2}+\frac{6T}{l}y_2-\frac{3T}{l}y_1 = 0 \quad ...(38)$$

Adding equations (37) and (38)

$$m\frac{d^2}{dt^2}(y_1+y_2)+\frac{3T}{l}(y_1+y_2) = 0 \quad ...(39)$$

Subtracting equations (37) and (38)

$$m\frac{d^2}{dt^2}(y_1-y_2)+\frac{9T}{l}(y_1-y_2) = 0 \quad ...(40)$$

Whence the solutions of equations (39) and (40) are

$$y_1+y_2 = a_1\cos\sqrt{\frac{3T}{ml}}t+b_1\sin\sqrt{\frac{3T}{ml.}}t \quad ...(41)$$

and $$y_1-y_2 = a_2\cos\sqrt{\frac{9T}{ml}}t+b_2\sin\sqrt{\frac{9T}{ml}}t \quad ...(42)$$

From equations (41) and (42)

$$y_1 = \frac{a_1}{2}\cos\sqrt{\frac{3T}{ml}}t+\frac{b_1}{2}\sin\sqrt{\frac{3T}{ml}}t+\frac{a_2}{2}\cos\sqrt{\frac{9T}{ml}}t+\frac{b_2}{2}\sin\sqrt{\frac{9T}{ml}}t \quad ...(43)$$

and $$y_2 = \frac{a_1}{2}\cos\sqrt{\frac{3T}{ml}}t+\frac{b_1}{2}\sin\sqrt{\frac{3T}{ml}}t-\frac{a_2}{2}\cos\sqrt{\frac{9T}{ml}}t-\frac{b_2}{2}\sin\sqrt{\frac{9T}{ml}}t \quad ...(44)$$

The constants a_1, b_1, a_2 and b_2 are determined from boundary conditions. Thus the masses will vibrate with frequencies.

$\frac{1}{2\pi}\sqrt{\frac{3T}{ml}}$ and $\frac{1}{2\pi}\sqrt{\frac{9T}{ml}}$ which depends on the signs of y_1 and y_2 when they begin to vibrate.

4.3. STIFFNESS COUPLED SYSTEM BY SPRINGS

Consider two masses m_1 and m_2 slide in a frictionless groove being attached by rigid supports by springs of stiffness constants s_1 and s_2. If y_1 and y_2 are the displacement of masses m_1 and m_2 at the instant t then the equations of motion with coupled system are

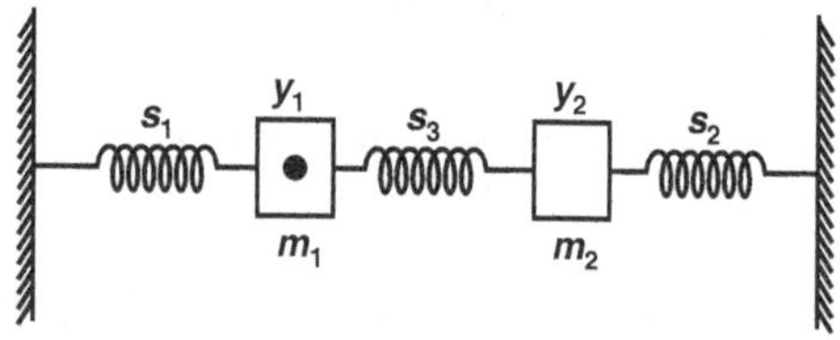

Fig. 4.

$$m_1\frac{d^2y_1}{dt^2}+s_1y_1+s_3(y_1-y_2) = 0 \quad ...(45)$$

and $$m_2\frac{d^2y_2}{dt^2}+s_2y_2+s_3(y_2-y_1) = 0 \quad ...(46)$$

Additional force $-s_3(y_1 - y_2)$ and $-s_3(y_2 - y_1)$ arises due to coupling. From equations (45) and (46) we get

$$m_1\frac{d^2y_1}{dt^2}+(s_1+s_3)y_1-s_3y_2 = 0 \quad ...(47)$$

and $$m_2\frac{d^2y_2}{dt^2}+(s_2+s_3)y_2-s_3y_1 = 0 \quad ...(48)$$

Putting $s_1 + s_3 = \mu_1$, $s_2 + s_3 = \mu_2$ and $-s_3 = \mu_{12}$ the equations (47) and (48) are the same from like equations (1) and (2). The solution of the equations will be like equations (29) and (30)

SOLVED PROBLEMS

1. Two particles of the same mass are attached to a weightless spring of length l with fixed ends stretched under a tension T. Find the normal frequencies of the vibration of the masses if they are attached at distances $\left(\frac{l}{3}\right)$ and $\left(\frac{2l}{3}\right)$ from one end. Assume that the tension in the string remains constant during the vibration.

This type of problem is worked out in thc Art. 4.2 as a special case.

The equations (37) and (38) are

$$m\frac{d^2y_1}{dt^2}+\frac{6T}{l}y_1-\frac{3T}{l}y_2 = 0 \quad ...(37)$$

and $$m\frac{d^2y_2}{dt^2}+\frac{6T}{l}y_2-\frac{3T}{l}y_1 = 0 \quad ...(38)$$

Adding equations (37) and (38) we have

$$m\frac{d^2}{dt^2}(y_1+y_2)+\frac{3T}{l}(y_1+y_2) = 0 \quad ...(39)$$

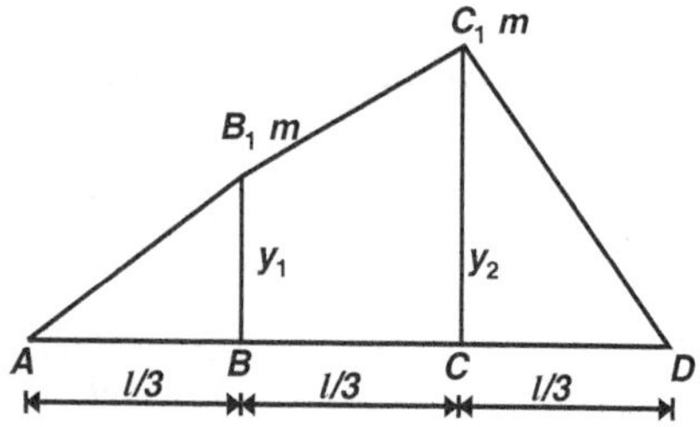

Fig. 4.

Subtracting equations (37) and (38) we have

$$m\frac{d^2}{dt^2}(y_1-y_2)+\frac{9T}{l}(y_1-y_2) = 0 \quad ...(40)$$

Whence the solution of equations (39) and (40) are

$$y_1 + y_2 = a_1 \cos\sqrt{\frac{3T}{ml}}t + b_1 \sin\sqrt{\frac{3T}{ml}}t \quad ...(41)$$

and
$$y_1 + y_2 = a_2 \cos\sqrt{\frac{9T}{ml}}t + b_2 \sin\sqrt{\frac{9T}{ml}}t \quad ...(42)$$

From equations (41) and (42) we get

$$y_1 = \frac{a_1}{2}\cos\sqrt{\frac{3T}{ml}}t + \frac{b_1}{2}\sin\sqrt{\frac{3T}{ml}}t + \frac{a_2}{2}\cos\sqrt{\frac{9T}{ml}}t + \frac{b_2}{2}\sin\sqrt{\frac{9T}{ml}}t \quad ...(43)$$

$$y_2 = \frac{a_1}{2}\cos\sqrt{\frac{3T}{ml}}t + \frac{b_1}{2}\sin\sqrt{\frac{3T}{ml}}t - \frac{a_2}{2}\cos\sqrt{\frac{9T}{ml}}t - \frac{b_2}{2}\sin\sqrt{\frac{9T}{ml}}t \quad ...(44)$$

So the normal frequencies of vibration of masses are

$$\frac{1}{2\pi}\sqrt{\frac{3T}{ml}} \text{ and } \frac{1}{2\pi}\sqrt{\frac{9T}{ml}}$$

2. Two masses m_1 and m_2 are connected by a massless spring of spring constant s. The system is free to oscillate along the length of the spring. Show that the system oscillates with a frequency $n = \frac{1}{2\pi}\sqrt{\frac{s}{\mu}}$.

where
$$\mu = \mu = \frac{m_1 m_2}{m_1 + m_2}.$$

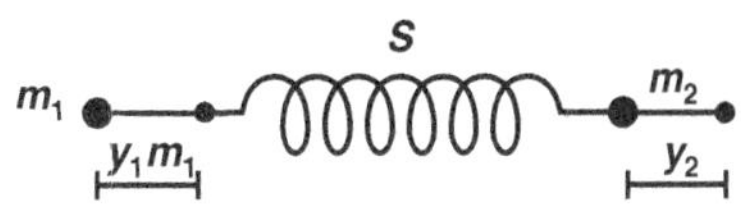

Fig. 5.

The Fig. 5 shows the displacement of the masses m_1 and m_2 from equilibrium position. Here the displacements are y_1 and y_2. The force exerted on the two masses are equal in magnitude and opposite in direction. So the equations of motion are

$$m_1\frac{d^2 y_1}{dt^2} = s(y_2 - y_1) \quad ...(45)$$

$$m_2\frac{d^2 y_2}{dt^2} = -s(y_2 - y_1) \quad ...(46)$$

Multiplying equation (45) by m_2 and equation (46) by m_1 and subtracting

$$m_1 m_2 \frac{d^2}{dt^2}(y_1 - y_2) = s(m_1 + m_2)(y_2 - y_1) \quad ...(47)$$

Putting $y_1 - y_2 = y$ in equation (47) we get

$$\frac{d^2 y}{dt^2} = -\frac{s}{\frac{m_1 m_2}{m_1 + m_2}}y = -\frac{s}{\mu}y$$

where
$$\mu = \frac{m_1 m_2}{m_1 + m_2}$$

or
$$\frac{d^2 y}{dt^2} = -\frac{s}{\mu}y \quad ...(48)$$

The equation (48) is the equation of S.H.M. So the frequency of oscillation is

$$n = \frac{1}{2\pi}\sqrt{\frac{s}{\mu}}$$

3. The equations of motion of two coupled oscillations are

$$\ddot{q}_1 + \sqrt{2}q_2 + 4q_1 = 0 \qquad ...(49)$$

$$\ddot{q}_2 + \sqrt{2}q_1 + 5q_2 = 0 \qquad ...(50)$$

Find normal frequencies and the ratio of amplitudes of the two normal modes.

[Burd. U. 2005]

Let $q_1 = Ae^{pt}$ and $q_2 = Be^{pt}$

where A and B are amplitudes.

$$\therefore \quad \ddot{q}_1 = Ap^2e^{pt} \text{ and } \ddot{q}_2 = Bp^2e^{pt}$$

Now substituting the values of $\ddot{q}_1$ and $\ddot{q}_2$ in equations (49) and (50) we get

$$Ap^2e^{pt} + \sqrt{2}Be^{pt} + 4Ae^{pt} = 0$$

or $$(p^2+4)A + \sqrt{2}B = 0 \qquad ...(51)$$

Similarly from equation (50) we get

$$\sqrt{2}A + (p^2+5)B = 0 \qquad ...(52)$$

From equations (51) and (52) non-zero solution of A and B solving equations (51) and (52) we get

$$(p^2 + 4)(p^2 + 5) - 2 = 0 \text{ or } (p^2 + 3)(p^2 + 6) = 0$$

or $$p^2 = -3 \text{ or } p^2 = -6$$

So the normal frequencies are $\omega_1 = \sqrt{3}$ or $\omega_2 = \sqrt{6}$

Putting the values of $p^2 = -3$ in equation (51) we get

$$A + \sqrt{2}B = 0$$

or $$\frac{A}{B} = -\sqrt{2}$$

and putting $p^2 = -6$ we get $\dfrac{A}{B} = \dfrac{1}{\sqrt{2}}$

The ratio of amplitudes of two normal modes is $(-\sqrt{2})$ or $\left(\dfrac{1}{\sqrt{2}}\right)$.

REVIEW QUESTIONS

1. Derive the expression of the displacement of two vibratory systems when coupled together. What is loose and tight coupling?
2. Two loads of masses m_1 and m_2 are attached to a string making the segment equal. Find the displacement of load at a given time.
3. Write the equation of motion of two masses coupled together and connected to two rigid supports by three springs of the same stiffness constant. Derive the characteristic equation and the frequencies of the normal modes of vibration. Write the general equations for the displacement of the masses in terms of the normal modes.
4. Give the characteristics of the in phase and out of phase mode of vibration of two identical coupled simple pendulums (oscillators).
5. How will you distinguish coupled oscillation from forced oscillation?

PROBLEMS

1. Potassium chloride is a diatomic molecule. The two atoms Potassium and chlorine are coupled by electrostatic forces which may be represented by a tiny massless spring. A potassium chloride molecule has a natural vibration frequency = 1.14×10^{13} Hz. Calculate the interatomic force constant. Mass of K atom = 39 a.m.u (1 a.m.u = 1.67×10^{-27} kg)

   [**Ans.** = 198.85 Nm^{-1}]

2. Two identical oscillators are coupled by a spring of stiffness constant s'. Calculate the frequencies and the ratio of the amplitudes for the normal modes and also the phase relation between the vibrations of the two oscillators in each mode.

   [**Ans.** $\sqrt{\frac{s}{m}}, \sqrt{\frac{s+2s'}{m}}$ and $\frac{A}{B} = \pm 1$]

3. Two identical particles each of mass M and lying on a smooth horizontal platform are connected to two identical massless springs of spring constant s. Determine the characteristic frequency for normal mode of oscillation.

   [**Ans.** $\frac{1}{2\pi}\sqrt{\frac{2s}{M}}$]

5 CHAPTER FOURIER ANALYSIS

5.1. FOURIER'S THEOREM AND FOURIER SERIES

Composition of any Number of S.H.M. Commensurate Frequencies

The vibration being in the same direction that is along the same straight line gives rise to Synthesis and Analysis of complex vibration. This leads to the development of Fourier's theorem which states that any single valued finite and continuous periodic function or has a finite number of finite discontinuities can be represented as a summation of simple harmonic vibrations having frequencies which are multiples of the frequency of the periodic function have a finite number of maxima a and minima. The conditions are known as Dirichlet's conditions.

Case I. Any finite periodic motion can be expressed as a sum of a series of Simple Harmonic Motions of commensurate frequencies

$$y_1 = a\sin \omega t$$

$$y_2 = b\cos \omega t$$

So $$y = y_1 + y_2 = a\sin \omega t + b\cos \omega t$$

The time displacement curve is drawn for each case and algebraic sum of the ordinate at any point gives the values of the ordinate at that point. These points are joined together and the curve so obtained represents the resultant time displacement curve of the two component vibration Fig. 1(*a*).

Case II. Let $$y_1 = a\sin \omega t$$

$$y_2 = \frac{a}{2}\sin 2\omega t$$

$$y = a \sin\omega t + \frac{a}{2}\sin 2\omega t$$

We obtain the resultant curve in the similar ways as shown in Fig. 1(b).

Case III. Let $$y_1 = a\sin \omega t$$

$$y_2 = \frac{a}{2}\sin 2\omega t$$

$$y_3 = \frac{a}{3}\sin 3\omega t$$

$$\therefore \quad y = y_1 + y_2 + y_3 = a\sin \omega t + \frac{a}{2}\sin 2\omega t + \frac{a}{3}\sin 3\omega t$$

The resultant curve is obtained in the similar way as above Fig. 1(*c*).

Thus a series

$$y = a_1\sin \omega t + a_2\sin 2\omega t + a_3\sin 3\omega t + ... + a_s \sin s\omega t$$
$$+ b_0 + b_1 \cos \omega t + b_2\cos 2\omega t + b_3\cos 3\omega t + ... + b_s \cos s\omega t$$

$$= \sum_{s=1}^{s} a_s \sin s\omega t + \sum_{s=1}^{s} b_s \cos s\omega t + b_0 \quad ... (1)$$

represents a resultant vibration of component S.H.M., b_0 represents the position of line of reference from which the displacement y is measured, that is, it measures the displacement of the axis of the curve from the coordinate axis.

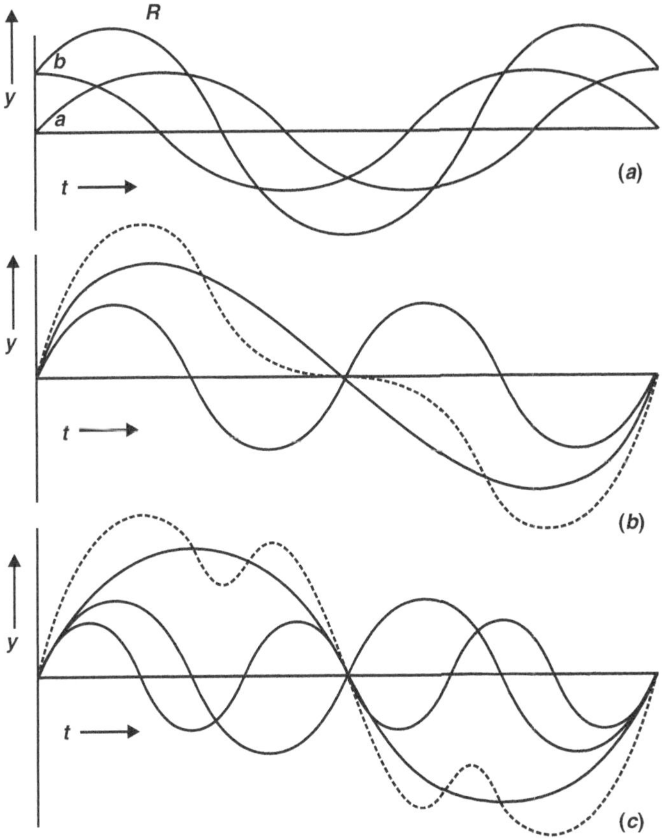

Fig. 1.

Fourier's series [Equation (1)] helps to obtain the component curve if the resultant curve is supplied. The method is known as Fourier Analysis. This is done by evaluating the constants in the Fourier series.

In accoustical vibrations the following are the conditions (*i*) the displacement be the single valued and continuous (*ii*) the displacement must always have a single finite value and has a finite number of maximum and minimum values and finite number of discontinuities and (*iii*) $\int_0^T /y/\, dt$ is finite. The above three conditions are called Dirichlet's conditions.

From equation (1) it is clear that the displacement repeats itself in a cycle when ωt changes by 2π. Hence the periodic time $T = \frac{2\pi}{\omega}$.

5.2. EVALUATION OF THE CONSTANTS OF THE FOURIER SERIES

To find b_0 integrate both sides of equation (1) from $t = 0$ to $t = T$ where T is the periodic time $\left(T = \frac{2\pi}{\omega}\right)$.

$$\int_0^T y\,dt = b_0\int_0^T dt + \sum_s \int_0^T a_s \sin s\omega t\,dt + \sum_s \int_0^T b_s \cos s\omega t\,dt \quad \text{... (2)}$$

or $$\int_0^T y\,dt = b_0 T$$

or $$b_0 = \frac{1}{T}\int_0^T y\,dt \quad \text{... (3)}$$

As $$a_s \int_0^T \sin s\omega t\; dt = \frac{a_s}{s\omega}[-\cos s\omega t]_0^T$$

$$= -\frac{a_s}{s\omega}[\cos s\omega T - \cos 0]$$

$$= -\frac{a_s}{s\omega}[1-1] = 0$$

and $$b_s \int_0^T \cos s\omega t\; dt = \frac{bs}{s\omega}[\sin s\omega t]_0^T$$

$$= \frac{bs}{s\omega}[\sin s\omega T - \sin 0] = 0 = 0$$

To find a_s multiply both sides of equation (1) by sin $s\omega t$ and integrate from $t = 0$ to $t = T$.

$$\int_0^T y \sin s\omega t\; dt = \int_0^T (a_1 \sin \omega t + a_2 \sin 2\omega t + ... + a_s \sin s\omega t)$$

$$.\sin s\omega t\; dt + \int_0^T (b_1 \cos \omega t + b_2 \cos 2\omega t + ... + b_s \cos s\omega t).\sin s\omega t\; dt + \int_0^T b_0 \sin s\omega t\; dt$$

As $$\int_0^T \sin k\omega t \sin s\omega t\; dt = 0 \text{ when } k \neq s$$

$$= \frac{T}{2} \text{ when } k = s$$

and $$\int_0^T \cos k\omega t \sin s\omega t\; dt = 0 \text{ for all values of } k$$

$\therefore$ $$\int_0^T y \sin s\omega t\; dt = a_s \frac{T}{2}$$

and $$\int_0^T b_0 \sin s\omega t\; dt = 0 = 0 \text{ or } a_s = \frac{2}{T}\int_0^T y \sin s\omega t\; dt \quad \text{... (4)}$$

Multiplying both sides of equation (1) by cos $s\omega t$ and integrating from $t = 0$ to $t = T$.

$$\int_0^T y\cos s\omega t\,dt = \int_0^T (a_1 \sin\omega t + a_2 \sin 2\omega t + \ldots + a_s \sin s\omega t)$$

$$\cos s\omega t\,dt + \int_0^T (b_1 \cos\omega t + b_2 \cos 2\omega t + \ldots + bs\cos s\omega t).\cos s\omega t\,dt$$

As $\int_0^T \sin k\omega t \cos s\omega t\,dt = 0$ for all values of k and s

and $\int_0^T \cos k\omega t\cos s\omega t\,dt = 0$ when $k \neq s$

$$= \frac{T}{2} \text{ when } k = s$$

and $\int_0^T \cos s\omega t\,dt = 0$

$\therefore$ $\int_0^T y\cos s\omega t\,dt = b_s \frac{T}{2}$

or $b_s = \frac{2}{T}\int_0^T y\cos s\omega t\,dt$ ... (5)

5.3. EVEN AND ODD FUNCTION

An even function like cos x whose graph for negative x is a reflection in the y-axis of its graph for positive x as shown in Fig. 2. or even function of $f(x)$

$$f(x) = f(-x) \quad \ldots (6)$$

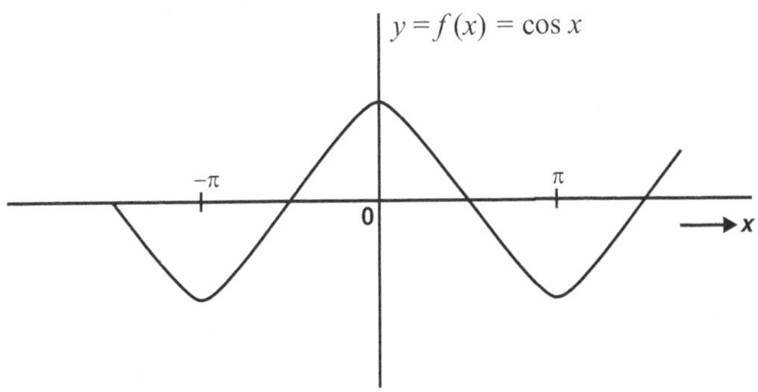

Fig. 2.

An odd function is like sin x for which the values of $f(x)$ and $f(-x)$ are negative of each other as shown in Fig. 3. For odd function of $f(x)$

$$f(x) = -f(-x) \quad \ldots (7)$$

From equation (5)

$$b_s = \frac{2}{T}\int_0^T y\cos s\omega t\,dt$$

is even function and from equation (4)

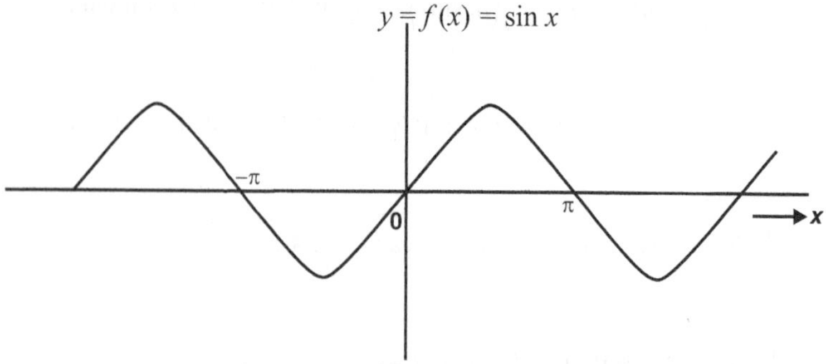

Fig. 3.

$$a_s = \frac{2}{T}\int_0^T y \sin s\omega t \, dt$$

is an odd function.

5.4. SOME EXAMPLES

The theorem is applied both for analysis and synthesis of complex curves. To find the component vibrations from the resultant one, is known as analysis of the curve by the Fourier's theorem. To find the resultant vibration from the component vibration is known as synthesis.

Example: The nature of the displacement curve is given by

$$y = C \text{ from } t = 0 \text{ to } t = \frac{T}{2} \quad \text{... (8)}$$

$$y = 0 \text{ from } t = \frac{T}{2} \text{ to } t = T \quad \text{... (9)}$$

From equation (3) (Art. 5.2) we get

$$b_0 T = \int_0^T y\, dt = \int_0^{T/2} y dt + \int_{T/2}^{T} y dt$$

or
$$b_0 T = \int_0^{T/2} C\, dt$$

or
$$b_0 = \frac{C}{2} \quad \text{... (10)}$$

This means that the axis of the curve is in the line $y = \frac{C}{2}$

From equation (4) [Art. 5.2]

$$a_s = \frac{2}{T}\int_0^T y \sin s\omega t\, dt$$

$$= \frac{2}{T}\left[\int_0^{T/2} y \sin s\omega t\, dt + \int_{T/2}^{T} y \sin s\omega t\, dt\right]$$

$$= \frac{2}{T} C \int_0^{T/2} \sin s\omega t\, dt$$

$$= \frac{2}{T}\frac{C}{s\omega}[-\cos s\omega t]_0^{T/2}$$

$$= \frac{C}{s\pi}[1-\cos s\pi]$$

As $(1 - \cos s\pi) = 2$ when s is odd

$= 0$ when s is even

$\therefore$ $a_s = \dfrac{2C}{s\pi}$ when s is odd

$= 0$ when s is even ... (11)

From eq. (5) with the help of equations (8) and (9)

$$b_s = \frac{2}{T}\int_0^T y \cos s\omega t \, dt$$

$$= \frac{2}{T}\frac{C}{s\omega}[\sin s\omega t]_0^{T/2} = 0 \quad \text{... (12)}$$

Thus the cosine terms disappear as they have zero values.

Hence
$$y = \frac{C}{2} + \frac{2C}{\pi}\left(\sin \omega t + \frac{1}{3}\sin 3\omega t + \delta\frac{1}{5}\sin 5\omega t + ... +\right) \quad \text{... (13)}$$

Hence square type time displacement curve analysed into some odd sine terms, even sine terms and cosine terms disappear is shown in Fig. 4.

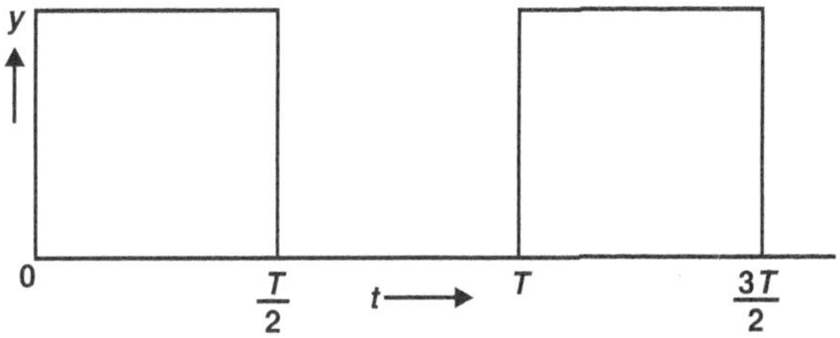

Fig. 4.

Synthesis. It is interesting to draw the above curve with the help of a few terms of the above trigonometric series to show how their sum approaches the value of y.

In the accompanying set of Fig. 5, the horizontal axis represents time axis and vertical axis the displacement. The heavy curve (a) represented by the thick line is given by the first term of the series. In (b) thin lower curve is given by the second term of the series and the thick curve above is given by the sum of two terms. In (c) the thin lower curve is given by the third term and the thick one above is given by the sum of three terms. Similarly in (d) the thin curve is given by the fourth term and thick one is given by the sum of four terms.

In (e) the curve represents the sum of 21 terms. Thus we see that the gradual increase of terms approximates to a straight line and square type time displacement curve is perfectly produced.

Case II. *The displacement is linear with time axes.*

$y = C$ at $t = 0$ and $y = 0$ at $t = T$

so
$$y = C\left(1-\frac{t}{T}\right)$$

when time axis passes through the lowest point of the periodic curve. From Art. 5.2., equations (3), (4) and (5)

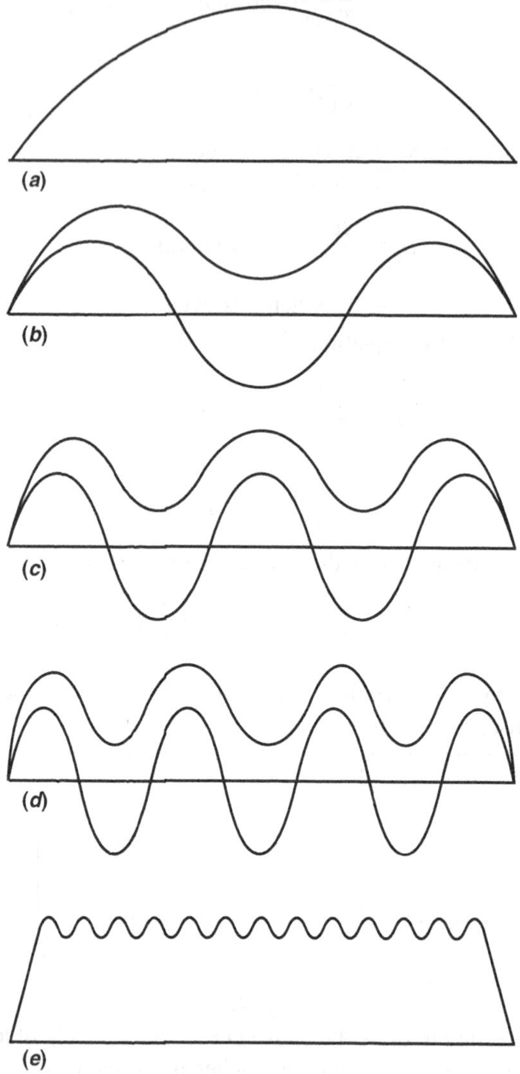

Fig. 5

$$b_0 = \frac{1}{T}\int_0^T y\,dt = \frac{C}{T}\int_0^T \left(1-\frac{t}{T}\right)dt = \frac{C}{2}$$

which gives the coordinate axes of the curve.

$$b_s = \frac{2C}{T}\int_0^T \left(1-\frac{t}{T}\right)\cos s\omega t\,dt = 0$$

$$a_s = \frac{2C}{T}\int_0^T \left(1-\frac{t}{T}\right)\sin s\omega t\,dt = \frac{2C}{\pi s}$$

$$\therefore \qquad y = \frac{2C}{\pi}\left(\sin\omega t + \frac{1}{2}\sin 2\omega t + \frac{1}{3}\sin 3\omega t + \ldots +\right) + \frac{C}{2}$$

The sum of the successive terms are shown as follows:

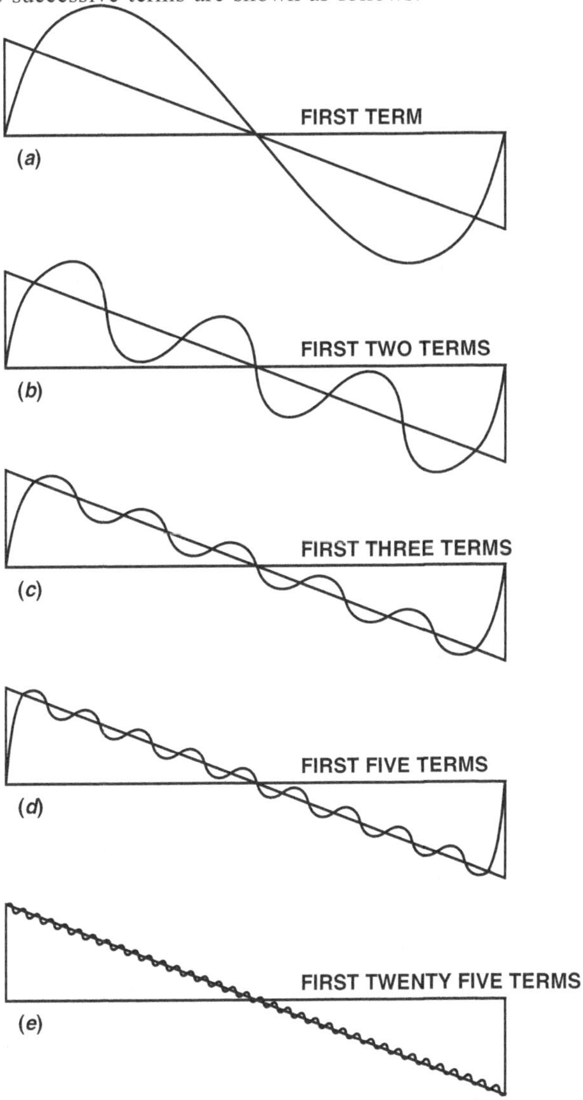

Fig. 6.

Case III. *Two straight lines meeting at* $\frac{T}{2}$ *and* $y = C$. *It passes through* $t = 0$ *and* $t = T$.

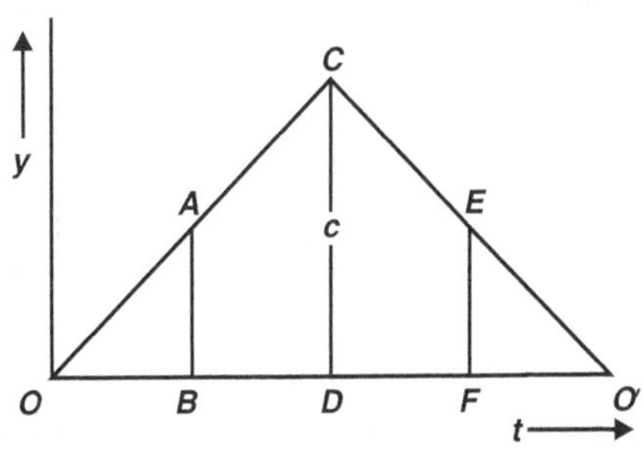

Fig. 7.

In Fig. 7. $OD = \ t = \dfrac{T}{2} \ \ CD = C$

so $$\frac{AB}{OD} = \frac{y}{t} = \frac{DC}{OD} = \frac{2C}{T}$$

or between $t = 0$ and $t = \dfrac{T}{2}$ $y = \dfrac{2Ct}{T}$

Similarly between $t = \dfrac{T}{2}$ to $t = T$ $y = \dfrac{2C}{T}(T - t) = 2C\left(1 - \dfrac{t}{T}\right)$

Hence
$$b_0 = \frac{1}{T}\left[\int_0^{T/2} \frac{2C}{T} t dt + \int_{T/2}^{T} \frac{2C}{T}\left(1 - \frac{t}{T}\right) dt\right]$$
$$= \frac{1}{T}\left[\left(\frac{2Ct^2}{2T}\right)_0^{T/2} + \frac{2C}{T}(t)_{T/2}^{T} - \frac{2C}{T}\left(\frac{t^2}{2}\right)_{T/2}^{T}\right]$$
$$= \frac{C}{2}$$
$$b_s = \left[\int_0^{T/2} \frac{2C}{T} t \cos s\,\omega t\, dt + 2C \int_{T/2}^{T} (1 - \frac{t}{T} \cos s\,\omega t) dt\right]$$
$= 0$ for even values of s

$= -\dfrac{2C}{\pi s^2}$ for odd values of s.
$$a_s = \frac{2}{T}\left[\frac{2C}{T}\int_0^{T/2} t \sin s\,\omega t dt + 2C \int_{T/2}^{T} \left(1 - \frac{t}{T}\right) \sin s\omega t\, dt\right]$$
$= 0$

Thus all sine terms disappear, only odd terms of cosine remains and the series becomes
$$y = \frac{C}{2} - \frac{4C}{\pi^2}\left\{\cos \omega t + \frac{1}{3^2}\cos 3\omega t + \frac{1}{5^2}\cos 5\omega t + ...\right\}$$

The resultant is obtained by adding the terms of the series. The series is rapidly convergent, so a few terms in the series will closely represent.

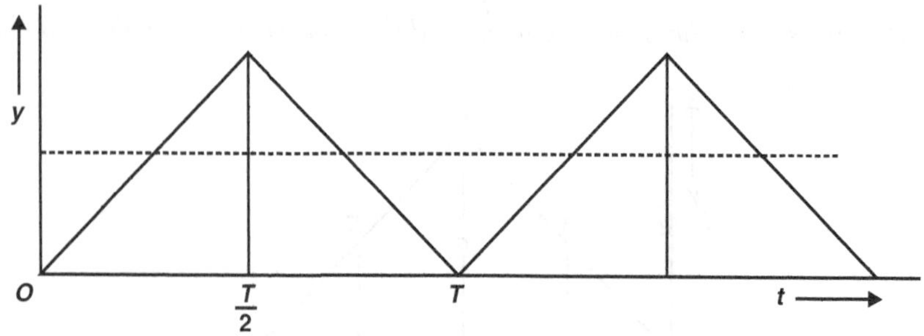

Fig. 8.

Fourier's theorem is applied for analysis and synthesis of not only simple types of periodic vibrations but also indicates the complexity of calculation when the curve between O and T is very irregular. In such cases the analysis by calculation is a long and tedious process. O.H. Taylor (Physical Review Vol. 6 Page 300, 1915) took trouble to prepare a chart of analysis which greatly minimises the trouble of the work. Instruments known as Harmonic Analyser has also been devised by many workers to minimise the trouble of analysis, as this Fourier series has wide application in many branches of Physics.

5.5. COMPLEX FORM OF FOURIER SERIES

Fourier series can be expressed by equation (1) as

$$y = \sum_{s=1}^{\infty} a_s \sin s\omega t + \sum_{s=1}^{\infty} b_s \cos s\omega t + b_0 \qquad \text{... (1)}$$

The sine and cosine can be expressed as

$$\sin s\omega t = \frac{1}{2i}\left(e^{is\omega t} - e^{-is\omega t}\right) \qquad \text{... (13)}$$

and

$$\cos s\omega t = \frac{1}{2}\left(e^{is\omega t} + e^{-is\omega t}\right) \qquad \text{... (14)}$$

Substituting equations 13 (a) and (14) in equation (1) wc get

$$y = \sum_{s=1}^{\infty} a_s \frac{e^{is\omega t} - e^{-is\omega t}}{2i} + \sum_{s=1}^{\infty} b_s \frac{e^{is\omega t} + e^{-is\omega t}}{2} + b_0$$

$$= \sum_{s=1}^{\infty} \frac{(b_s - ia_s)}{2} e^{is\omega t} + \sum_{s=1}^{\infty} \left(\frac{b_s + ia_s}{2}\right) e^{-is\omega t} + b_0 \qquad \text{... (15)}$$

Put

$$\frac{b_s - ia_s}{2} = C_s,$$

$$\frac{b_s + ia_s}{2} = C_{-s},$$

and $b_0 = C_0$ in equation (15) we get

$$y = \sum_{s=1}^{\infty} C_s e^{is\omega t} + \sum_{s=1}^{\infty} C_{-s} e^{-is\omega t} + C_0$$

$$= \sum_{s=-\infty}^{\infty} C_s e^{is\omega t} \qquad \text{... (16)}$$

The equation (16) is referred to as exponential form of Fourier series.

C_0 and C_s can be obtained as follows

$$C_0 = \frac{1}{T}\int_0^T y \, dt,$$

and

$$C_s = \frac{1}{T}\int_0^T y e^{-is\omega t} dt \qquad \text{... (17)}$$

This equation for C_s holds irrespective of whether s is positive, negative or zero.

5.6. FOURIER EXPANSION OF FUNCTION OTHER THAN TIME

If the Fourier function may be expanded not involving time and may be expanded in independent variable x such that $y = f(x)$ is expanded in the range $x = -l$ to $x = +l$ or $x = 0$ to $x = 2l$

We put $\omega = \frac{2\pi}{2l} = \frac{\pi}{l}$, then the equation (1) is given by

$$f(x) = \sum_{s=1}^{\infty} a_s \sin\frac{s\pi x}{l} + \sum_{s=1}^{\infty} b_s \cos\frac{s\pi x}{l} + b_0 \quad \text{... (18)}$$

The constants are

$$b_0 = \frac{1}{2l}\int_{-l}^{+l} f(x)dx$$

$$a_s = \frac{1}{l}\int_{-l}^{+l} f(x)\sin\frac{s\pi x}{l}dx$$

$$b_s = \frac{1}{l}\int_{-l}^{+l} f(x)\cos\frac{s\pi x}{l}dx$$

If the interval $(-\pi, \pi)$the constants are

$$b_0 = \frac{1}{2\pi}\int_{-\pi}^{\pi} f(x)dx$$

$$a_s = \frac{1}{\pi}\int_{-\pi}^{+\pi} f(x)\sin\frac{s\pi x}{l}dx$$

$$b_s = \frac{1}{\pi}\int_{-\pi}^{\pi} f(x)\cos\frac{s\pi x}{l}dx$$

5.7. (A) FOURIER TRANSFORMS

If function $f(x)$ of variable x the Fourier transform of function $f(x)$ as defined by $F(k)$ is given by

$$F(k) = \frac{1}{\sqrt{2\pi}}\int_{-\infty}^{\infty} e^{-ikx} f(x)dx \quad \text{... (19)}$$

The Fourier integral theorem says that if a function $f(x)$ satisfies the Dirichlet condition on every finite interval and if $\int_{-\infty}^{\infty} |f(x)|\, dx$ is finite then $f(x)$ can be written in the form of complex Fourier integral as

$$f(x) = \frac{1}{2\pi}\int_{-\infty}^{\infty} dk \int_{-\infty}^{\infty} f(t)e^{ik(x-t)}dt \quad \text{... (21)}$$

Rearranging the equation (21) we have

$$f(x) = \frac{1}{\sqrt{2\pi}}\int_{-\infty}^{\infty} dke^{ikx}\left[\frac{1}{\sqrt{2\pi}}\int_{-\infty}^{\infty} e^{-ikt} f(t)dt\right] \quad \text{... (22)}$$

$$= \frac{1}{\sqrt{2\pi}}\int_{-\infty}^{\infty} dke^{ikx} F(k) \quad \text{... (23)}$$

where $$F(k) = \frac{1}{\sqrt{2\pi}}\int_{-\infty}^{\infty} e^{-ikt} f(t)dt$$

The equation (22) is often referred to as Eourier inversion formula.

(b) Fourier Sine Transforms

We define $f_s(x)$ and $F_s(k)$ a pair of Fourier sine transforms representing odd functions by equations

$$f_s(x) = \sqrt{\frac{2}{\pi}} \int_0^\infty F_s(k) \sin kx \, dk \quad \text{... (23)}$$

where
$$F_s(k) = \sqrt{\frac{2}{\pi}} \int_0^\infty f_s(t) \sin kt \, dt \quad \text{... (24)}$$

(c) Fourier Cosine Transforms

We define $f_s(x)$ and $F_s(k)$ a pair of cosine transforms representing even functions by the equations

$$f_s(x) = \sqrt{\frac{2}{\pi}} \int_0^\infty F_s(k) \cos kx \, dk \quad \text{... (25)}$$

and
$$F_s(k) = \sqrt{\frac{2}{\pi}} \int_0^\infty f(t) \cos kx \, dt \quad \text{... (26)}$$

(d) Properties of Fourier Transforms

(*i*) Linear property: If $F(k)$ and $G(k)$ are Fourier transforms of $f(x)$ and $g(x)$ respectively then

$$F\,[\alpha f(x) + \beta g(x)] = \alpha F(k) + \beta G(k) \quad \text{... (27)}$$

where α and β are constants

(*ii*) Change of scale property. If $F(k)$ is the complex Fourier transform of $f(x)$, then

$$F\,[f(\alpha x)] = \frac{1}{\alpha} F\left(\frac{k}{a}\right), \alpha \neq 0 \quad \text{... (28)}$$

(*iii*) Shifting property: If $F(k)$ is the complex Fourier transform of $f(x)$ then

$$F[f(x-a)] = e^{ika} F(k) \quad \text{... (29)}$$

5.8. THE DIRAC DELTA FUNCTION AND FOURIER TRANSFORMS

The Fourier integral representation of a function

$$F(k) = \frac{1}{\sqrt{2\pi}} \int_{-\infty}^{\infty} e^{ikx} f(x) dx \quad \text{... (22)}$$

$$= \frac{1}{\sqrt{2\pi}} \int_{-\infty}^{\infty} dx \left[\frac{1}{\sqrt{2\pi}} \int_{-\infty}^{\infty} du \, f(u) e^{-iku} \right] e^{ikx}$$

$$= \int_{-\infty}^{\infty} f(u) du \left[\frac{1}{2\pi} \int_{-\infty}^{\infty} dx \, e^{i\,k(x-u)} \right]$$

$$= \int_{-\infty}^{\infty} du \, f(u) \delta(u-x) \quad \text{... (29)}$$

where
$$\delta(u-x) = \frac{1}{2\pi} \int_{-\infty}^{\infty} dx \, e^{ik(x-u)} \quad \text{... (30)}$$

and has the property

$$\delta(u - x) = 1 \quad x = u$$
$$= 0 \quad x \neq u \qquad \text{... (31)}$$

$$\int_{-\infty}^{\infty} f(u)\delta(u - x)\, du = f(x) \qquad \text{... (32)}$$

$$\int_{-\infty}^{\infty} \delta(x)\, dx = 1$$

and $$f(x)\ \delta(x - a) = f(a)\ \delta(x - a)$$

SOLVED PROBLEMS

1. Express $f(x) = x$ as a Fourier series in the interval $-\pi < x < \pi$. *[C.U. 95, H.P.U. 2003]*

The Fourier series can be expressed as

$$y = \sum_{s=1}^{\infty} a_s \sin sx + \sum_{s=1}^{\infty} b_s \cos sx + b_0$$

where

$$b_0 = \frac{1}{2\pi}\int_{-\pi}^{\pi} x\, dx = \frac{1}{2\pi}\left[\frac{x^2}{2}\right]_{-\pi}^{\pi} = 0$$

$$a_s = \frac{1}{\pi}\int_{-\pi}^{\pi} x \sin sx\, dx$$

$$= \frac{1}{\pi}\left\{\left[-\frac{x\cos sx}{s}\right]_{-\pi}^{\pi} + \int_{-\pi}^{\pi}\frac{\cos sx}{s}\, dx\right\}$$

$$= -\frac{1}{\pi s}\left[\pi\cos s\pi - (-\pi)\cos s\pi\right] + 0$$

$$= -\frac{2}{s}\cos s\pi$$

and $$b_s = \frac{1}{\pi}\int_{-\pi}^{\pi} x\cos sx\, dx$$

$$= \frac{1}{\pi}\left\{\left[\frac{x\sin sx}{s}\right]_{-\pi}^{\pi} - \int_{-\pi}^{\pi}\frac{\sin sx}{s}\, dx\right\}$$

$$= \frac{1}{\pi}\left[\frac{1}{s^2}\cos s\pi\right]_{-\pi}^{\pi}$$

$$= 0$$

Hence $$y = \sum_{s=1}^{\infty} -\frac{2\cos s\pi}{s}\sin sx$$

$$= 2\left[\sin x - \frac{\sin 2x}{2} + \frac{\sin 3x}{3} + \ldots\right]$$

2. Find a Fourier series to represent $(x - x^2)$ from

$x = -\pi$ to $x = \pi$ [Andhra U. 1998]

Let $$y = x - x^2 = \sum_{s=1}^{s=\infty} a_s \sin sx + \sum_{s=1}^{s=\infty} b_s \cos sx + b_0$$

$\therefore$ $$b_0 = \frac{1}{2\pi}\int_{-\pi}^{\pi} y\, dx$$

$$= \frac{1}{2\pi}\int_{-\pi}^{\pi}(x - x^2)\, dx$$

$$= \frac{1}{2\pi}\left[\frac{x^2}{2} - \frac{x^3}{3}\right]_{-\pi}^{\pi}$$

$$= \frac{1}{2\pi}(-)\frac{2\pi^3}{3} = -\frac{\pi^2}{3}$$

$$a_s = \frac{1}{\pi}\int_{-\pi}^{\pi} y \sin sx\, dx$$

$$= \frac{1}{\pi}\int_{-\pi}^{\pi}(x - x^2)\sin sx\, dx$$

$$= \frac{1}{\pi}\left[(x - x^2)(-)\frac{\cos sx}{s} - (1 - 2x)(-)\frac{\sin sx}{s^2} + (-2)\frac{\cos sx}{s^3}\right]_{-\pi}^{\pi}$$

$$= (-2)\frac{(-1)^s}{s}$$

and $$b_s = \frac{1}{\pi}\int_{-\pi}^{\pi} y \cos sx\, dx$$

$$= \frac{1}{\pi}\int_{-\pi}^{\pi}(x - x^2)\cos sx\, dx$$

$$= \frac{1}{\pi}\left[(x - x^2)\frac{\sin sx}{s} - (1 - 2x)(-)\frac{\cos sx}{s^2} + (-2)\frac{(-\sin sx)}{s^3}\right]_{-\pi}^{\pi}$$

As $\cos s\pi = (-1)^s$, $$b_s = -\frac{4(-1)^s}{s^2}$$

$$y = x - x^2$$

$$= -\frac{\pi^2}{3} + \sum_{s=1}^{\infty}\frac{-2(-1)^s}{s}\sin sx + \sum_{s=1}^{\infty}\frac{-4(-1)^s}{s^2}\cos sx$$

$$= -\frac{\pi^2}{3} + 2\left[\sin x - \frac{\sin 2x}{2} + \frac{\sin 3x}{3} - \ldots\right] + 4\left[\cos x - \frac{\cos 2x}{4} + \frac{\cos 3x}{9} - \ldots\right]$$

3. The function $f(x) = x^2$ is defined within the interval $-\pi \le x \le \pi$ and outside it is periodic *i.e.*, (*i*) Show whether the function is even or odd within $-\pi \le x \le \pi$ (*ii*) Expand $f(x)$ in a Fourier series to show

$$f(x) = \frac{\pi^2}{3} + \sum_{n=1}^{\infty} \frac{4}{n^2} \cos nx$$

(*iii*) Use the above result to show that

$$\sum_{n=1}^{\infty} \frac{1}{n^2} = \frac{\pi^2}{6}$$

[C. U. 1997, H.P.U. 1996]

Let $\qquad y = x^2$

(*i*) Now $f(-x) = (-x)^2 = x^2 = f(x)$

hence the function $f(x)$ is even

(*ii*) The Fourier constants are determined as

$$b_0 = \frac{1}{2\pi}\int_{-\pi}^{\pi} y\,dx = \frac{1}{2\pi}\int_{-\pi}^{\pi} x^2\,dx = \frac{1}{2\pi}\left[\frac{x^3}{3}\right]_{-\pi}^{\pi} = \frac{\pi^2}{3}$$

$$b_s = \frac{1}{\pi}\int_{-\pi}^{\pi} y\cos sx\,dx$$

$$= \frac{1}{\pi}\int_{-\pi}^{\pi} x^2 \cos sx\,dx$$

$$= \frac{1}{\pi}\left\{\left[\frac{x^2 \sin sx}{s}\right]_{-\pi}^{+\pi} - \int_{-\pi}^{+\pi} 2x\frac{\sin sx}{s}\,dx\right\}$$

$$= \frac{4}{s^2}(-1)^s$$

Since $f(x)$ is an even function the sine terms will be absent

$$\therefore \qquad f(x) = \frac{\pi^2}{3} + \sum_{s=1}^{\infty} \frac{4}{s^2}(-1)^s \cos sx$$

(*iii*) Putting $x = \pi$

$$f(\pi) = \pi^2 = \frac{\pi^2}{3} + 4\sum_{s=1}^{\infty}\frac{1}{s^2}$$

$$\therefore \qquad 4\sum_{s=1}^{\infty}\frac{1}{s^2} = \left(\pi^2 - \frac{\pi^2}{3}\right) = \frac{2}{3}\pi^2$$

$$\therefore \qquad \sum_{s=1}^{\infty}\frac{1}{s^2} = \frac{\pi^2}{6}$$

4. Find the Fourier transform of

$$f(x) = \begin{matrix} 1 \text{ for } |x| < 1 \\ 0 \text{ for } |x| > 1 \end{matrix}$$

and hence evaluate $\displaystyle\int_0^{\infty} \frac{\sin x}{x}\,dx$

The Fourier transform of $f(x) = F[f(x)] = \int_{-\infty}^{\infty} f(x) e^{ikx} dx$

$$= \int_{-1}^{+1} 1\, e^{ikx} dx$$

$$= \left[\frac{e^{ikx}}{ik}\right]_{-1}^{+1} = \frac{e^{ik} - e^{-ik}}{ik} = 2\frac{\sin k}{k}$$

For $k = 0$ the Fourier transform $f(k) = 2$

Now using Fourier inversion formula we get

$$f(x) = \frac{1}{2\pi}\int_{-\infty}^{\infty} F(k) e^{-ikx} dk$$

$$= \frac{1}{2\pi}\int_{-\infty}^{\infty} \frac{2\sin k}{k} e^{-ikx} dk = \begin{matrix} 1 \text{ for } |x| < 1 \\ 0 \text{ for } |x| > 1 \end{matrix}$$

Putting $x = 0$

$$\int_{-\infty}^{\infty} \frac{\sin k}{k} dk = \pi$$

$$\therefore \quad \int_{0}^{\infty} \frac{\sin k}{k} dk = \frac{\pi}{2}$$

REVIEW QUESTIONS

1. What is Fourier series? Explain with suitable examples the application of the series.
2. Explain how Fourier theorem is applied to analysis and synthesis of complex vibration.
3. State Fourier's theorem regarding the expansion of a periodic function. How are Fourier coefficients evaluated?
4. What are the limitations under which a Fourier series expansion is possible?
5. What are Dirichlet's conditions for a function to be piece-wise regular in a given interval?
6. What are odd functions and even functions? Why is the Fourier series for one limited to cosine terms and for the other to the sine terms?
7. (*a*) What are Fourier transforms? Explain.
   (*b*) Define infinite Fourier sine and cosine transforms.
8. Apply Fourier theorem to analyse a saw tooth wave.

[N.H. Bahuguna Garhwal University 2010]

PROBLEMS

1. Obtain the Fourier series expansion of the following function with period 2π.

$$f(x) = x(\pi - x) \quad 0 \le x \le \pi$$
$$= x(\pi + x) \quad -\pi \le x$$

Hence prove that

$$1 - \frac{1}{3^3} + \frac{1}{5^3} - \frac{1}{7^3} + \ldots = \frac{\pi^3}{32}$$

[**Ans.** $b_0 = 0,\ a_s = \frac{4}{\pi s^3}(1 - \cos s\pi),\ b_s = 0$

For odd s $\quad a_s = \dfrac{8}{\pi s^3}$ and for even s $a_s = 0$

and $\quad f(x) = \dfrac{8}{\pi}\Sigma\dfrac{1}{s^3}\sin nx$

$$f\left(\frac{\pi}{2}\right) = \frac{\pi}{2}\left(\pi - \frac{\pi}{2}\right) = \frac{8}{\pi}\left[\sin\frac{\pi}{2} + \frac{1}{3^3}\sin\frac{3\pi}{2} + ...\right]$$

or $\quad 1 - \dfrac{1}{3^3} + \dfrac{1}{5^3} .... = \dfrac{\pi^3}{32}\Bigg]$

2. Analyse with the help of Fourier's theorem a square periodic wave given by

$$y = A \text{ (constant) for } 0 \le t \le \frac{T}{2}$$

$$\le 0 \text{ for } \frac{T}{2} \le t \le T$$

What harmonics are present in the expansion? [C.U. 1988]

$$\left[\textbf{Ans. } y = \frac{A}{2} + \frac{2A}{\pi}\left(\sin\omega t + \frac{1}{3}\sin 3\omega t + ...\right)\right]$$

3. Find the Fourier series for $f(x)$ in the interval $(-\pi, \pi)$ when

$$f(x) = (\pi + x) \text{ for } -\pi < x < 0$$
$$= \pi - x \text{ for } 0 < x < \pi$$

[**Ans.** $b_0 = \dfrac{\pi}{2}$ $\quad a_s = \dfrac{4}{\pi s^2}$ when s is odd and 0 when s is even and $b_s = 0$.

$$f(x) = \frac{\pi}{2} + \frac{4}{\pi}\sum_{s=1,3}^{\infty}\frac{1}{s^2}\cos sx\Bigg]$$

4. Find the Fourier transform of the function

$$f(x) = |1 - x^2| \qquad |x| < 1$$
$$= 0 \qquad |x| > 1$$

$$\textbf{Ans. } \frac{1}{2\pi}\left\{-\frac{4}{k^3}(k\cos k - \sin k)\right\}$$

5. Find the Fourier transform of the function

$$f(x) = \begin{cases} 1 & |x| < a \\ 0 & |x| > a \end{cases}$$

[**Ans.** $\quad F(k) = \sqrt{\dfrac{2}{\pi}}\dfrac{\sin ka}{k}$ for $k \neq 0$

and $= \sqrt{\dfrac{2}{\pi}}\, a$ for $k = 0$]

6
CHAPTER

Anharmonic (Asymmetric) Vibration and Combinational Tones

6.1. LARGE DISPLACEMENT VIBRATION

In the case of simple harmonic motion the amplitude is considered small and the restoring force is found always proportional to the displacement only. But in the case of motions with large displacement the restoring force, without being proportional to the displacement, may also involve square and higher power terms of the displacement. Such can be represented, by the equation of motion,

$$\frac{d^2x}{dt^2}+\omega^2x+ax^2+bx^3+.... = 0 \qquad ... (1)$$

where x is the displacement at any time t, a, b, ... are small constants and ω is the angular frequency of vibration. For small amplitude terms ax^2, bx^3.... can be neglected with reference to ω^2x. It is easily seen from the above equation (1) that the higher terms with even power of x give correction in the force which are asymmetric because the force arising out of the terms will not change sign of x. Higher terms with odd powers will give rise to symmetric vibration. The vibrations of nonlinear vibration are referred to as anharmonic vibration.

6.2. THE ASYMMETRIC VIBRATION IN WHICH THE FORCE OF RESTITUTION CONTAINS x^2 TERMS

Let us consider the simple case where the force of restitution contains x^2 term. The equation of such a motion is given by

$$m\frac{d^2x}{dt^2}+\mu x+\mu_1x^2 = 0$$

or

$$\frac{d^2x}{dt^2}+\omega^2x+ax^2 = 0 \qquad ... (2)$$

where $\omega^2=\frac{\mu}{m}$ and $a=\frac{\mu_1}{m}$

The equation is solved by successive approximation when a is very small compared to other terms the equation (2) becomes

$$\frac{d^2x}{dt^2}+\omega^2x = 0 \qquad ... (3)$$

The solution of equation (3) is

$$x = A\cos(\omega t+\alpha) \qquad ... (4)$$

or

$$ax^2 = aA^2\cos^2(\omega t+\alpha)$$

$$= \frac{aA^2}{2}\{1+\cos 2(\omega t+\alpha)\} \qquad ... (5)$$

Writing D for $\frac{d}{dt}$ and D^2 for $\frac{d^2}{dt^2}$ the equation (2) can be written with the help of equation (4) as

$$(D^2 + \omega^2)x = -\frac{aA^2}{2} - \frac{aA^2}{2}\cos 2(\omega t + \alpha)$$

or
$$x = -\frac{aA^2}{2(D^2 + \omega^2)} - \frac{aA^2}{2(D^2 + \omega^2)}\cos 2(\omega t + \alpha) \quad ...(6)$$

$$\left[\frac{d}{dt}\cos 2(\omega t + \alpha) = D\cos 2(\omega t + \alpha) = -2\omega \sin 2(\omega t + \alpha)\right.$$

$$\frac{d^2}{dt^2}\cos 2(\omega t + \alpha) = D^2 \cos 2(\omega t + \alpha) = D(-2\omega \sin 2(\omega t + \alpha)$$

$$= -(2\omega)^2 \cos 2(\omega t + \alpha)$$

or
$$\left. D^2 = -(2\omega)^2 \right]$$

Writing $-(2\alpha)^2$ for D^2 in the second term and zero for D^2 in the constant term then the equation (6) becomes

$$x = -\frac{aA^2}{2\omega^2} - \frac{aA^2}{-2(-4\omega^2 + \omega^2)}\cos 2(\omega t + \alpha)$$

or
$$x = -\frac{aA^2}{2\omega^2} + \frac{aA^2}{6\omega^2}\cos 2(\omega t + \alpha) \quad ... (7)$$

The total solution is given by equations (4) and (7) as

$$x = -\frac{aA^2}{2\omega^2} + A\cos(\omega t + \alpha) + \frac{aA^2}{6\omega^2}\cos 2(\omega t + \alpha) \quad ... (8)$$

The constant term is responsible for asymmetry and the system vibrates also with a new frequency, double that of the normal frequency along with the natural vibration. The amplitude of the new vibration is very small as *'a'* is very small.

Higher Order Correction: If *'a'* is not sufficiently small so that ax^2 term cannot be neglected, we need higher order of approximations for corrections.

Now with the help of equation (2) we have

$$\frac{d^2x}{dt^2} + \omega^2 x = -ax^2$$

Substituting the value of x obtained by first approximation from equation (8) the equation (2) becomes

$$\frac{d^2x}{dt^2} + \omega^2 x = -a\left\{-\frac{aA^2}{2\omega^2} + A\cos(\omega t + \alpha) + \frac{aA^2}{6\omega^2}\cos 2(\omega t + \alpha)\right\}^2$$

$$= \gamma_0 + \gamma_1 A\cos(\omega t + \alpha) + \gamma_2 \cos 2(\omega t + \alpha)$$
$$+ \gamma_3 \cos 3(\omega t + a) + \gamma_4 \cos 4(\omega t + \alpha) \quad ... (9)$$

where
$$\gamma_0 = -\left(\frac{3}{8}a^3 A^4 + \frac{aA^2}{2}\right)$$

$$\gamma_1 = \frac{3}{2}a^2 A^2$$

$$\gamma_2 = -\left(\frac{a^3A^4}{2}+\frac{aA^2}{2}\right)$$

$$\gamma_3 = \frac{a^2A^3}{2}$$

$$\gamma_4 = -\frac{a^3A^4}{8}$$

The second term $\gamma_1\ A\cos(\omega t + \alpha)$will give rise to resonance and will make the displacement infinite, which is not the case here so the solution is made by putting

$$A\cos\ (\omega t + \alpha) = x$$

Hence the equation (9) becomes

$$\frac{d^2x}{dt^2}+\omega_1^2 x = \gamma_0 + \gamma_2 \cos2(\omega t + \alpha) + \gamma_3 \cos3(\omega t + \alpha) + \gamma_4 \cos4(\omega t + \alpha) \qquad \text{... (10}$$

where $$\omega^2{}_1 = \omega^2 - \frac{3}{2}a^2A^2 = \omega^2 - \gamma_1 \qquad \text{... (11)}$$

Thus owing to asymmetry the frequency of the system is slightly reduced. In first order approximation there was no appreciable change of natural frequency noticeable. This change becomes pronounced with second order approximation. Assumption $x = A\cos\ (\omega t + \alpha)$ made at the beginning when there is asymmetry is not justified but a change in frequency which may be produced due to asymmetry might have been taken into consideration leading the approximate solution $x = A\cos\ (\omega_1 t + \alpha)$ instead of $A\cos\ (\omega t + \alpha)$.

The equation (10) may be written as

$$(D^2 + \omega^2{}_1)x = \gamma_0 + \gamma_2 \cos2(\omega_1 t + \alpha) + \gamma_3 \cos3(\omega_1 t + \alpha) + \gamma_4 \cos4(\omega_1 t + \alpha)$$

or $$x = \frac{\gamma_0}{\omega_1^2}+\gamma_2\frac{\cos 2(\omega_1 t+\alpha)}{\omega_1^2-4\omega_1^2}+\gamma_3\frac{\cos 3(\omega_1 t+\alpha)}{\omega_1^2-9\omega_1^2}+\gamma_4\frac{\cos 4(\omega_1 t+\alpha)}{\omega_1^2-16\omega_1^2}$$

or $$x = \frac{\gamma_0}{\omega_1^2}-\frac{\gamma_2}{3\omega_1^2}\cos 2(\omega_1 t+\alpha)-\frac{\gamma_3}{8\omega_1^2}\cos 3(\omega_1 t+\alpha)-\frac{\gamma_4}{15\omega_1^2}\cos 4(\omega_1 t+\alpha) \qquad \text{... (12)}$$

The total solution is obtained with the substitution of the values of constants γ and is given by

$$x = A\cos(\omega_1 t+\alpha)-\left(\frac{3}{8}a^3A^4+\frac{aA^2}{2}\right)\frac{1}{\omega_1^2}+\frac{1}{3\omega_1^2}\left(\frac{a^3A^4}{2}+\frac{aA^2}{2}\right)$$

$$\cos 2(\omega_1 t+\alpha)-\frac{1}{8\omega_1^2}\frac{a^2A^3}{2}\cos 3(\omega_1 t+\alpha)+\frac{1}{15\omega_1^2}\frac{a^3A^4}{8}\cos 4(\omega_1 t+\alpha) \qquad \text{... (13)}$$

where $\omega_1$is given by equation (11)

Here also the constant term represents the asymmetry as before.

6.3. THE SYMMETRIC VIBRATION IN WHICH THE FORCE OF RESTITUTION CONTAINS x^3 TERM

The equation of motion is written in the form

$$\frac{d^2x}{dt^2}+\omega^2x+bx^3 = 0 \qquad \text{... (14)}$$

or

$$\frac{d^2x}{dt^2}+\omega^2x = -bx^3 \qquad \text{... (15)}$$

But

$$\frac{d^2x}{dt^2}+\omega^2x = 0$$

has a solution of the form

$$x = A\cos(\omega t+\alpha) \qquad \text{... (16)}$$

The equation (15) with the help of equation (16) is written as

$$\frac{d^2x}{dt^2}+\omega^2x = -bA^3\cos^3(\omega t+\alpha)$$

or

$$\frac{d^2x}{dt^2}+\omega^2x = -\frac{bA^3}{4}\{\cos 3(\omega t+\alpha)+3\cos(\omega t+\alpha)\}$$

By equation (16)

$$\frac{d^2x}{dt^2}+\left(\omega^2+\frac{3}{4}bA^2\right)x = -\frac{bA^3}{4}\cos 3(\omega t+\alpha) \qquad \text{... (17)}$$

or

$$\frac{d^2x}{dt^2}+\omega_1^2x = -\frac{bA^3}{4}\cos 3(\omega t+\alpha) \qquad \text{... (18)}$$

where

$$\omega_1^2 = \omega^2+\frac{3}{4}bA^2 \qquad \text{... (19)}$$

Here also we find due to asymmetry the frequency is altered from natural frequency ω to ω_1. So the assumption $x = A\cos(\omega t+\alpha)$ is not strictly valid but $x = A\cos(\omega_1 t+\alpha)$ is more legitimate where ω_1 differs from ω.

So the equation (18) is written as

$$\frac{d^2x}{dt^2}+\omega_1^2x = -\frac{bA^3}{4}\cos 3(\omega_1 t+\alpha)$$

or

$$x = \frac{bA^3}{4}\cdot\frac{1}{8\omega_1^2}\cos 3(\omega_1 t+\alpha) \qquad \text{... (20)}$$

The complete solution is

$$x = A\cos(\omega_1 t+\alpha)+\frac{bA^3}{4}\cdot\frac{1}{8\omega_1^2}\cos 3(\omega_1 t+\alpha) \qquad \text{... (21)}$$

The solution does not contain any constant term, hence the vibration is symmetrical about the position of rest. Presence of higher power term produces a departure of the magnitude of normal frequency from ω to ω_1 given by (19).

6.4. COMBINATIONAL TONES

It was observed by two violinists, Sorge in Germany (1745) and Tarline in Italy (1754) independently that when two notes are sounded together, a third note is heard often having frequency equal to the difference of the frequencies of two initial tones. This tone is called difference tone.

Koning tried to explain the difference tone by identifying the same with the formation of beats. Beats are formed by the superposition of two notes having very small difference of frequencies. Number of beats formed per second is equal to the difference of frequencies of two notes. But when the rapidity of beats is more than 16 per second, ear will recognise a separate note having the frequency equal to the difference of frequencies. As the combination is only realised by ear as suggested by the theory, the phenomenon is entirely subjective.

Meanwhile it was observed that a summation tone having the frequency equal to the sum of the frequencies of the original tones is emitted. The observation of summation tone cannot be explained by Konig's beat theory, so it was discarded.

Helmholtz defined combination tones as the series of new notes which are heard when two pure strong tones of frequencies p and q act on the same system. The combination tones have frequencies $p - q$, $p + q$ and in general $\alpha p + \beta q$, where α and β are integers. Tones having frequency $p - q$ is called first order difference tone and that having frequency $p + q$ is called first order summation tone. Frequencies as $p - 2q$, $2p - q$, $p + 2q$ .... etc., are called higher order combination tones.

Helmholtz then put forward the intensity theory of combination tones by assuming asymmetry system under double forcing.

The equation of motion of asymmetric system is given by

$$m\frac{d^2x}{dt^2} + \mu_1 x + \mu_2 x^2 = F \sin pt + G \sin qt \qquad \text{... (22)}$$

or

$$\frac{d^2x}{dt^2} + \omega^2 x + ax^2 = f \sin pt + g \sin qt \qquad \text{... (23)}$$

where

$$\omega^2 = \frac{\mu_1}{m}, \; a = \frac{\mu_2}{m} \text{ (small quantity)}$$

$$f = \frac{F}{m} \text{ and } g = \frac{G}{m}$$

ω represents angular frequency of the system. Neglecting x^2 term as 'a' is very small at the first hand we have from equation (23)

$$x = \frac{f}{D^2 + \omega^2}\sin pt + \frac{g}{D^2 + \omega^2}\sin qt$$

$$= \frac{f}{\omega^2 - p^2}\sin pt + \frac{g}{\omega^2 - q^2}\sin qt \qquad \text{... (24)}$$

where D stands for $\frac{d}{dt}$ and D^2 for $\frac{d^2}{dt^2}$. and $-p^2$ is written for D^2 in the first term and $-q^2$ for D^2 in the second term.

Substituting the value of x from equation (24) in equation (23) we have

$$(D^2 + \omega^2)x = -a\left\{\frac{f}{\omega^2 - p^2}\sin pt + \frac{g}{\omega^2 - q^2}\sin qt\right\}^2 + f \sin pt + g \sin qt$$

$$= -a\left\{\frac{f^2}{(\omega^2 - p^2)}\sin^2 pt + \frac{g^2}{(\omega^2 - q^2)^2}\sin^2 qt + \frac{2fg}{(\omega^2 - p^2)(\omega^2 - q^2)}\sin pt \sin qt\right\} + f\sin pt + g\sin qt$$

$$= -a\left\{\frac{f^2}{2(\omega^2 - p^2)^2} + \frac{g^2}{2(\omega^2 - q^2)^2} - \frac{f^2}{2(\omega^2 - p^2)^2}\cos 2pt\right.$$

$$- \frac{g^2}{2(\omega^2 - q^2)^2}\cos 2qt + \frac{2fg}{(\omega^2 - p^2)(\omega^2 - q^2)}\cos(p - q)t$$

$$\left. - \frac{2fg}{(\omega^2 - p^2)(\omega^2 - q^2)}\cos(p + q)t\right\} + f\sin pt + g\sin qt \qquad \text{... (25)}$$

Now operating each term of equation (25) by $\frac{1}{D^2 + \omega^2}$

we have

$$x = -a\left\{\frac{f^2}{2(\omega^2 - p^2)\omega^2} + \frac{g^2}{2(\omega^2 - q^2)\omega^2} - \frac{f^2\cos 2pt}{2(\omega^2 - p^2)(\omega^2 - 4p^2)}\right.$$

$$- \frac{g^2}{2(\omega^2 - p^2)(\omega^2 - 4g^2)} + \frac{fg\cos(p - q)t}{(\omega^2 - p^2)(\omega^2 - q^2)\left\{\omega^2 - (p - q)^2\right\}}$$

$$\left. - \frac{fg\cos(p + q)t}{(\omega^2 - p^2)(\omega^2 - g^2)\left\{\omega^2 - (p + q)^2\right\}}\right\} + \frac{f}{\omega^2 - p^2}\sin pt + \frac{g}{\omega^2 - q^2}\sin qt \qquad \text{... (26)}$$

The first two terms in equation (26) are constant, existence of which make the vibration an asymmetric one.

The vibration having frequencies $2p$, $2q$, $(p - q)$ and $(p + q)$ over and above the frequencies p and q are actually present.

Thus Helmholtz showed that the summation tone of frequency $(p + q)$ and difference tone of frequency $(p - q)$ are audible. In order to explain asymmetry of vibration, he suggested that the tympanum of the ear is pressed inside, as the wave of high amplitude falls on one side of the drum, it is pressed inward and the vibration is asymmetric. So asymmetry is objective in character, as it does not depend upon the special construction of ear or tympanum but it depends upon the mode of excitation of tympanum and large amplitude.

The analysis shows that apart from the fundamentals, second harmonics of each of the component tones, tones of frequencies equal to sum and difference of the frequencies of the component tones, and higher order combinational tones will appear when higher power terms are included in the equation (22).

Edser and Rüker devised an ingenious experiment and confirmed the objective nature of the combinational tones.

A tuning fork was used as a harmonic oscillator. Two notes were emitted by a double siren. The notes emitted were condensed by a conical horn. The tuning fork was placed at the throat of the horn. The frequencies of the component notes were adjusted so that the sum was equal to the natural frequency of the fork. A mirror of Michelson's interferometer was attached to one of the prong of the tuning fork. Interference bands were produced when the prong was without any motion. The fork was used as a resonator to one of the combination tone. It is difficult to excite the tuning fork by resonance, but arrangement was made so sensitive that even the movement of the prong by 10^{-4} of an inch (1 inch = 2.54 cm) altered the optical length of one of the rays, could cause the shift of fringes. The shift of fringes can be easily detected by photographing the fringes by moving plate. This proves the objective nature of the summation tone.

Similar experiment could be carried out for difference tones, when a fork of frequency equal to the difference of frequencies of the component tones is used, which will prove the objective nature of the difference tone.

Here in both cases the fork is vibrated with a small amplitude so it cannot be regarded as asymmetric system as suggested by Helmholtz.

To explain the anomally it is assumed by Weitzmann that the air molecules within the siren act as an asymmetric system which discards the hypothesis of asymmetric vibration of tympanum. The difference tone is found louder than the summation tone.

From the equation (26) the amplitude of different notes can be tabulated in Table I.

The ratio of the amplitude of difference tone to that of summation tone is given by

$$\frac{G}{F} = \frac{\omega^2 - (p+q)^2}{\omega^2 - (p-q)^2}$$

which is less than 1; this leads to the conclusion that the difference tone has smaller amplitude. So the above theory fails to explain how the difference tone is louder than summation tone.

Table 1.

Angular Frequency	*Amplitude*
p	$A = \frac{f}{\omega^2 - p^2}$
q	$B = \frac{g}{\omega^2 - q^2}$
Displaced zero	$C = \frac{a}{2\omega^2}\left(A^2 + B^2\right)$
$2p$	$D = -\frac{aA^2}{2(\omega^2 - 4p^2)}$
$2q$	$E = -\frac{aB^2}{2(\omega^2 - 4q^2)}$
$p + q$	$F = -\frac{aAB}{\omega^2 - (p+q)^2}$
$p - q$	$G = -\frac{aAB}{\omega^2 - (p-q)^2}$

SOLVED PROBLEMS

1. A tuning fork of frequency 96 Hz is acted on by two harmonic tone one of which has frequency 256 Hz. If the tuning fork responds to the difference frequency find the frequency of the other tone.

If p be the desired frequency then $p = (256 + 96)$ Hz
$= 352$ Hz

or $p = (256 - 96)$ Hz
$= 160$ Hz

2. An asymmetric nonlinear vibration is subjected simultaneously to two harmonic tones of frequencies 80 Hz and 64 Hz respectively. If the resonant frequency is very high give all the frequencies present in the forced vibration. Neglect higher order of the combinational tone.

If p and q are the two existing frequencies, then

(*i*) $p = 80$ Hz and $q = 64$ Hz

(*ii*) The frequencies $2p = 160$ Hz and $2q = 128$ Hz

(*iii*) The frequencies $p + q = 144$ Hz and $p - q = 16$ Hz

(*iv*) For a steady component displacement is zero

REVIEW QUESTIONS

1. Give the theory of combinational tone and show how you prove its objective reality. How far Helmholtz theory explains the combinational tones?
2. Explain theoretically the beat tone. How does it differ from difference tone of combination vibration. The difference tone is found louder than summation tone — Explain.
3. What are combination tones? Give a theory to show that they are produced as a result of double forcing of a nonlinear system. [Burd. U. 1996]
4. How do combination tones originate? Are they subjective or objective? Give the reason for your answer.
5. Write down the forcing equation for (*i*) asymmetric vibration (*ii*) symmetric vibration.

CHAPTER 7

WAVE MOTION

7.1. WAVE MOTION IN AN ELASTIC MEDIUM

The motion of a particle under different conditions has been considered. But when we have got a number of particles linked together by elastic chains, it is not possible to displace any one of them without disturbing the other. The motion imparted to one is communicated through the chains to others and they in turn are set into motion. An elastic medium consists of very fine particles with elastic connection between them. If we impart periodic motion to any particle, its energy of motion will be communicated to the adjacent particles surrounding it, which will thus execute similar periodic motions and in doing so will in their turn transfer their energy to the neighbouring particles surrounding them and so on. The motion imparted to the particle will move uniformly to all directions from particle to particle with a definite velocity. The propagation of a disturbance of this kind due to transfer of energy from particle to particle is called wave motion.

There are two types of wave motion (*i*) transverse wave motion (*ii*) longitudinal wave motion.

In a transverse wave motion the particles of a medium vibrate at right angles about the mean position to the direction of propagation of the wave.

Longitudinal wave motion is that wave motion in which the particles of the medium vibrate about their mean position in the same direction in which the wave is propagated.

There are waves in which no material medium is necessary viz. radiowaves, light waves, heat radiation, x-rays, γ-rays and in general all these waves are called electromagnetic waves. Now we shall discuss ourselves to elastic waves in material media.

Wavelength of a wave is the minimum distance along the line of propagation between two particles of the medium vibrating in the same phase. The phase difference between two particles separated by a distance of one wavelength is 2π. It is the distance through which the energy transference advances through the medium in the interval when the source completes one oscillation. If the distance through which the energy is propagated *i.e.*, wave is propagated in unit time 'c', the velocity of propagation of the wave and n be the frequency of vibration of the source, then the wavelength $\lambda = \frac{c}{n}$ or $c = n\lambda$ where the frequency of the wave is the number of waves passing through a point in one second. The time period (T) of a wave is the time taken by a single wave to pass over a point in the medium. The frequency of the wave is given by $n = \frac{1}{T}$. The S.I. unit of wavelength is metre, frequency Hertz (Hz) and time period second.

7.2 GENERAL EQUATION OF WAVE MOTION

In a wave each particle executes simple harmonic motion and the motion is transferred from particle to particle with a definite velocity.

The motion of particle executing S.H.M. is represented by the equation

$$y = a\sin(\omega t - \alpha) \qquad \text{... (1)}$$

where y is the displacement of the particle at any time t (vide Art.2.3) or

$$y = a\cos(\omega t - \alpha) \qquad \text{... (2)}$$

where a is the amplitude, ωt is the phase and α is the epoch.

The epoch α of a particle depends on x, its distance from the origin measured along X-axis, the line of advance of the waves. The phase changes continuously along X-axis at such a rate that x increases through λ, the phase changes through 2π.

Therefore $$\frac{\alpha}{2\pi} = \frac{x}{\lambda} \quad \text{or} \quad \alpha = \frac{2\pi}{\lambda}x$$

and $$\omega t = 2\pi nt = \frac{2\pi ct}{\lambda}, \quad \lambda = cT$$

where n is the frequency and c velocity. Substituting these values the equations (1) and (2) may be represented as

$$y = a\sin\frac{2\pi}{\lambda}(ct - x) \quad \text{... (3)}$$

or $$y = a\cos\frac{2\pi}{\lambda}(ct - x) \quad \text{... (4)}$$

which advances along positive direction *i.e.* forward wave but

$$y = a\sin\frac{2\pi}{\lambda}(ct + x) \quad \text{... (5)}$$

or $$y = a\cos\frac{2\pi}{\lambda}(ct + x) \quad \text{... (6)}$$

advances along negative directions *i.e.*, backward wave.

The equations (3), (4), (5) and (6) advance respectively in positive or negative direction in the medium are said to be Progressive Wave.

7.3 VELOCITY AND ACCELERATION OF A PARTICLE IN A PROGRESSIVE WAVE

The velocity of a particle at a place at a time t is obtained by differentiating its displacement with regard to time.

Displacement is given by equation (3) as

$$y = a\sin\frac{2\pi}{\lambda}(ct - x) \quad \text{... (3)}$$

Velocity $$\frac{dy}{dt} = \frac{2\pi c}{\lambda}a\cos\frac{2\pi}{\lambda}(ct - x) \quad \text{... (7)}$$

But $$\frac{dy}{dx} = -\frac{2\pi}{\lambda}a\cos\frac{2\pi}{\lambda}(ct - x) \quad \text{... (8)}$$

From equations (7) and (8)

$$\frac{dy}{dt} = -c\frac{dy}{dx} \quad \text{... (9)}$$

Thus the velocity of the particle in the wave, at any point x at the time t is equal to $-c$ times the slope of the wave displacement curve at the point.

Differentiating velocity $\frac{dy}{dt}$ in equation (7) we get acceleration

$$\frac{d^2y}{dt^2} = -\left(\frac{2\pi c}{\lambda}\right)^2 a\sin\frac{2\pi}{\lambda}(ct - x)$$

$$= -\left(\frac{2\pi c}{\lambda}\right)^2 y$$

$$= -\omega^2 y \quad \text{... (10)}$$

as $$\omega = 2\pi n = \frac{2\pi}{T} = \frac{2\pi c}{\lambda}$$

where n is the frequency, T the time period, this shows that the particle executing S.H.M. From equation (8) we get

$$\frac{d^2y}{dx^2} = -\left(\frac{2\pi}{\lambda}\right)^2 a\sin\frac{2\pi}{\lambda}(ct - x)$$

or $$\frac{d^2y}{dx^2} = -\frac{\omega^2}{c^2}y \qquad \text{... (11)}$$

From equations (10) and (11)

$$\frac{d^2y}{dt^2} = c^2\frac{d^2y}{dx^2} \qquad \text{... (12)}$$

Thus the acceleration of the particle is equal to c^2 times the curvature of the wave displacement curve. The equation (12) represents the differential equation of the wave motion.

The general solution of equation (12) is

$$y = f(ct - x) + g(ct + x)$$

where f and g are any two functions. f represents plane wave moving in the positive x-direction with velocity c *i.e.,* forward wave and g represents a plane wave moving in the negative x-direction *i.e.*, backward wave.

Note : The energy of the progressive wave is calculated in Art. 17.1.

7.4 DIFFERENT TYPES OF PROGRESSIVE WAVE

There are two types of waves:

(*i*) Progressive longitudinal wave in which every particle vibrates along the direction of propagation of the wave. They are same in amplitude but differ in phase with distance.

(*ii*) Progressive transverse wave in which every particle vibrates in a direction perpendicular to the direction of propagation of the wave. They have the same in amplitude but differ in phase with distance.

The motion by smart push at one end of a spiral spring, suspended from a wooden frame horizontally by bifilar suspension is an example of longitudinal wave. The rims move to and fro along the axis of the spring producing compressions (C) and rare-fraction (R) in Fig. 1 which will travel along the length of the spring. Here the motion of individual turns is along the direction of propagation. So the wave is called longitudinal wave.

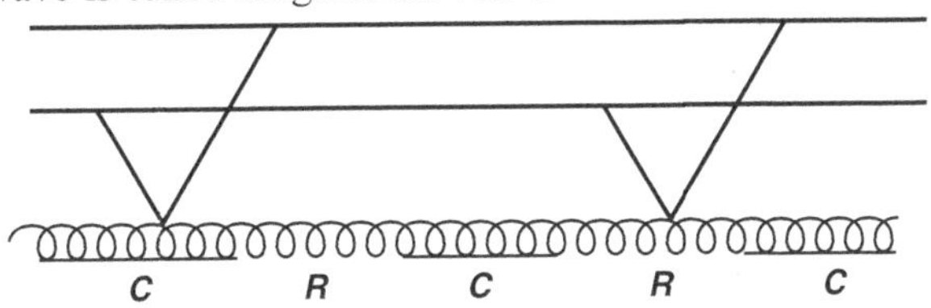

Fig. 1.

Water waves and rope waves are transverse. Alternate dipping and rising of a small rod at any point on the surface of still water will generate ripples which propagates onwards uniformly in all directions in form of circles of alternate crests and troughs of gradually increasing radius. Every particle of water executes S.H.M. about their mean position of rest at right angles to the line of advance. A cork floating on water which was at rest originally, will rise and fall as waves advance through it.

A thick rope say a rubber tubing suspended from one end, the other end of which is moved a little at right angles to the length with given horizontal periodic motion will cause the disturbance in the form of sinusoidal wave to travel along the length upwards. Every particle of the rope will execute to and fro motion perpendicular to the length of the rope without advancing in the forward direction.

The motion of stringed musical instruments are the examples of transverse wave. The total length of a crest and a trough, is called the wavelength of the transverse wave. The distance between two consecutive crests or consecutive trough is called the wavelength of the transverse wave.

7.5. PROPAGATION OF SOUND WAVE THROUGH AIR

The source of sound is always a vibrating body. All vibrating bodies do not perform a single S.H.M. A tuning fork is ordinarily the only source which produces sound by performing a single S.H.M. of a fixed frequency.

When a fork vibrates and motion of one of the prongs is outwardly for the first time at the beginning will compress the layer of air immediately in front of it. When the prong moves back to the extreme opposite position, the compressed layer expands and in doing so, pushes the layer in front of it and comes backwards in the space in which pressure has been released due to motion backwards. The compression will successively travel outwards from layer to layer from the prong. By this time, the prong swing back and start a journey leftward. As such, it creates a partial vacuum behind causing a rarefaction of the layer in contact with it. This rarefied layer tends to return to its normal condition and causes a rarefaction of the adjoining layer. Thus a pulse of rarefaction travels outwards.

During the vibration of the prong alternate compression (condensation) and rarefaction are created in the medium which travels outwards from the source. A compression and rarefraction constitute a longitudinal wave progressive in the outward direction. The distance between successive compression or rarefraction is called a wavelength of longitudinal sound wave.

7.6. DIFFERENTIAL EQUATION FOR WAVES IN THREE DIMENSION

The differential equation for the wave in three dimensions is of the form

$$\frac{\partial^2 \psi}{\partial t^2} = c^2 \nabla^2 \psi \qquad \text{... (13)}$$

where ∇^2 is called the Laplacian operator.

The operator ∇^2 has the following forms in three coordinate systems

(1) Rectangular Cartesian co-ordinates (x, y, z)

$$\nabla^2 \psi = \frac{\partial^2 \psi}{\partial x^2} + \frac{\partial^2 \psi}{\partial y^2} + \frac{\partial^2 \psi}{\partial z^2} \qquad \text{... (14)}$$

(2) Spherical polar co-ordinates (r, θ, φ)

$$\nabla^2 \psi = \frac{1}{r^2}\frac{\partial}{\partial r}\left(r^2 \frac{\partial \psi}{\partial r}\right) + \frac{1}{r^2 \sin\theta}\frac{\partial}{\partial \theta}\left(\sin\theta \frac{\partial \psi}{\partial \theta}\right) + \frac{1}{r^2 \sin^2\theta}\frac{\partial^2 \psi}{\partial \varphi^2} \qquad \text{... (15)}$$

(3) Cylindrical co-ordinates (r, θ, z)

$$\nabla^2 \psi = \frac{1}{r^2}\frac{\partial}{\partial r}\left(r^2 \frac{\partial \psi}{\partial r}\right) + \frac{1}{r^2}\frac{\partial^2 \psi}{\partial \theta^2} + \frac{\partial^2 \psi}{\partial z^2} \qquad \text{... (16)}$$

(A) Solution of Wave Equation in Cartesion Co-ordinates

The equation (13) in cartesian co-ordinate system takes the form

$$\frac{\partial^2 \psi}{\partial x^2} + \frac{\partial^2 \psi}{\partial y^2} + \frac{\partial^2 \psi}{\partial z^2} = \frac{1}{c^2}\frac{\partial^2 \psi}{\partial t^2} \qquad \text{... (17)}$$

Solution of equation (17) in Cartesian co-ordinate system by the method of separation of variable is given below:

Let us assume $$y(x, y, z, t) = X(x)\, Y(y)\, Z(z)\, T(t) \qquad \text{... (18)}$$

where X is a function of x alone Y is a function of y alone and Z is a function of z and T is a function of t alone.

The equation (17) with the help of equation (18) takes the form

$$YZT\frac{\partial^2 X}{\partial x^2} + XZT\frac{\partial^2 Y}{\partial y^2} + XYT\frac{\partial^2 Z}{\partial z^2} = \frac{XYZ}{c^2}\frac{\partial^2 T}{\partial t^2} \qquad \text{... (19)}$$

Dividing throughout by $XYZT$ the equation (19) takes the form

$$\frac{1}{X}\frac{\partial^2 X}{\partial x^2} + \frac{1}{Y}\frac{\partial^2 Y}{\partial y^2} + \frac{1}{Z}\frac{\partial^2 Z}{\partial z^2} = \frac{1}{c^2 T}\frac{\partial^2 T}{\partial t^2} \qquad \text{... (20)}$$

Since x, y, z and t are independent of each other, each term must be set equal to a constant and we can write

$$\left.\begin{aligned} \frac{1}{X}\frac{d^2 X}{dx^2} &= -k_x^2 \\ \frac{1}{Y}\frac{d^2 Y}{dy^2} &= -k_y^2 \\ \frac{1}{Z}\frac{d^2 Z}{dz^2} &= -k_z^2 \end{aligned}\right\} \qquad \text{... (21)}$$

where k^2_{x}, k^2_{y}, k^2_{z} are constants.

The equation (20) with the help of equation (21) takes the form $\frac{1}{c^2 T}\frac{\partial^2 T}{\partial t^2} = -(k_x^2 + k_y^2 + k_z^2)$

or $$\frac{\partial^2 T}{\partial t^2} = (K^2 c^2 T) \qquad \text{... (22)}$$

where $K^2 = k^2_{x}, k^2_{y}, k^2_{z}$

The solution of equations (21) and (22) can be written as

$$\left.\begin{aligned} X &= A_1 \cos k_x x + B_1 \sin k_x x \\ Y &= A_2 \cos k_y y + B_2 \sin k_y y \\ Z &= A_3 \cos k_z z + B_3 \sin k_z z \\ T &= A_4 \cos Kct + B_4 \sin Kct \end{aligned}\right\} \qquad \text{... (23)}$$

The solution of equation (17) takes the form as

$$\begin{aligned} y(x, y, z, t) = {} & (A_1 \cos k_x x + B_1 \sin k_x x)(A_2 \cos k_y y + B_2 \sin k_y y) \\ & (A_3 \cos k_z z + B_3 \sin k_z z)\,(A_4 \cos Kct + B_4 \sin Kct) \end{aligned} \qquad \text{... (24)}$$

It is convenient to write the equation (24) in exponential form as

$$\psi(x, y, z, t) = A \exp\,[i(k_x x + k_y y + k_z z + Kct + \varphi)]$$

where φ is the phase

or $$\psi(x, y, z, t) = A \exp\,[i(\bar{k}\,.\bar{r} + Kct + \varphi] \qquad \text{... (25)}$$

but $Kc = \frac{2\pi}{\lambda}c = \frac{2\pi}{T}$ Therefore the equation (25) can be written as

$$\psi(x, y, z, t) = A \exp\,[i(\bar{k}\,.\bar{r} + \omega t + \varphi)] \qquad \text{... (26)}$$

(B) SOLUTION OF WAVE EQUATION IN SPHERICAL WAVE WHICH DEPENDS ON THE RADIAL DISTANCE AND TIME.

For spherical wave the displacement ψ depends only on r and t then the equation (13) takes the form

$$\frac{1}{r^2}\frac{\partial}{\partial r}\left(r^2\frac{\partial \psi}{\partial r}\right) = \frac{1}{c^2}\frac{\partial \psi}{\partial t^2} \qquad \text{... (27)}$$

We substitute $$\psi(r, t) = \frac{u(r,t)}{r}$$

$$\therefore \quad \frac{1}{r^2}\frac{\partial}{\partial r}\left(r^2\frac{\partial \psi}{\partial r}\right) = \frac{1}{r^2}\frac{\partial}{\partial r}\left(r\frac{\partial u}{\partial r} - u\right) = \frac{1}{r}\frac{\partial^2 u}{\partial r^2}$$

The equation (27) becomes

$$\frac{1}{r}\frac{\partial^2 u}{\partial r^2} = \frac{1}{c^2 r}\frac{\partial^2 u}{\partial t^2}$$

or $$\frac{\partial^2 u}{\partial r^2} = \frac{1}{c^2}\frac{\partial^2 u}{\partial t^2} \qquad \text{... (28)}$$

Let us assume $$u(r, t) = R(r)\, T(t) \qquad \text{... (29)}$$

where R is a function of r alone and T is a function of t only

The equation (28) with the help of equation (29)

$$T\frac{d^2 R}{dr^2} = \frac{R}{c^2}\frac{d^2 T}{dt^2}$$

or $$\frac{1}{R}\frac{d^2 R}{dr^2} = \frac{1}{c^2 T}\frac{d^2 T}{dt^2} \qquad \text{... (30)}$$

Since r and t are independent of each other then each term must be set equal to a constant then

$$\frac{1}{R}\frac{d^2 R}{dr^2} = -k^2 \qquad \text{... (31)}$$

The solution of equation (31) is

$$R = (A_1 \cos kr + B_1 \sin kr) \qquad \text{... (32)}$$

and $$\frac{1}{c^2 T}\frac{d^2 T}{dt^2} = -k^2$$

or $$\frac{d^2 T}{dt^2} = -k^2 c^2 T \qquad \text{... (33)}$$

The solution of equation (33) is

$$T = (A_2 \cos kct + B_2 \sin kct) \qquad \text{... (34)}$$

$$\therefore \quad \psi(r, t) = \frac{u(r,t)}{r} = \frac{1}{r}(A_1 \cos kr + B_1 \sin kr)(A_2 \cos kct + B_2 \sin kct)$$

$$= \frac{1}{r}(A_1 \cos kr + B_1 \sin kr)(A_2 \cos \omega t + B_2 \sin \omega t) \qquad \text{... (35)}$$

as $\omega = kc$

The equation (35) can also be written as

$$\psi(r, t) = \frac{A}{r} \exp [i(kr + \omega t + \varphi)] \quad \text{... (36)}$$

The factor $\frac{1}{r}$ in equation (36) implies that the amplitude decreases inversely as r.

(C) SOLUTION OF THE WAVE EQUATION IN SPHERICAL POLAR CO-ORDINATES

The equation (13) with the help of equation (15) in spherical polar co-ordinates.

$$\frac{1}{r^2}\frac{\partial}{\partial r}\left(r^2\frac{\partial \psi}{\partial r}\right) + \frac{1}{r^2 \sin\theta}\frac{\partial}{\partial \theta}\left(\sin\theta\frac{\partial \psi}{\partial \theta}\right) + \frac{1}{r^2 \sin^2\theta}\frac{\partial^2 \psi}{\partial \varphi^2} = \frac{1}{c^2}\frac{\partial^2 \psi}{\partial t^2} \quad \text{... (37)}$$

Let us assume

$$y(r, \theta, \varphi, t) = R(r)\, \Theta(\theta)\, \phi(\varphi)\, T(t) \quad \text{... (38)}$$

Since r, θ, φ and t are independent of each other on substituting the equation (38) in equation (37) the equation (37) takes the form

$$\Theta\, \phi\, T \frac{\partial^2 R}{\partial r^2} + 2\, \Theta\, \phi\, T \frac{\partial R}{\partial r} + \frac{\cot\theta}{r^2} R\, \phi\, T \frac{\partial \Theta}{\partial \theta} + \frac{R\phi T}{r^2}\frac{\partial^2 \Theta}{\partial \theta^2} + \frac{R\Theta T}{r^2 \sin^2\theta}\frac{\partial^2 \phi}{\partial \varphi^2} = \frac{R\Theta\phi}{c^2}\frac{\partial^2 T}{\partial t^2}$$

Dividing both sides by $R\theta\phi T$

$$\frac{1}{R}\frac{\partial^2 R}{\partial r^2} + \frac{2}{Rr}\frac{\partial R}{\partial r} + \frac{1}{r^2\Theta}\frac{\partial^2 \Theta}{\partial \theta^2} + \frac{\cot\theta}{r^2\Theta}\frac{\partial \theta}{\partial \theta} + \frac{1}{r^2 \sin^2\theta\phi}\frac{\partial^2 \phi}{\partial \varphi^2} = \frac{1}{c^2 T}\frac{\partial^2 T}{\partial t^2} \quad \text{... (39)}$$

Since r, θ, φ and t are independent of each other each term of equation (39) must be separately equal to a constant and we can write

$$\frac{1}{c^2 T}\frac{d^2 T}{dt^2} = -k^2 \quad \text{... (40)}$$

$$\frac{1}{\phi}\frac{d^2 \phi}{d\varphi^2} = -m^2 \quad \text{... (41)}$$

and

$$\frac{1}{\Theta}\frac{d^2 \Theta}{d\theta^2} + \frac{\cot\theta}{\Theta}\frac{d\Theta}{d\Theta} - \frac{m^2}{\sin^2\theta} = -l(l + 1) \quad \text{... (42)}$$

Now the equation (39) can be written as

$$\frac{1}{R}\frac{d^2 R}{dr^2} + \frac{2}{Rr}\frac{dR}{dr} + \frac{m^2}{r^2 \sin^2\theta} - \frac{l(l+1)}{r^2} - \frac{m^2}{r^2 \sin^2\theta} = -k^2$$

or

$$\frac{1}{R}\frac{d^2 R}{dr^2} + \frac{2}{Rr}\frac{dR}{dr} = \frac{l(l+1)}{r^2} - k^2 \quad \text{... (43)}$$

The solution of equation (40) and (41) are

$$T = Ae^{ikct} \quad \text{... (44)}$$

$$\phi = A_1 e^{im\varphi} \quad \text{... (45)}$$

The equation (42) can be written as

$$\frac{d^2\Theta}{d\theta^2} + \cot\theta\frac{d\Theta}{d\theta} + \left\{l(l+1) - \frac{m^2}{\sin^2\theta}\right\}\Theta = 0$$

or

$$\frac{1}{\sin\theta}\frac{d}{d\theta}\left(\sin\theta\frac{d\Theta}{d\theta}\right) + \left\{l(l+1) - \frac{m^2}{\sin^2\theta}\right\}\Theta = 0 \quad \text{... (46)}$$

Put $$x = \cos\theta$$

$$\frac{dx}{d\theta} = -\sin\theta$$

or $$\frac{d}{dx} = -\frac{1}{\sin\theta}\frac{d}{d\theta} \quad \text{... (47)}$$

From equation (47) the equation (46) becomes

$$\frac{d}{dx}\left\{(1-x^2)\frac{d\Theta}{dx}\right\} + \left\{l(l+1) - \frac{m^2}{1-x^2}\right\}\Theta = 0 \quad \text{... (48)}$$

The equation (48) is associated Legendre equation and its solution can be given as

$$\Theta = A_l P_l^m(x) + B_l Q_l^m(x)$$

or $$\Theta = A_l P_l^m(\cos\theta) + B_l Q_l^m(\cos\theta) \quad \text{... (49)}$$

If θ has to be finite when $x \to 0$ $B_1 = 0$ so the equation (49) can be written as

$$\theta = A_l P_l^m(\cos\theta) \quad \text{... (50)}$$

where $P_l^m(\cos\theta)$ is associated Legendre function.

The equation (43) can be written as

$$\frac{d^2R}{dr^2} + \frac{2}{r}\frac{dR}{dr} + \left\{k^2 - \frac{l(l+1)}{r^2}\right\}R = 0 \quad \text{... (51)}$$

Putting $R = r^{-1/2}u$ the equation (51) reduces to

$$\frac{d^2u}{dr^2} + \frac{1}{r}\frac{du}{dr} + \left\{k^2 - \frac{\left(l+\frac{1}{2}\right)}{r^2}\right\}u = 0 \quad \text{... (52)}$$

Putting $y = kr$ in equation (52) the equation (52) takes the form

$$\frac{d^2u}{dy^2} + \frac{1}{y}\frac{du}{dy} + \left\{1 - \frac{\left(l+\frac{1}{2}\right)^2}{k^2}\right\}u = 0 \quad \text{... (53)}$$

The equation (53) is Bessel's differential equation and its solution is

$$u = C J_n(l) + DJ_{-n}(l)$$

where $n = l + \frac{1}{2}$ and $J_n(kl)$ is called Bessel's function of nth order and first kind.

Therefore the total solution of the spherical wave equation is

$$\psi(r, \theta, \varphi, t) = r^{-1/2}\,[C J_n(l)\, DJ_{-n}(l)]\, A_l P_l^m(\cos\theta).\, A_1 e^{im\varphi}\, Ae^{i\omega t} \quad \text{... (54)}$$

where $$kc = \omega$$

(D) SOLUTION OF THE WAVE EQUATION IN CYLINDRICAL CO-ORDINATES

The equation (13) with the help of equation (16) in cylindrical coordinate system

$$\frac{1}{r^2}\frac{\partial}{\partial r}\left(r\frac{\partial\psi}{\partial r}\right) + \frac{1}{r^2}\frac{\partial^2\psi}{\partial\theta^2} + \frac{\partial^2\psi}{\partial z^2} = \frac{1}{c^2}\frac{\partial^2\varphi}{\partial t^2} \quad \text{... (55)}$$

Put $$\psi = R(r)\,\Theta(\theta)\,Z(z)\,T(t) \quad \text{... (56)}$$

where r, θ, z and t are independent of each other.

Substituting equation (56) in equation (55) and dividing both sides by $R\Theta ZT$ we get

$$\frac{1}{R}\frac{\partial^2 R}{\partial r^2} + \frac{1}{Rr}\frac{\partial R}{\partial r} + \frac{1}{r^2\Theta}\frac{\partial^2 \Theta}{\partial \theta^2} + \frac{1}{Z}\frac{\partial^2 Z}{\partial z^2} = \frac{1}{c^2 T}\frac{\partial^2 T}{\partial t^2} \quad \text{... (57)}$$

Since r, θ, z and t are independent of each other, each term of equation (57) must be separately equal to a constant and we can write

$$\frac{1}{Z}\frac{\partial^2 Z}{\partial z^2} = -k^2$$

Then

$$Z = c_1 e^{ikz} \quad \text{... (58)}$$

$$\frac{1}{\theta}\frac{\partial^2 \theta}{\partial \theta^2} = -l^2$$

$$\theta = c_2 e^{il\theta} \quad \text{... (59)}$$

$$\frac{1}{T}\frac{\partial^2 T}{\partial t^2} = -c^2\lambda^2$$

$$T = c_3 e^{ic\lambda t} \quad \text{... (60)}$$

The equation (57) becomes

$$\frac{1}{R}\left(\frac{\partial^2 R}{\partial r^2} + \frac{1}{r}\frac{\partial R}{\partial r}\right) - \frac{l^2}{r^2} - k^2 = -\lambda^2$$

or

$$\frac{\partial^2 R}{\partial r^2} + \frac{1}{r}\frac{\partial R}{\partial r} + \left(\alpha^2 - \frac{l^2}{r^2}\right) R = 0 \quad \text{... (61)}$$

where

$$k^2 - \lambda^2 = -\alpha^2$$

Putting $x = \alpha r$ in equation (61) takes the form

$$\frac{d^2 R}{dx^2} + \frac{1}{x}\frac{dR}{dx} + \left(1 - \frac{l^2}{x^2}\right) R = 0 \quad \text{... (62)}$$

The solution of equation (62) is

$$R = A\, J_l\,(ar) + B\, Y_l\,(\alpha r) \quad \text{... (63)}$$

$$Y_l\,(\alpha r) \to \infty \text{ when } r \to 0, \qquad \therefore B = 0$$

The complete solution of equation (55) is

$$\psi(r, \theta, z, t) = A\, J_l\,(\alpha r)\, c_2 e^{il\theta}\, c_1 e^{ikz}\, c_3 e^{ic\lambda t} \quad \text{... (64)}$$

7.7 (a) SOLVED PROBLEMS

1. A plane wave travelling through a medium is given by $\psi(x, t) = 0.02 \sin(2\pi t - 0.04\,\pi x)$ where ψ and x are in metres and t in seconds. Find the wavelength and frequency.

[Burd. U. 2001]

Given

$$\psi(x, t) = 0.02 \sin(2\pi t - 0.04\,\pi x)$$

We know that the general equation of wave motion may be represented as

$$\psi(x, t) = a \sin\frac{2\pi}{\lambda}(ct - x)$$

Comparing above two equations we get

$$\frac{2\pi c}{\lambda} = \frac{2\pi}{T} \text{ and}$$

we know $c = n\lambda$

$\therefore$ $n = 1$ Hz

and $$\frac{2\pi}{\lambda} = 0.04\pi$$

$\therefore$ $$\lambda = \frac{2}{0.04} \text{ metre}$$

$$= 50 \text{ metres}$$

2. A wave of frequency 400 Hz is travelling with a velocity 800 m/sec. How far are two points situated whose displacement differ in phase by $\frac{\pi}{4}$? *[P.U. 1996]*

Given the frequency of the wave $n = 400$ Hz

Velocity of the wave $c = 800$ m/sec

$\therefore$ Wavelength of the wave $\lambda = \frac{c}{n} = \frac{800}{400}$ m

or $\lambda = 2$ m

Phase difference between two points at a distance x apart

$$\phi = \frac{2\pi}{\lambda}x$$

For the displacement to differ in phase by $\frac{\pi}{4}$

$$\phi = \frac{\pi}{4} = \frac{2\pi}{\lambda}x = \frac{2\pi}{2}x$$

or $x = 0.25$ metre

3. A transverse harmonic wave of amplitude 0.02 m is generated at one end ($x = 0$) of along horizontal string by a tuning fork of frequency 500 Hz. At a given instant of time, the displacement of the particle at $x = 0.1$ m is 0.005 m and that of the particle at $x = 0.2$ m is 0.005 m. Calculate the wavelength and the velocity of the wave. Obtain the equation of the wave assuming that the wave is travelling along the positive x-direction and that the end $x = 0$ is at equilibrium position at $t = 0$.

The general equation of this wave is given by

$$y(x, t) = a\sin\frac{2\pi}{\lambda}(ct - x)$$

Here $a = 0.02$ m

Again when $x = 0.1$ m $y = -0.005$ m

$\therefore$ $$-0.005 = 0.02\sin\frac{2\pi}{\lambda}(ct - 0.1)$$

or $$\sin\frac{2\pi}{\lambda}(ct - 0.1) = -\frac{0.005}{0.02} \text{ m} = -0.25 \text{ m}$$

$\therefore$ Phase $$\phi_1 = \frac{2\pi}{\lambda}(ct - 0.1) = \sin^{-1}(-0.25)$$

Again $x = 0.2 \text{ m} \quad y = 0.005 \text{ m}$

$$\therefore \quad 0.005 = 0.02 \sin \frac{2\pi}{\lambda}(ct - 0.2)$$

$$\therefore \text{ Phase} \quad \phi_2 = \frac{2\pi}{\lambda}(ct - 0.2) = \sin^{-1}\frac{0.005}{0.02} = \sin^{-1}(0.25)$$

Phase difference $\Delta\phi = \phi_1 - \phi_2$

$$= \sin^{-1}(-0.25) - \sin^{-1}(0.25)$$
$$= \pi^c$$

But
$$\Delta\phi = \frac{2\pi}{\lambda}\Delta x$$

i.e.,
$$\pi = -\frac{2\pi}{\lambda}(0.1 - 0.2)$$

or
$$\lambda = 0.2 \text{ metre}$$

Frequency $n = 500$ Hz (given)

$\therefore$ The velocity $c = n\lambda = 500 \times 0.2$ metre

$$= 100 \text{ metre/sec}$$

The equation of the wave is given by

$$y = 0.02 \sin\left[\frac{2\pi}{0.2}(100t - x)\right]$$
$$= 0.02 \sin [10\pi (100t - x)]$$

4. Solve the diffusion equation $\frac{\partial c}{\partial t} = D\frac{\partial^2 c}{\partial x^2}$ where c is the concentration of diffusing substance and D is the diffusion constant.

To solve the diffusion equation $\frac{\partial c}{\partial t} = D\frac{\partial^2 c}{\partial x^2}$ ... (1)

We choose $c = X(x)\, T(t)$ [1.1.]

where x is a function of x only and T is a function of t only.

Substituting the above value of c in diffusion equation (1) and dividing both sides by XT we get

$$\frac{1}{T}\frac{dT}{dt} = \frac{D}{X}\frac{d^2X}{dx^2}$$

or
$$\frac{1}{DT}\frac{dT}{dt} = \frac{1}{X}\frac{d^2X}{dx^2} = -\lambda^2 \quad \text{(say)}$$

$$\therefore \quad \frac{1}{T}\frac{dT}{dt} = -\lambda^2 D \quad \text{... (2)}$$

and
$$\frac{1}{X}\frac{d^2X}{dx^2} = -\lambda^2 D \quad \text{... (3)}$$

The solution of equations (2) and (3) are

$$T = A_1 e^{-\lambda^2} Dt \quad \text{... (4)}$$
$$X = A \sin \lambda x + B \cos \lambda x \quad \text{... (5)}$$

The solution of equation (1) with the help of equations (1.1), (4) and (5) is

$$c = (A \sin \lambda x + B \cos \lambda x) A_1 e^{-\lambda^2 Dt} \quad \text{... (6)}$$

5. The equation of a plane sound wave is

$$s = 6.0 \times 10^{-6} \cos(1900t + 5.72\,x) \text{ m}$$

Find the velocity of the wave. [C.U. 2013]

Here $s = 6.0 \times 10^{-6} \cos(1900t + 5.72\,x)$

∴ The velocity of the wave is

$$V = \frac{ds}{dt} = -6.0 \times 10^{-6} \times 1900 \sin(1900t + 5.72\,x) \text{ m/sec.}$$

$$= -1.14 \times 10^{-2} \sin(1900t + 5.72\,x) \text{ m/sec.}$$

6. A progressive harmonic wave is respresented by

$$y(x, t) = a \sin(0.5x - 10t)$$

where x is in metre and t is second. Obtain the wave velocity. [C.U. 2008]

Here $y(x, t) = a \sin(0.5x - 10t)$

∴ wave velocity $V = \dfrac{dy}{dt} = -10a \cos(0.5x + 10t)$ metre/sec.

REVIEW QUESTIONS

1. Derive the general differential equation of the wave motion. What are the different types of progressive waves?
2. (a) What do you mean by wave motion?
   (b) What are transverse and longitudinal waves?
3. Derive an equation for the progressive wave of wavelength λ and amplitude a moving with a velocity c in the positive x-direction.
4. Write down the wave equation in three dimensions in (a) Cartesian coordinate (b) cylindrical coordinates and (c) spherical polar co-ordinate.
5. Solve the three dimensions wave equations in
   (a) rectangular cartesian co-ordinates
   (b) cylindrical co-ordinates
   (c) spherical polar co-ordinates

PROBLEMS

1. If a wave of frequency 500 Hz is travelling with a velocity of 200 m/sec, then find the change in phase at a given point in a space between a time interval of 10^{-3} sec and also find the path difference between two points which differ in phase by $\frac{\pi}{2}$ radian. [P.U. 1993]

   [**Ans.** Phase change = π radian, Path difference = 0.1 m]
2. A simple harmonic wave travelling along x-axis is given by $y = 5 \sin 2\pi\,(0.2t - 0.5x)$ m. Calculate the amplitude, frequency, wavelength, wave velocity, particle velocity and particle acceleration. [P.U. 1992]

   [**Ans.** Amplitude = 5 m, frequency = 0.2 Hz, wavelength = 2 m, wave velocity = 0.4 m/sec, particle velocity = $2\pi \cos 2\pi\,(0.2t - 0.5x)$ m/sec and particle acceleration = $-0.8\pi^2 \sin 2\pi\,(0.2t - 0.5x)$ m/sec^2]

3. A transverse harmonic wave of amplitude 0.01 m is generated at one end ($x = 0$) of a long horizontal string by a tuning fork of frequency 500 Hz. At a given instant of time the displacement of the particle at $x = 0.1$ m is – 0.005 m and that of the particle at $x = 0.2$ m is 0.005 m. Calculate the wavelength and wave velocity. Obtain the equation of the wave assuming that the wave is travelling along the x-direction and that the end $x = 0$ is at the equilibrium position at $t = 0$.

[**Ans.** Wavelength = 0.2 m, velocity = 100 m/sec and $y = 0.01 \sin(1000\ \pi t - 10\pi x)$]

4. The rectilinear flow of heat in one dimension is

$$\frac{d\theta}{dt} = h\frac{d^2\theta}{dx^2}$$

where θ is the excess of temperature of the surface over that of the surroundings and h is a constant. Solve the above equation by the method of separation of variables.

[**Ans.** $\theta = (A \sin \lambda x + B \cos \lambda x).\ A_1 e^{-\lambda^2 ht}$ where λ is a constant.]

CHAPTER 8

Reflection, Refraction, Interference, Diffraction, Stationary Wave, Beats, Phase velocity and Group velocity, Doppler effect

8.1. INTRODUCTION

Sound wave is a longitudinal wave. It can be reflected, refracted, interfered and diffracted under appropriate conditions. The question of polarisation in case of longitudinal vibration does not arise as it is only possible in case of transverse wave where the particles vibrate in all possible directions perpendicular to the direction of propagation, but in case of longitudinal wave the particles vibrate in one direction along the line of propagation of waves. Sound waves can form stationary waves, beats in appropriate condition.

8.2. REFLECTION OF WAVES

Sound waves play an important role in number of phenomena such as echoes, reverberation in buildings. It also plays part in speaking tube and horn and in formation of stationary waves.

When a sound wave meets an obstacle its behaviour is determined by the dimension of the obstacle, compared to the wavelength of the wave. When the obstacles are large compared with the wavelength of the wave, the wave is reflected from the obstacle in the same medium. Sound waves are generally of large wavelength so very large obstacle is required to produce reflection and to produce shadow behind.

Sound wave follows the laws of reflection and refraction and Huygens principle as is given in the case of light waves.

If the dimension of the obstacle is comparable to the wavelength of the wave the waves bend round the corner and encroaches into shadow appreciably. This phenomenon is called diffraction.

If the obstacle is sufficiently smaller compared to the wavelength of the wave, the particle behaves as a source of new waves spreading out in all directions. This is known as scattering of waves.

8.3. CHANGE OF PHASE BY REFLECTION

The velocities of the vibrating particles are reversed and suffer a change of phase π when a wave meets a rigid wall or a denser medium. The pressure waves are reflected back without change of phase at the surface of reflection. This will be found in case of closed organ pipe. The state of compression or rarefaction will remain unchanged in form in the case of longitudinal waves. As the displacement of the vibrating particles are reversed in direction so in case of transverse wave the crest is reflected as trough and trough as crest.

In case of reflection from a gas surface, that is, when the wave meets a yielding wall or a lighter medium, the resultant velocity is maximum and remains unchanged in direction. The change of phase which takes place is zero in velocity wave and π in pressure wave. In compressional wave compression is reflected as rarefaction and rarefaction as compression. In transverse wave crest is reflected as crest and trough as a trough.

8.4. ECHO

It is heard by reflection of sound wave from an extended reflector like a big wall or cliff and is distinguished from the original.

The human ear is unable to distinguish two sounds unless the time interval between them is 0.1 second. So when a sound produces echo, the time interval between the echo and the original sound must be at least 0.1 second. The original sound must go to the reflector and come back. If the distance of the reflector be '*s*' metre it will travel a distance 2*s* metre. Sound travels 332 metres per second. So $s = \frac{332 \times 0.1}{2}$ metres = 16.6 metres should be the minimum distance of the reflector. Again a man ordinarily requires $\frac{1}{5}$ sec to utter a monosyllabic word. So there must be at least $\frac{1}{5}$ second interval between the utterance of monosyllabic and reception of sound. Accordingly the distance of the reflector should be at least $\frac{1}{2} \times 332 \times \frac{1}{5}$ metres = 33.2 metres from the source.

In a room of ordinary size the interval between the source and its echo is so short that (less than 0.1 sec) we hear a continuous rolling of sound due to multiple reflection, persisting for some time even after the source is withdrawn. This phenomenon is known as reverberation. This will be discussed in detail hereafter.

Whispering Gallery. St. Paul's Cathedral in London has a whispering gallery, in which any person who whispers along the wall on one side can be distinctly heard at any one of the galleries due to successive reflection provided the listener places his ear near the wall. Whisper is inaudible at any point away from the wall. The gallery has a circular form round the base, inside a hemispherical dome. Rayleigh gives a theory which states that whisper is audible with abnormal loudness not only at the point diametrically opposite to the whisperer but also where whispers seem to creep round the gallery horizontally along that towards which the whisperer faces. Raman carried out the experiment with high frequency source and sensitive flame (1922) and his observations confirmed by Rayleigh's conclusions. The complete explanation seems yet wanting.

8.5. TECHNICAL APPLICATION OF SOUND WAVE REFLECTION

The depth of sea from a moving ship by the help of echoes of large importance.

There are many ways of finding the depth. They are divided into two groups:

(*a*) Low frequency sonic system

(*b*) High frequency supersonic system

(a) Low Frequency Sonic System

The method adopted by British Admiralty. The old method of Fessender has been improved by taking short intense and heavily damped sound impulse, so that, received echo is the only sound heard. The source of sound is the steel diaphragm which is struck by an electromagnetic hammer. The hammer is kept aside by electromagnet except when circuit is broken. This happens when an insulated bar fixed on revolving metal drum passes through connecting brushes. A damped high frequency sound wave generated by the diaphragm is reflected from the bottom of the sea and is picked up by hydrophone on the other side of the ship. The observer's telephone is short circuit except for a short period. The phase of the short circuited except for a short period. The phase of the short listening period can be set at will anywhere between 'zero position' where diaphragm is struck by the hammer and the 'full scale' which is one complete revolution of the drum. The depth is measured as a fraction of the time for one revolution. The brushes connected to the phones can be moved through any angle by means of a hand-wheel. They are moved through a correct angle so that the telephone comes into operation at the moment when the reflected sound is received by the hydrophone. The hand wheel is generally calibrated directly in feet (1 foot = 0.3042 m) or fathoms.

(b) High Frequency Supersonic System

It consists of Langevin's high frequency piezo-electric crystal oscillator which includes a sandwich of quartz plates connected between steel plates such that the mass of the crystal and steel plates was forced to vibrate as a whole in a desired frequency say 30 to 50 kc/sec.

A high voltage spark oscillatory circuit tuned to the desired frequency, controls the sandwiched crystal oscillator. A single train of damped high frequency wave is emitted from the crystal transmitter. The crystal is kept in contact with the surface of the sea water. The damped signal travels to the sea bed and is reflected and returns to the quartz oscillator. The crystal acts as its own detector both of the emitted and reflected ultrasonic wave. The time interval between emission and reception of the waves is directly recorded by oscillogram in which the small deflection of the spot of light at right angles to the previous calibrated time base, once during the transmission of the impulse and again during echo impulse. Knowing the velocity of sound in sea water with the help of recorded time, depth of the sea bed is calculated.

8.6. REFRACTION OF SOUND WAVE

Refraction is a general property of the wave motion. As sound travels in wave it is expected that when it falls on the surface of separation of two media, where there is difference in density and hence difference in velocity of propagation of sound, there will be a change in the direction of propagation in the two media, except when the incidence is normal to the surface of separation.

If i be the angle of incidence in the first medium and r be the angle of refraction in the second medium, as in the case of light.

$$\frac{\sin i}{\sin r} = {}_1\mu_2 = \frac{c_1}{c_2} \quad \text{... (1)}$$

where ${}_1\mu_2$ is acoustical refractive index of the second medium with respect to the first and c_1 and c_2 are the velocities of the sound wave in the first and second medium.

Like light it is also totally internally reflected when travels from denser to rarer medium making the angle of incidence greater than the critical angle.

It is possible to converge the sound by convex lens of say carbon dioxide of lens being bounded by thin rubber membranes. The focal length f of a lens is given by

$$\frac{1}{f} = ({}_1\mu_2 - 1)\left(\frac{1}{r_1} - \frac{1}{r_2}\right) = \left(\frac{c_1}{c_2} - 1\right)\left(\frac{1}{r_1} - \frac{1}{r_2}\right)$$

r_1, r_2 are the radii of curvature of two surfaces.

8.7. INTERFERENCE OF SOUND WAVE

Interference is the inherent property of the wave motion. This is due to superposition of the wave of the same period and meeting at points in a medium. When they meet at the same phase they combine to give a large displacement which is equal to the sum of their amplitudes. But when they meet in opposite phase the amplitude becomes less being equal to the difference of their amplitude. They will annihilate and silence is produced when their amplitudes are equal.

Let the two waves travelling along the same direction be represented by

$$y_1 = a\sin\frac{2\pi}{\lambda}(ct - x) \quad \text{... (1)}$$

$$y_2 = b\sin\left\{\frac{2\pi}{\lambda}(ct - x) + \alpha\right\} \quad \text{... (2)}$$

α is the difference in phase, a and b are the amplitudes, λ is the wavelength and c is the velocity of wave propagation.

By the principle of superposition

$$
\begin{aligned}
y &= y_1 + y_2 \\
&= a \sin \frac{2\pi}{\lambda}(ct - x) + b \sin \left\{ \frac{2\pi}{\lambda}(ct - x) + \alpha \right\} \\
&= a \sin \phi + b(\sin \theta + \alpha) \qquad \text{... (3)}
\end{aligned}
$$

where θ stands for $\frac{2\pi}{\lambda}(ct - x)$

or $\quad y = (a + b \cos \alpha) \sin \theta + b \sin \alpha \cos \theta = A \sin (\theta + \varepsilon)$

where $\quad a + b \cos \alpha = A \cos \varepsilon,\ b \sin \alpha = A \sin \varepsilon$

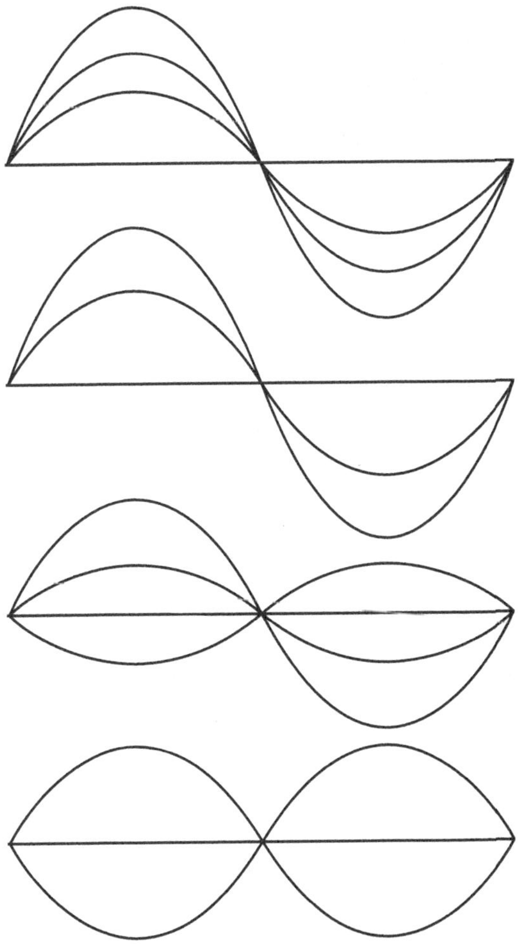

Fig. 1.

Squaring and adding

$$A = \sqrt{a^2 + b^2 + 2ab \cos \alpha}$$

and

$$\varepsilon = b \sin \left\{ \frac{2\pi}{\lambda}(ct - x) + \alpha \right\}$$

or $$y = A\sin\left\{\frac{2\pi}{\lambda}(ct - x) + \varepsilon\right\}$$

when $\alpha = 0, 2\pi, 4\pi...$ even multiple of π

$A = a + b$ if $a = b$, $A = 2a$ and

$\alpha = \pi, 3\pi, 5\pi...$ odd multiple of π

$A = a - b$ if $a = b$, $A = 0$

Thus when phase difference is the multiple of 2 π that path difference is the multiple of λ two waves will reinforce each other and intense sound is heard. When the phase difference is odd multiple of π, that is, path difference $\frac{\lambda}{2}$ or its odd multiple, two waves will destroy each other. When amplitudes are different very feeble sound will be heard. No sound will be heard when the amplitudes are same. They are represented graphically in Fig. 1. Interference of sound wave can be directly demonstrated by Quenche's experiment Fig. 1(*a*).

Quenche's Experiment. Experimental demonstration of interference of sound wave B_1BCC_1 is a continuous hollow U-tube with openings at A and E. Another U-tube $LMNO$ is fitted to the open end BC of the U-tube by sliding it over the limbs AB_1 and EC_1.

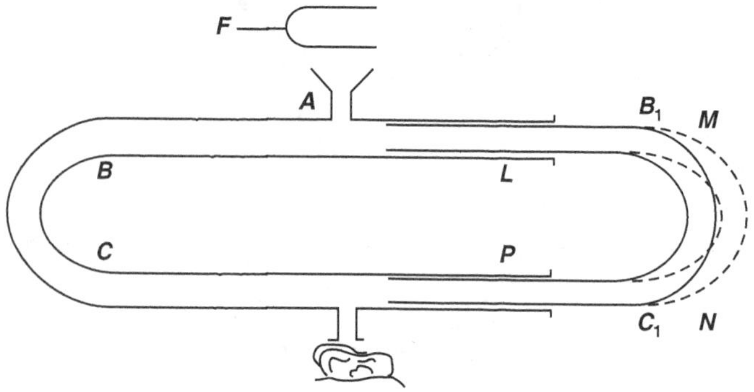

Fig. 1(*a*).

The U-tube $LMNP$ is pushed towards A and E up to a maximum when the path $ABCE$ becomes equal to the path AB_1MNC_1E within the pipe. A vibrating tuning fork is placed near A as in figure, ear or an ear trumpet is placed at E. A quite loud sound will be received.

Then the movable tube is gradually drawn outwards and at a certain position of the tube it will be found that sound at E is inaudible from the known frequency n of the tuning fork the wavelength of sound can be known and the measurement will show that, at that moment the path along AB_1MNPE is greater by $\frac{\lambda}{2}$. The tube $LMNP$ is further drawn so that the difference in path is λ when maximum sound will be heard.

In this way we shall get alternately intensified and no sound with the change of path of $LMNP$. The experiment gives a direct evidence of interference of sound wave.

8.8. BEATS

Composition of Two Waves Slightly Different Frequencies, Velocity of Propagation Being Same.

If the two component waves that travel along the same direction are not very strictly but very nearly of equal wavelength the resulting wave motion presents peculiarities. It is not harmonic, the resultant amplitude is greater in some places than in others so that the actual wave consists of condensation and rarefaction which is intensified in some region while diminished in other place. The sound heard fluctuates in loudness. This phenomenon of destruction and reinforcement of two sounds together is called Beats.

Let the displacements y_1 and y_2 at a place due to waves having same amplitudes are given by

$$y_1 = a\sin\frac{2\pi}{\lambda_1}(ct-x) = a\sin 2\pi\left(n_1 t - \frac{x}{\lambda_1}\right) \quad \text{... (5)}$$

as $$\frac{c}{\lambda_1} = n_1$$

and $$y_2 = a\sin\frac{2\pi}{\lambda_2}(ct-x) = a\sin 2\pi\left(n_2 t - \frac{x}{\lambda_2}\right) \quad \text{... (6)}$$

and $$n_2 = \frac{c}{\lambda_2}$$

where n_1 is the frequency of one and $n_1 = n_1 + v$ of the other. The difference in frequency is v of two waves, which produces the difference in wavelength

The resultant wave

$$y = y_1 + y_2 = a\sin\frac{2\pi}{\lambda_1}(ct-x) + a\sin\frac{2\pi}{\lambda_2}(ct-x)$$

$$= 2a\sin\frac{2\pi}{2}\left(\frac{1}{\lambda_2}+\frac{1}{\lambda_1}\right)(ct-x).\cos\frac{2\pi}{2}\left(\frac{1}{\lambda_1}-\frac{1}{\lambda_2}\right)(ct-x) \quad \text{... (7)}$$

But $$\frac{1}{\lambda_1}-\frac{1}{\lambda_2} = \frac{1}{\lambda}-\frac{1}{\lambda+d\lambda} = \frac{d\lambda}{\lambda^2}$$

$$\frac{1}{\lambda_1}+\frac{1}{\lambda_2} = \frac{1}{\lambda}+\frac{1}{\lambda+d\lambda} = \frac{2}{\lambda}$$

or $$y = 2a\cos\frac{2\pi}{\lambda'}(ct-x)\sin\frac{2\pi}{\lambda}(ct-x) \quad \text{... (8)}$$

where $$\lambda' = \frac{2\lambda^2}{d\lambda}$$

The wave of wavelength λ advances with periodic vibration of amplitude which has a wave character of wavelength $\lambda' = \frac{2\lambda^2}{d\lambda}$ advances with the same velocity c as the original wave. The resultant displacement is given in Fig. 5. in Art. 2.9.

The distance between two consecutive maxima or minima is

$$\frac{\lambda'}{2} = \frac{\lambda^2}{d\lambda} = \frac{\lambda_1\lambda_2}{\lambda_2-\lambda_1} \quad \text{... (9)}$$

The time interval between two maxima or two minima is given by

$$\frac{\text{distance}}{\text{velocity}} = \frac{\lambda_1\lambda_2}{c(\lambda_2-\lambda_1)} \quad \text{... (10)}$$

Hence the frequency of the occurrence of beats is

$$\frac{c(\lambda_2-\lambda_1)}{\lambda_1\lambda_2} = c\left(\frac{1}{\lambda_1}-\frac{1}{\lambda_2}\right) = n_1 - n_2 \quad \text{... (11)}$$

Thus beats produce the periodic fluctuation of loudness of sound.

8.9. PHASE VELOCITY AND GROUP VELOCITY

Composition of trains of waves differing slightly in frequencies *i.e.*, in wavelength and travelling with different velocities.

The compression and rarefactions of longitudinal wave or crest and trough of a transverse wave advances with the same velocity in a medium or in other words the advance of phase takes place with the same velocity, this velocity of advance is known as phase velocity.

The equation of a plane progressive wave is given by

$$y = a\sin\frac{2\pi}{\lambda}(ct-x) = a\sin(\omega t - kx) \quad \text{... (12)}$$

where $\omega = \frac{2\pi}{\lambda}c = 2\pi n$ as $c = n\lambda$ and $k = \frac{2\pi}{\lambda}$

$$\therefore \quad \frac{\omega}{k} = \frac{2\pi n}{\frac{2\pi}{\lambda}} = n\lambda = c$$

c is called the phase velocity of the wave. The phase velocity is the velocity with which a plane progressive wave front travels forward and which has a constant phase given by $(\omega t - kx)$. Thus

$$\omega t - kx = \text{constant}$$

Differentiating with respect to time t, we get

$$\omega - k\frac{dx}{dt} = 0$$

or $\frac{dx}{dt} = \frac{\omega}{k} = c$, c the phase velocity of the wave.

In the resultant curve of a beat note there are maxima at regular interval of space. In between two maxima there is a minimum. Propagation of maxima and minima of the resultant wave including a group of waves takes place with the same velocity as that of the component waves.

But there will be a change of the type of the resultant wave if the velocity changes with wavelength. The velocity of advance of the central maximum is called group velocity.

The displacement due to two waves of equal amplitudes, different in wavelengths λ_1 and λ_2 different in velocities c_1 and c_2 are given by

$$y_1 = a\sin\frac{2\pi}{\lambda_1}(c_1t - x) \quad \text{... (13)}$$

$$y_2 = a\sin\frac{2\pi}{\lambda_2}(c_2t - x) \quad \text{... (14)}$$

The resultant displacement is given by

$$y = y_1 + y_2 = 2a\sin 2\pi.\frac{1}{2}\left\{\left(\frac{c_1}{\lambda_1}+\frac{c_2}{\lambda_2}\right)t - \left(\frac{1}{\lambda_1}+\frac{1}{\lambda_2}\right)x\right\}\cos 2\pi.\frac{1}{2}\left\{\left(\frac{c_1}{\lambda_1}-\frac{c_2}{\lambda_2}\right)t - \left(\frac{1}{\lambda_1}-\frac{1}{\lambda_2}\right)x\right\} \quad \text{... (15)}$$

If λ and c be the mean wavelength and velocity

$$\frac{c_1}{\lambda_1}+\frac{c_2}{\lambda_2} = \frac{2c}{\lambda} \quad \text{and} \quad \frac{1}{\lambda_1}+\frac{1}{\lambda_2} = \frac{2}{\lambda}$$

$$\frac{1}{\lambda_2}-\frac{1}{\lambda_1} = d\left(\frac{1}{\lambda}\right) = -\frac{d\lambda}{\lambda^2} = -\frac{1}{\frac{\lambda^2}{d\lambda}}$$

or
$$\frac{1}{\lambda_1}-\frac{1}{\lambda_2} = =\frac{1}{\frac{\lambda^2}{d\lambda}}$$

and
$$\frac{c_1}{\lambda_1}-\frac{c_2}{\lambda_2} = d\left(\frac{c}{\lambda}\right)=\frac{cd\lambda-\lambda dc}{\lambda^2}=\frac{c-\lambda\frac{dc}{d\lambda}}{\frac{\lambda^2}{d\lambda}}$$

The equation (15) becomes
$$y = 2a\cos\frac{2\pi}{\left(\frac{2\lambda^2}{d\lambda}\right)}\left\{\left(c-\lambda\frac{dc}{d\lambda}\right)t-x\right\}.\sin\frac{2\pi}{\lambda}(ct-x) \quad ...(16)$$

The above equation represents the periodic variation of amplitude of the original wave as a result of superposition. This periodic variation possesses wave character of wavelength $2\lambda^2/d\lambda$ advancing with a velocity given by $C_g = c-\lambda\frac{dc}{d\lambda}$, C_g is called the group velocity of the wave. If $\frac{dc}{d\lambda} = 0$, $C_g = c$ and the group velocity and phase velocity are equal as found in the previous case Art. 8.7.

8.10. STATIONARY OR STANDING WAVE

Stationary waves are formed in any medium when two exactly similar wave trains having the same wavelength, velocity and amplitude travel along the same straight line in a medium in opposite direction.

By the principle of superposition of waves, the resultant displacement of a particle of the medium is equal to the algebraic sum of the displacement due to component waves if the amplitudes are small.

The formation of stationary waves most conveniently will be discussed by drawing the space displacement curves of the component waves and finding their resultant at particular instants.

Since the space displacement curves of transverse and longitudinal waves are similar the figures are same in both kinds of waves.

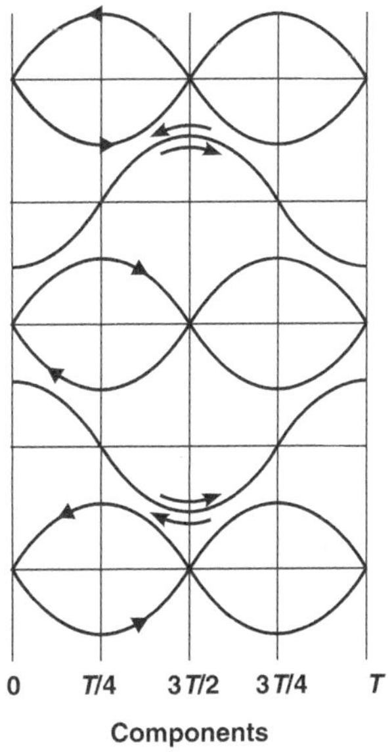

Fig. 2.

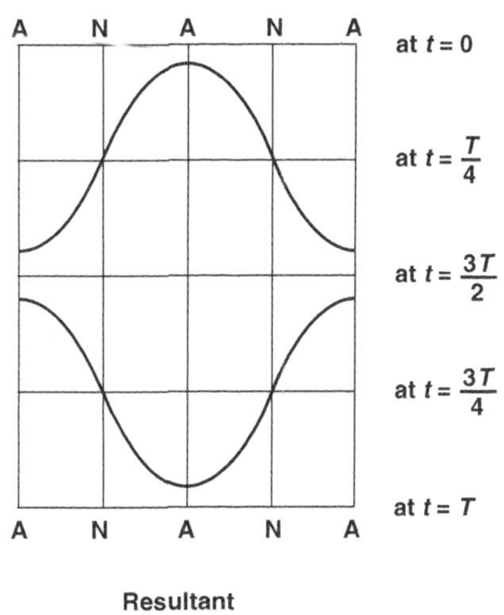

Fig. 3.

Fig. 2 represents two similar train of waves moving in the opposite direction through the same medium and the resultant is shown in Fig. 3 at five different instants $0, \frac{T}{4}, \frac{T}{2}, \frac{3T}{4}, T$ where T represents the time period of the vibration of the particles.

Let the two waves be represented by

$$y_1 = a\sin\frac{2\pi}{\lambda}(ct - x) \qquad \text{... (17)}$$

$$y_2 = a\sin\frac{2\pi}{\lambda}(ct + x) \qquad \text{... (18)}$$

Resultant

$$y = y_1 + y_2$$

$$= a\sin\frac{2\pi}{\lambda}(ct - x) + a\sin\frac{2\pi}{\lambda}(ct + x)$$

$$= 2a\sin\frac{2\pi}{\lambda}ct.\cos\frac{2\pi}{\lambda}x \qquad \text{... (19)}$$

The slope of the curve or strain

$$\frac{dy}{dx} = -\frac{4\pi a}{\lambda}\sin\frac{2\pi}{\lambda}ct.\sin\frac{2\pi}{\lambda}x \qquad \text{... (20)}$$

The velocity of the vibrating particles

$$\frac{dy}{dt} = \frac{4\pi a}{\lambda}\cos\frac{2\pi}{\lambda}ct.\cos\frac{2\pi}{\lambda}x \qquad \text{... (21)}$$

Acceleration

$$\frac{d^2y}{dt^2} = -\frac{8\pi^2c^2}{\lambda^2}a\cos\frac{2\pi x}{\lambda}\sin\frac{2\pi}{\lambda}ct \qquad \text{... (22)}$$

From equation (20)

$$\frac{d^2y}{dt^2} = -\frac{8\pi^2}{\lambda^2}a\cos\frac{2\pi}{\lambda}x\sin\frac{2\pi}{\lambda}ct \qquad \text{... (23)}$$

or

$$c^2\frac{d^2y}{dx^2} = -\frac{8\pi^2c^2}{\lambda^2}a\cos\frac{2\pi}{\lambda}x\sin\frac{2\pi}{\lambda}ct \qquad \text{... (24)}$$

From equation (22) and (24) we can write

$$\frac{d^2y}{dt^2} = c^2\frac{d^2y}{dx^2} \qquad \text{... (25)}$$

This is the characteristic equation of a stationary wave.

When $\cos\frac{2\pi}{\lambda}x = 0$; gives $y = 0$, $\frac{dy}{dt} = 0$

$\frac{dy}{dx}$ is maximum at places given by $\frac{2\pi}{\lambda}x = (2m+1)\frac{\pi}{2}$

where m is an integer.

or

$$x = \left(\frac{m}{2} + \frac{1}{4}\right)\lambda$$

Thus there are points at regular intervals, which are never displaced. They are called *Nodes.* The distance between two consecutive nodes is $\frac{\lambda}{2}$.

The displacement will be maximum or minimum at places given by

$$\cos\frac{2\pi}{\lambda}x = \pm 1$$

or $$\frac{2\pi}{\lambda}x = m\pi$$

where m is an integer or zero

or $$x = \frac{m\lambda}{2}$$

at instants given by

$$\sin\frac{2\pi}{\lambda}ct = \pm 1$$

or $$t = (2m+1)\frac{\pi}{4}$$

Thus there are points at regular intervals, where the particles periodically attain their maximum value $2a$ or a minimum value $-2a$ where strain $\frac{dy}{dx} = 0$ and velocity is maximum.

These are called *Antinodes*. The distance between two consecutive antinodes is $\frac{\lambda}{2}$ and between a node and next antinode is $\frac{\lambda}{4}$.

In a stationary wave

(*i*) All the particles are in a state of vibration except where there are nodes which are at rest.

(*ii*) All the particles between two consecutive nodes have the same phase but differ in amplitude.

(*iii*) The particles on the opposite sides of each node move in opposite direction hence they are in opposite phase.

8.11. DISTINCTION BETWEEN PROGRESSIVE AND STATIONARY WAVES

Progressive wave	Stationary wave
1. A progressive wave is a single train of waves which proceeds along a certain direction in a medium.	1. It is produced by superposition of two trains of wave moving in opposite direction.
2. In a progressive wave each particle will excute S.H.M of same period and amplitude with different phase depending on the position of the particle.	2. In a stationary wave each particle in the medium with certain exception executing S.H.M with same period and phase but with different amplitude depending on the position of the particle.
3. In a progressive wave there is no point where the particles will remain permanently at rest and there is no point where the particle will be permanently vibrating with maximum amplitude.	3. In case of stationary wave there are points where the particle is at rest which are called *Nodes*. The point between two nodes where the particle move with maximum amplitude are called *Antinodes*.
4. The waveform in the case of progressive wave proceeds with definite velocity through the medium.	4. The waveform in stationary wave remains stationary in the medium, without changing its type or advancing in any direction of the medium.
5. The instantaneous form of a progressive wave repeats itself at certain interval of length in the medium known as wavelength.	5. The instantaneous form of at stationary wave repeats itself at certain interval of length in the medium known as *wavelength*.
6. Each particle in the progressive wave undergoes similar change of pressure and density.	6. There is maximum or minimum variation of pressure and density in the case of stationary wave.
7. There is a transmission of energy across every plane.	7. There is no flow of energy across any plane. This is due to the fact that condensation and velocity curve differ in phase.

8.12 ACOUSTIC SHADOW AND DIFFRACTION OF SOUND

It is observed that sound bends appreciably round any obstacle, while light appears to be linear. In case of fine obstacle such as the edge of a razor blade or the end of a needle, light bends round the appropriate obstacle. It is due to the fact that wavelength of light is very short while that of sound is more than million times larger.

The region behind the obstacle placed before the sound source where sound is absent is known as *Acoustic shadow* from optical analogy.

The bending of waves round the obstacle and their encroachment within the geometrical shadow is known as Diffraction of sound wave.

The acoustical shadow is not so sharp as the optical shadow. In order to produce sharp acoustic shadow, the wavelength of sound wave should be short, that is, its frequency should be high and the obstacle should be of moderate size.

The sound wave can be diffracted by suitable acoustic grating similar to optical grating. Grating for this purpose has been designed and constructed by using glass rods spaced at about one centimeter apart. Source of sound is the sound generated by spark discharge of a condenser, which produces sound wave of a few millimeter in length.

8.13 (A) DOPPLER EFFECT

It is a common experience that an observer standing on a railway platform, the frequency of the note emitted by the whistle of an approaching locomotive engine appear higher and passes away, the frequency appears decreasing continuously with the increasing distance. The apparent change of frequency of the waves emitted due to the relative motion of the source and observer is known as *Doppler effect* and the principle which gives the explanation of the effect is called *Doppler principle.*

1. Stationary Source, Observer in Motion

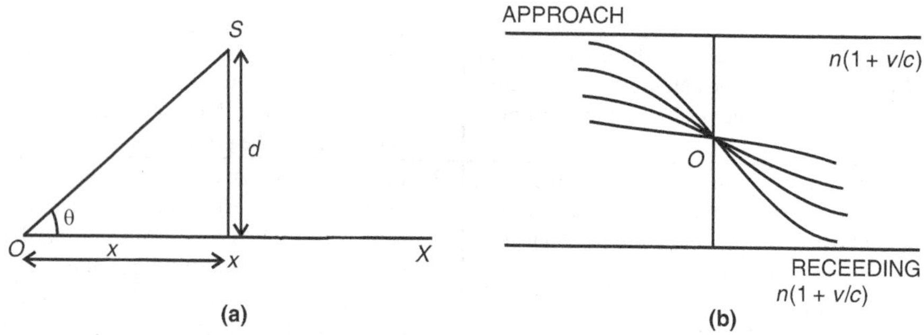

Fig. 4.

Let S be the stationary source and O is the position of the observer moving along OX with a velocity v, distance SX equal to d from the source OS makes an angle θ with OX. θ increases as O approaches X and becomes 90° when comes to X. The observer approaches S with a velocity $v \cos \theta$, the apparent velocity of sound waves at O is $c + v \cos \theta$. If n be the actual frequency of the source and n_1 the apparent frequency of the source on arrival at O then

$$\frac{c}{n} = \frac{c + v\cos\theta}{n_1} = \lambda_1$$

i.e.,

$$n_1 = \frac{c+v\cos\theta}{c} n$$

$$= \left(1+\frac{v}{c}\cos\theta\right) n \qquad \text{... (25)}$$

But from geometry

$$\cos\theta = \frac{x}{\sqrt{x^2+d^2}}$$

so

$$n_1 = \left[1+\frac{v}{c}\frac{x}{(x^2+d^2)^{1/2}}\right] n \qquad \text{... (26)}$$

Thus it is found n_1 is a function of x for different values of d. A set of curves showing the variation of the magnitude of n_1 for different values of d is given in Fig. 4. From equation (26)

$$n_1 - n = \frac{nv}{c}\frac{x}{(x^2+d^2)^{1/2}} \qquad \text{... (27)}$$

$$= \pm\frac{nv}{c}$$

when $x >> d$ and

$$\frac{dn}{dx} = 0$$

As the observer passes x the value of $(n_1 - n)$ changes sign and at $x = 0$

$$\frac{dn_1}{dx} = \frac{nv}{c}\frac{1}{(x^2+d^2)^{1/2}} - \frac{nv}{c}\frac{x^2}{(x^2+d^2)^{3/2}}$$

$$= \frac{nv}{cd} \qquad \text{when } x = 0$$

2. The Source is in Motion, Observer is Stationary

In this case, we interchange the position of the source and the observer in the Fig. 4 (*a*). O is the source moving with velocity u along x direction. S is the position of the observer. But in the case of apparent change of frequency as observed by the observer at S is due to motion of the source towards the observer. The observer will receive the wave of wavelength

$$\lambda_1 = \frac{c-u\cos\theta}{n} \text{ instead of } \frac{c}{n} = \lambda$$

and hence

$$n_1 = \frac{c}{\lambda_1} = \left(\frac{c}{c-u\cos\theta}\right) n$$

$$= n\left[\frac{c-u\cos\theta+u\cos\theta}{c-u\cos\theta}\right]$$

$$= n\left[1+\frac{u\cos\theta}{c-u\cos\theta}\right]$$

or

$$n_1 - n = n\frac{u\cos\theta}{c-u\cos\theta} \qquad \text{... (28)}$$

when x is large compared to d, $\cos\theta \to 1$

and $$n_1 - n = \frac{nu}{c-u} \qquad \text{... (29)}$$

gives $n_1 > n$ when $\theta = \frac{\pi}{2}$ gives $n_1 = n$; when the moving source passes away the observer $\theta = \pi$, $\cos\theta = -1$ gives

$$n_1 - n = \frac{-nu}{c+u} \text{ gives } n_1 < n$$

3. Observer, Source and Wind are moving along the same direction

Let the motion of the source, observer and wind be from left to right.

Let the velocity of sound be c, of the wind ω,of the source be v_s and observer be v_0. The source S_1 gives out n waves in one second which

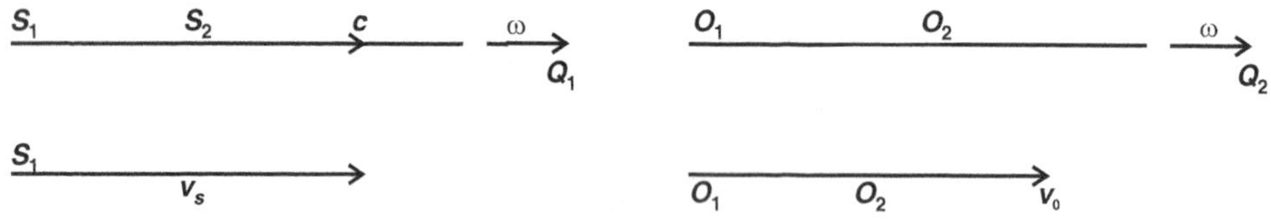

Fig. 5.

spread over the length $S_1Q_1 = c + \omega$. By this time the source has moved to S_2, naturally n waves will be crowded in the length $S_2Q_1 = c + \omega - v_s$. The new wavelength λ_1 is

$$\frac{c+\omega-v_s}{n}$$

Again the wave which reaches the observer at O_1 at certain instant will move to Q_2 in one second, so $Q_1Q_2 = c + \omega$. The observer has moved in the mean time to O_2, so the number of waves n_1 say will be crowded over the length $c + \omega - v_0$ so, n_1 is given by

$$n_1 = \frac{c+\omega-v_0}{\lambda_1} = \frac{c+\omega-v_0}{c+\omega-v_s}n \qquad \text{... (30)}$$

as perceived by the observer.

If the motion of the source, observer or the wind be not in the direction from the source to the observer then the component of the motion along the direction is to be taken instead of u_s, v_0 or ω in the above calculation.

Case I If $v_s = v_0$ or $v_s = v_0 = 0$ we get from equation (30) $n_1 = n$

Case II If $v_s > v_0$ the source approaches the observer gives $n_1 > n$; $v_s < v_0$ the source recedes the observer $n_1 < n$.

Case III If $v_s = 0$ the source is at rest

$$n_1 = \frac{c+\omega-v_0}{c+\omega}n \text{ gives } n_1 < n$$

when observer recedes.

If $v_s = 0$ the observer is moving in opposite direction which means v_0 is $-u_0$

$$n_1 = \frac{c+\omega+v_0}{c+\omega}n \text{ gives } n_1 > n$$

when observer approaches,
If $v_0 = 0$ the observer at rest

$$n_1 = \frac{c+\omega}{c+\omega-v_s}n \text{ gives } n_1 > n$$

the source approaches the observer
If $v_0 = 0$ the source is moving in opposite direction v_s is $-v_s$

$$n_1 = \frac{c+\omega}{c+\omega+v_s}n \text{ gives } n_1 < n$$

the source recedes from the observer.

(B) Limitation of Doppler's Principle

When the velocity of the source is greater than the velocity of sound that is the source is moving with supersonic velocity Doppler's principle is not applicable. It is applicable so long the relative velocity of the source and observer is less than the velocity of sound.

When a bullet is shot through air with velocity less than that of sound the concentrated sound waves move towards the observer with velocity of sound, which gains over the velocity of the bullet. But when velocity of the bullet is supersonic, the observer will hear three sounds.

(*i*) A click as the bullet flies past the observer

(*ii*) Next a buzzing sound produced with the formation of eddies by the motion of the bullet.

(*iii*) Finally the usual sound of explosion of gun-powder.

(C) Verification of Doppler's Principle

Doppler's principle is verified in a simple way in the laboratory by tightly fixing a whistle of moderately high frequency to a rubber tubing free end of which fitted to a blower. The whistle is blown steadily and whirled in a circle. To an observer some distance away, the frequency of the whistle is found to rise and fall apparently. Maximum change is noted when the whistle is moving towards or away from the observer.

The principle can also be verified from optical observation. The spectral lines of light from a star are displaced towards red end which indicates the increase of wavelength that is decrease of frequency, when the star recedes from the earth and when these lines are displaced towards the violet indicate the decrease of wavelength *i.e.*, increase of frequency, when the star approaches the earth.

Let λ be the wavelength of light when the star is stationary and λ' when in motion. When the source be moving with velocity v_s and observer is at rest $v_0 = 0$, we have

$$n' = \frac{c}{c-v_s}n$$

or $$\frac{c}{\lambda'} = \frac{c}{\lambda}\frac{c}{c-v_s}$$

or $$\lambda' = \frac{c-v_s}{c}\lambda$$

$\therefore$ $$\lambda - \lambda' = \frac{v_s}{c}\lambda$$

or $$v_s = \frac{(\lambda - \lambda')}{\lambda}.c$$

So the velocity of the star is found out by noting the change of wavelength in the spectral line.

SOLVED PROBLEMS

1. Calculate the velocity of sound in a gas in which the waves of wavelength 0.50 m and 0.505 m produce 6 beats per second.

Let c be the velocity of sound in the gas. Given wavelengths are

d $$\lambda_1 = 0.50\text{m}$$

and $$\lambda_2 = 0.505\text{m}$$

$$\therefore \quad \frac{c}{\lambda_1} - \frac{c}{\lambda_2} = 6$$

or $$\frac{c}{0.500} - \frac{c}{0.505} = 6$$

or $$c = c = \frac{6 \times 0.500 \times 0.505}{0.005}\text{m}$$

$$= 303 \text{ m/sec}$$

2. Two trains are approaching each other with speed 60 km/hr and 45 km/hr. A whistle of frequency 512 Hz is sounded by the first train. Calculate the frequency of the note heard by a listener in the second train (*i*) before and (*ii*) after the trains pass each other. (speed of the sound 332 m/sec) [Nag. U. 2001]

Velocity of the sound $c = 332$ m/sec

Velocity of the first train $v_s = 60$ km/h.

$$= \frac{60 \times 1000}{3600}\text{m/sec}$$

$$= \frac{50}{3}\text{m/sec}$$

Velocity of the second train (observer) = 45 km/h

$$= \frac{25}{2}\text{m/sec}$$

Case I. Before the trains pass each other *i.e.*, source and observer are moving towards each other so the apparent frequency

$$n' = \frac{c + v_0}{c - v_s}n$$

$$= \frac{332 + \frac{25}{2}}{332 - \frac{50}{3}}.512 \text{ Hz}$$

$$= 559.3 \text{ Hz}$$

Case II. After the trains pass each other *i.e.*, source and observer are moving away from each other. The apparent frequency

$$n'' = \frac{c-v_0}{c+v_s}n$$

$$= \frac{332-\frac{25}{2}}{332+\frac{50}{3}}.512\,\text{Hz}$$

$$= 469 \text{ Hz}$$

3. Two sinusoidal waves

$$y_1 = 0.03 \sin (7t - 10x) \text{ m}$$
$$y_2 = 0.03 \sin (5t - 8x)\text{m}$$

were superimposed. Calculate the group velocity *[GNDU 2003]*

We know the wave equation is

$$y = a\sin\frac{2\pi}{\lambda}(ct-x)$$

$$= a\sin(\omega t - kx)$$

as $$\frac{2\pi}{\lambda}c = \frac{2\pi}{T} = \omega \text{ and} \frac{2\pi}{\lambda} = k$$

From the given equation, we get

$$\omega_1 = 7 \quad k_1 = 10$$

and $$\omega_2 = 5 \quad k_2 = 8$$

$$c_1 = \frac{\omega_1}{k_1} = \frac{7}{10} = 0.7 \text{ m/sec m/sec.}$$

$$c_2 = \frac{\omega_2}{k_2} = \frac{5}{8} = 0.625 \text{ m/sec. m/sec.}$$

$$k_1 = \frac{2\pi}{\lambda_1} \quad \text{or} \quad \frac{1}{\lambda_1} = \frac{10}{2\pi}$$

and $$k_2 = \frac{2\pi}{\lambda_2} = 8 \quad \text{or} \quad \frac{1}{\lambda_2} = \frac{8}{2\pi}$$

Again we know $$\frac{c_1}{\lambda_1} + \frac{c_2}{\lambda_2} = \frac{2c}{\lambda}$$

or $$0.7.\frac{10}{2\pi} + 0.625.\frac{8}{2\pi} = \frac{2c}{\lambda}$$

or $$c = \frac{12}{2\pi}\lambda$$

Again $$\frac{1}{\lambda_1} + \frac{1}{\lambda_2} = \frac{2}{\lambda}$$

or $$\frac{10}{2\pi} + \frac{8}{2\pi} = \frac{2}{\lambda}$$

or $$\lambda = \frac{2\pi}{9}$$

$\therefore$ $$c = \frac{12}{2\pi}\cdot\frac{2\pi}{9} \text{ m/sec}$$

$$= \frac{4}{3}\text{m/sec m/sec}$$

We know $$c_g = c - \lambda\frac{dc}{d\lambda}$$

$$= \frac{4}{3} - \frac{2\pi}{9}\cdot\frac{0.7-0.625}{\frac{25}{8}-\frac{25}{10}}\text{m/sec. m/sec.}$$

$$= 1 \text{ m /sec}$$

REVIEW QUESTIONS

1. What is an echo. How is it formed? What are the practical applications of the phenomenon. Describe one such application.
2. Two sound waves having the same frequency and amplitude travel along the same direction what will be their resultant effect. How will this effect depend upon the phase of the wave.
3. What is interference of sound? How does it differ from that of the light?
4. What is beat? How is it formed? Give the analytical explanation of the formation of beat.
5. What is stationary wave? How is it formed? Distinguish a stationary wave from a progressive wave.
6. Define the terms wave velocity and group velocity. Find a relation connecting the two.
7. What is Doppler effect? Derive an expression for the change in the frequency of a sound wave when the source and the observer are both in motion in a stationary medium.
   [C.U. 1996]
8. Discuss the refraction, reflection and diffraction of sound waves.
   [Barkatullah University 2011]
9. Give an experiment to explain the reflection of sound.
   [Pandit Ravisankar University 2010]

PROBLEMS

1. A police car driving 15 m/sec sounds a siren emitting a note of frequency 550 Hz which is reflected back from a stationary object some distance ahead of the car. What will be the number of beats heard per second by an observer in the car? Assume the velocity of sound = *332 m/sec.* *[C.U. 1996]*
   [**Ans.** No. of beats = 52.05/sec]
2. Calculate the group velocity when two waves
   $$y_1 = 10 \sin(2\pi t - 5x)$$
   $$y_2 = 15 \sin(5\pi t - 5x)$$
   superimpose y_1 and y_2 are in metres. *[P.U. 1994]*
   [**Ans.** Group velocity = $.0.3\pi$ m/sec]
3. Sound waves from a vibrating body reach a point by two paths. There is a silence at the point when the path differ by 12 cm or 36 cm. If the velocity of sound be 330 metres/sec find the frequency of the fork.
   [**Ans.** Frequency = 1375 Hz]

9 CHAPTER

VELOCITY OF ACOUSTIC WAVES (LONGITUDINAL AND TRANSVERSE WAVE)

9.1. VELOCITY OF PLANE LONGITUDINAL WAVES IN A GAS

Imagine a longitudinal plane sound wave consisting of condensation and rarefaction travels through a medium of density ρ, to the right with a velocity c.

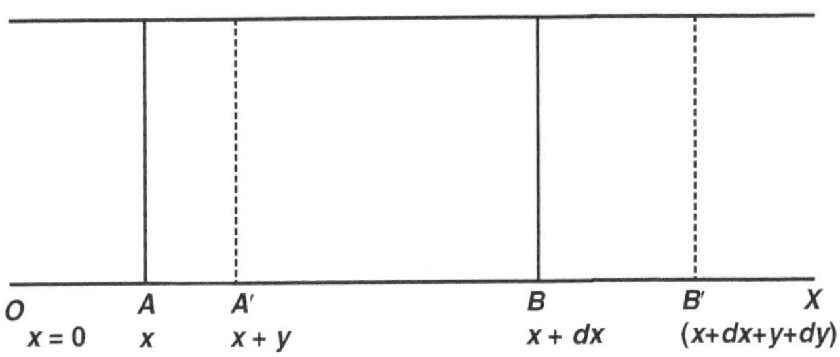

Fig. 1.

To obtain the steady conditions we may imagine that the medium itself is moving towards the left also with a velocity c, so that the condensation and rarefaction will remain localised with respect to the observer.

Let OX be the line of advance of the wave and let AB be a very thin layer of the medium at right angles to it and of unit cross-section. Let x be the distance of A from the origin O at any given instant, $x + dx$ be the distance of the layer B from the origin at the same instant (Fig. 1). The volume enclosed within $AB = dx$. Let y be the displacement from A to A' after time dt, $y + dy$ be the displacement of B to B' in the same time dt. Now the distance between A' and B' from the displaced position of A and B is

$$(x + dx + y + dy) - (x + y) = dx + dy$$

So that the volume enclosed is $(dx + dy)$

Hence the change of volume undergone by the original volume dx in time dt is

$$(dx + dy) - dx = dy$$

$\therefore$ Change of volume per unit volume

$$= \text{Volume strain} = \frac{dy}{dx} \qquad \text{... (1)}$$

Now the bulk modulus of the fluid

$$E = \frac{\text{Volume stress}}{\text{Volume strain}}$$

$$\therefore \qquad \text{Volume stress} = E.\frac{dy}{dx} = \text{Excess pressure } dp \qquad \text{... (2)}$$

If p is the excess of pressure on A and

$p+\frac{dp}{dx}dx$ on B, the change of pressure $= \left(p+\frac{dp}{dx}dx\right)-p$

$$= \frac{dp}{dx}dx$$

From equation (2) we can write

$$\frac{dp}{dx}dx = \frac{d}{dx}\left(E\frac{dy}{dx}\right)dx = E\frac{d^2y}{dx^2}dx$$

= Force per unit area = acoustic pressure = mass × acceleration

$$= \rho dx\frac{d^2y}{dt^2}$$

where ρ is the density of the fluid.

Hence
$$\rho\frac{d^2y}{dt^2} = E\frac{d^2y}{dx^2}$$

or
$$\frac{d^2y}{dt^2} = \frac{E}{\rho}\frac{d^2y}{dx^2} \quad \text{... (2)}$$

or
$$\frac{d^2y}{dt^2} = c^2\frac{d^2y}{dx^2} \quad \text{... (3)}$$

where $c^2=\frac{E}{\rho}$ where c is the velocity of wave propagation.

The general solution of the wave equation is given by

$$y = f(ct-x) + F(ct+x) \quad \text{... (3a)}$$

$$\frac{dy}{dt} = cf'(ct-x) + cF'(ct+x)$$

$$\frac{d^2y}{dt^2} = c^2 f''(ct-x) + c^2 F''(ct+x) \quad \text{... (4)}$$

Again
$$\frac{dy}{dx} = f'(ct-x) + F'(ct+x)$$

$$\frac{d^2y}{dx^2} = f''(ct-x) + F''(ct+x) \quad \text{... (5)}$$

Where $f'(ct+x)$, $F'(ct+x)$, $f''(ct-x)$ and $F''(ct+x)$ are the first and second order differential co-efficients of the functions.

Substituting equations (4) and (5) in equations (2) and (3)

$$c^2[f''\,ct-x+F''(ct+x)] = \frac{E}{\rho}[f''(ct-x)+F''(ct+x)]$$

or
$$c^2 = \frac{E}{\rho} \quad \text{... (6)}$$

or $$c = \sqrt{\frac{\gamma P}{\rho}}$$

This expression of the velocity of the longitudinal wave was given by Newton.

9.2. DISCUSSION ON THE VELOCITY OF SOUND THROUGH GASES

(i) Newton's Value

Newton assumed that the passage of the train of longitudinal wave through air is so slow that heat generated during compression or cold produced during rarefaction is dissipated or absorbed to or from the body of the gas respectively, so the process is isothermal.

Hence Boyle's law

$$PV = \text{constant}$$

holds. Differentiating both sides we get

$$Pdv + Vdp = 0$$

or $$P = \frac{dp}{\frac{-dv}{V}} = \frac{\text{change in pressure}}{\text{volume strain}}$$

$$= E \text{ the bulk modulus}$$

So the velocity $$c = \sqrt{\frac{E}{\rho}} = \sqrt{\frac{P}{\rho}}$$

Taking air at standard temperature and pressure $c = \sqrt{\frac{76 \times 13.6 \times 981}{0.001293}}$ cm/sec

$$= 280 \text{ metres per second}$$

The value so obtained is much less than the observed value which is 331.5 metres per second.

(ii) Laplace Correction

This discrepancy was explained by Laplace, nearly after a century. He pointed out that the compression and rarefaction in sound wave in a gas occur so quickly, so that heat generated during compression or cold produced during rarefaction cannot pass away or absorb heat to and from the surrounding. So changes are adiabatic and not isothermal.

Adiabatic gas equation is PV^{γ} = constant

where γ is the ratio of specific heats of the gas at constant pressure and constant volume.

Differentiating with respect to V we get,

$$\gamma P V^{\gamma-1} + V^{\gamma}\frac{dP}{dV} = 0$$

or $$\gamma P = -\frac{dP}{\frac{dV}{V}} = E \text{ the bulk modulus}$$

Hence $$c = \sqrt{\frac{\gamma P}{\rho}}$$

where γ = 1.41 for diatomic gas

For air at standard temperature and pressure

$$c = \sqrt{\frac{1.41 \times 76 \times 13.6 \times 981}{0.001293}} \text{ cm/sec}$$

$$= 331.5 \text{ metres/sec}$$

This is in well agreement with experimental value which is 331.5 metres per second.

(iii) Effect of Pressure and Temperature

Let c_0 be the velocity of sound at 0°C and pressure P_0 density ρ_0 such that

$$c_0 = \sqrt{\frac{\gamma P_0}{\rho_0}}$$

When pressure is P_1, temperature T_1 density is ρ_1 the velocity $c_1 = \sqrt{\frac{\gamma P_1}{\rho_1}}$

Hence
$$\frac{c_1}{c_0} = \sqrt{\frac{P_1}{\rho_1} \cdot \frac{\rho_0}{P_0}} = \sqrt{\frac{P_1 V_1}{P_0 V_0}} \text{ as } \rho = \frac{m}{V}$$

But from gas laws
$$\frac{P_1 V_1}{T_1} = \frac{P_0 V_0}{T_0}$$

or
$$\frac{P_1 V_1}{P_0 V_0} = \frac{T_1}{T_0}$$

Hence
$$\frac{c_1}{c_0} = \sqrt{\frac{T_1}{T_0}} = \sqrt{\frac{273+t}{273}} = \left(1 + \frac{t}{273}\right)^{1/2} \quad \text{... (7)}$$

as $T_1 = 273 + t$ where t in degree centigrade.

The velocity of sound is not affected by change of pressure. Let the pressure of air change from P_0 to P_1 and density from ρ_0 to ρ_1. Then according to Boyle's law $P_0/\rho_0 = P_1/\rho_1$ $\therefore$ $c = \sqrt{\frac{\gamma P_0}{\rho_0}} = \sqrt{\frac{\gamma P_1}{\rho_1}}$

Thus the velocity of sound in a gas does not depend upon the pressure to which the gas is subjected but it depends on temperature and increases with the rise of temperature as given by equation (7).

(iv) Effect of Moisture on Velocity of Sound

Let c_m be the velocity of moist air at temperature t°C and pressure P, c_d is the velocity of dry air at t°C and pressure 76 cm of mercury.

$$\therefore \quad c_m = c_m = \sqrt{\frac{\gamma p}{\rho_m}} \text{ and } c_d = \sqrt{\frac{\gamma 76}{\rho_d}}$$

$$\therefore \quad \frac{c_d}{c_m} = \sqrt{\frac{76 \rho_m}{P \rho_d}} \quad \text{... (8)}$$

ρ_m = mass of 1 cc of moist air at t°C, Pressure P = mass of 1 cc dry air at t°C and pressure $(P - f)$ + mass of 1 cc of moisture at t°C and pressure f

where f is the aqueous tension at t°C. But the mass of 1 cc of moisture is equal to $\frac{5}{8}$ of mass of 1 cc dry air. Again as 1 cc gas at pressure $(P - f)$ becomes $\frac{P-f}{76}$ cc at pressure 76 cm of mercury. So

$$\rho_m = \frac{P-f}{76}\rho_d + \frac{f}{76}\cdot\frac{5}{8}\rho_d$$

or $$\rho_m = \frac{\left(P-\frac{3}{8}f\right)\rho_d}{76}$$

Substituting the value of ρ_m in equation (8)

$$\frac{c_d}{c_m} = \sqrt{\frac{(P-\frac{3}{8}f)}{P}}$$

or $$c_d = c_m\sqrt{\left(1-\frac{3}{8}\frac{f}{P}\right)} = c_m\sqrt{\frac{P-0.378f}{P}} \quad \text{... (9)}$$

With the help of equation (7) the equation (9) becomes

$$c_0 = c_m\sqrt{\left(1-\frac{3}{8}\frac{f}{P}\right)\left(\frac{273}{273+t}\right)} = c_0\sqrt{\frac{P-.378}{P}\left(\frac{273}{273+t}\right)} \quad \text{... (10)}$$

Thus if the velocity of moist air is known at temperature t°C and pressure P, the velocity of dry air at 0°C and pressure 76 cm of mercury can be calculated with the help of equation (10).

(v) Effect of Wind Velocity

Wind is the medium through which the compressions and rarefactions travel. The velocity of sound in a wind will depend on the direction of wind blowing so it is different in different direction. The effective velocity of sound is the algebraic sum of the velocity of sound in still air and the component of velocity of wind in the direction in which sound travels that is $C_1 = c + v\cos\theta$ where C_1 is the effective velocity of the sound wave, v is the velocity of wind and θ is the angle between the direction in which wind blows and the direction in which sound waves travel.

When $\theta = 0$

$C_1 = c + v$, $\theta = 90°$ $C_1 = c$ and $\theta = \pi$, $C_1 = c - v$

(vi) Effect of Atmospheric Conditions

The velocity of sound is different in different layer at different heights of the atmospheric air with variation of temperature of different layers. A simple relation is worked out between the curvature of the path of sound and the temperature gradient of the layer of air in the atmosphere while the sound wave tends to travel nearly horizontally.

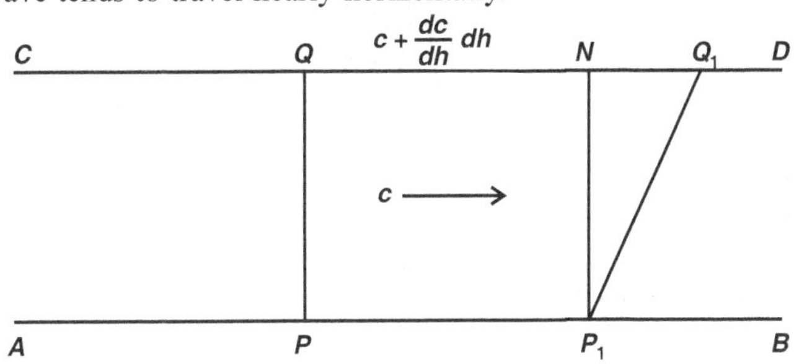

Fig. 2.

We consider a plane wave front PQ, P at a height h, Q is at a height $h + dh$. PQ is the position of the plane wave travelling horizontally, the plane is initially vertical. Let the velocity of sound at a height h be c and at a height $h + dh$ be $c+\frac{dc}{dh}dh$. The point P on the wave in time dt travels a distance $c\,dt$ and occupies the point P_1 while Q comes to Q_1 such that $QQ_1 = \left(c+\frac{dc}{dh}dh\right)dt$

Imagine a straight line drawn perpendicular on CD at N from P_1. The angle through which the wave form turns

$$= \frac{QQ_1 - PP_1}{P_1N} = \frac{\left(c + \frac{dc}{dh}dh\right)dt - cdt}{dh}$$

$$= \frac{dc}{dh}dt \qquad \text{... (11)}$$

The radius of curvature of the path R is given by

$$R = \frac{\text{Arc travelled}}{\text{Angle swept out}} = \frac{cdt}{\frac{dc}{dh}dt} = \frac{c}{\frac{dc}{dh}} \qquad \text{... (12)}$$

The velocity c changes with temperature given by

$$c = k\sqrt{T} \qquad \text{... (13)}$$

T represent, the absolute temperature, k is a constant. From eqn. (13)

$$\frac{dc}{dh} = \frac{h}{2\sqrt{T}} \cdot \frac{dT}{dh} \qquad \text{... (14)}$$

Hence eq. (12) with the help of (13) and (14)

$$R = 2T / \frac{dT}{dh}$$

When $\frac{dT}{dh}$ is negative the sound rays will be curved as in Fig. 3. A sound cannot be heard beyond a certain distance form the source. This distance depends upon the height of the source above the ground.

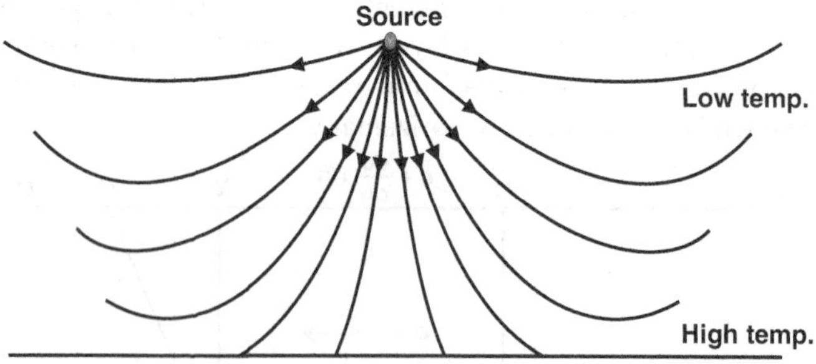

Fig. 3.

The velocity of wind is also different at different height. So wind will produce similar effect as above. Usually wind velocity is greater at higher level and $\frac{dc}{dh}$ will be negative on the windward side and positive on the opposite side *i.e.,* on leeward side.

On the windward side *i.e.,* on the right side of the source the velocity of sound gradually decreases with height as wind velocity gradually increases, hence dc/dh is negative.

On the left of the source the velocity of sound gradually increases with height, hence $\frac{dc}{dh}$ is positive.

Accordingly the rays are curved upward in the windward side and downwards on the leeward side, Fig. 4. An observer at ground level will receive less sound energy if he is on the right side of the source having less audibility of sound and will receive large share when he is on the left side, hence he will have good audibility of sound.

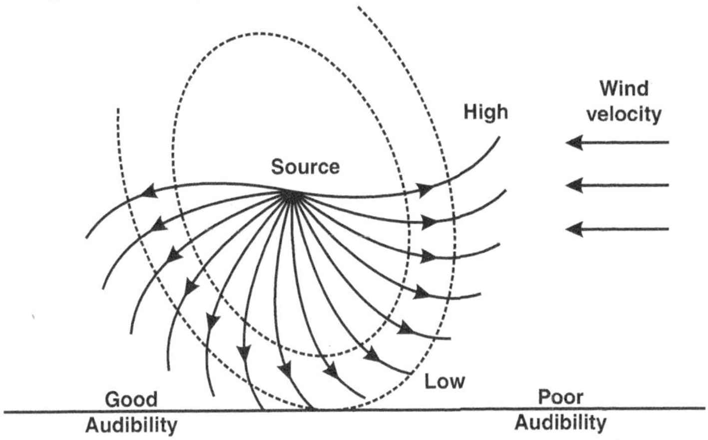

Fig. 4.

The gradient of temperature or wind velocity is responsible for curving the sound 'rays' in the upward direction as $\frac{dT}{dh}$ or $\frac{dv}{dh}$ is negative. These may be reversed in the upper atmosphere and sound is deflected down again. So sound will be audible at a great distance though it remains inaudible in the intermediate region. This intermediate region where sound remains inaudible is known as Zone of Silence.

9.3. VELOCITY OF LONGITUDINAL WAVE IN A SOLID MEDIUM

When a thin rod of large length is stretched in the direction of its length, there will be increase in length with contraction in cross-section and the rod is set in longitudinal vibration.

We consider such a rod whose length is large comparable with the diameter. Let ρ be the density, Y the Young's modulus of the material of the rod. Let α be the cross-section of the rod.

Consider an elementary slice AB in the rod bounded by two planes A and B at x and $x + dx$ respectively at right angles to the axis of the rod Fig. 5. Let these planes be displaced by the amount y and $y + dy$.

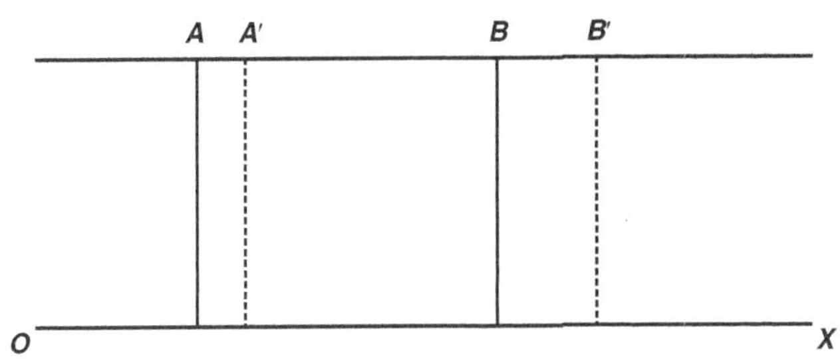

Fig. 5.

The distance between the planes A, B, which was initially dx becomes after displacement

$$(x + dx + y + dy) - (x + y) = dx + dy$$

So increase in length is dy and increase in length per unit length is $\frac{dy}{dx}$

If f is the total force acting on the area α the longitudinal stress $= \frac{f}{\alpha}$

So Young's modulur $Y = \dfrac{f/\alpha}{\dfrac{dy}{dx}}$

or
$$f = Y\alpha\frac{dy}{dx} \qquad \text{... (15)}$$

The force acting on the second face is
$$f + \frac{df}{dx}dx \qquad \text{... (16)}$$

Equation (16) with the help of equation (15)
$$\alpha Y\frac{dy}{dx} + \frac{d}{dx}\left(\alpha Y\frac{dy}{dx}\right)dx \qquad \text{... (17)}$$

The resultant force is given by the difference of expression (17) and (15)
$$\alpha Y\frac{dy}{dx} + \frac{d}{dx}\left(\alpha Y\frac{dy}{dx}\right)dx - \alpha Y\frac{dy}{dx}$$
$$= \alpha Y\frac{d^2y}{dx^2}dx \qquad \text{... (18)}$$

But the mass of the element dx is $\rho\alpha dx$ and its acceleration $\dfrac{d^2y}{dt^2}$

Hence
$$\rho\alpha\frac{d^2y}{dt^2}dx = \alpha Y\frac{d^2y}{dx^2}dx \qquad \text{... (19)}$$

or
$$\frac{d^2y}{dt^2} = \frac{Y}{\rho}\frac{d^2y}{dx^2} \qquad \text{... (20)}$$

or
$$\frac{d^2y}{dt^2} = c^2\frac{d^2y}{dx^2} \qquad \text{... (21)}$$

where $c^2 = \dfrac{Y}{\rho}$ where c is the velocity of wave propagation.

Equation (21) is a wave equation so solution should be of the form
$$y = f(ct - x) + F(ct + x)$$

Proceeding in the similar way vide Art.9.1 equations (3a), (4) and (5) we get velocity of longitudinal wave in a solid rod
$$c = \sqrt{\frac{Y}{\rho}}$$

Here lateral contraction or extension takes place which is not considered in the case of longitudinal vibration of a thin rod.

9.4. VELOCITY OF PLANE WAVE IN A BULK OF A SOLID

An elastic body be given a linear strain along the x-axis so that all the strains except along x-axis is zero, we assume, the wave is propagated along x-axis, the wave equation may be assumed to be of the form
$$\frac{d^2y}{dt^2} = \frac{Y_1}{\rho}\frac{d^2y}{dx^2} \qquad \text{... (22)}$$

where Y_1 stands for Axial modulus in place of Young's modulus Y as in the case of thin rod, where the velocity of propagation of wave $c=\sqrt{\frac{Y}{\rho}}$. Here lateral contraction or extension takes place which is not considered in the case of longitudinal vibration of a thin rod.

The axial modulus Y_1 is calculated as follows:

Let Y be the Young's modulus of the material of the medium, σ the Poisson's ratio and P is the stress acting along OX, the direction of propagation of the wave. P_1 is applied along the direction OY and similar stress P_1 is acting along OZ transverse to the direction of propagation of the wave. The strain produced along OX, OY, and OZ are given by

$$\frac{P}{Y}-\frac{\sigma P_1}{Y}-\frac{\sigma P_1}{Y} \quad \text{... (23)}$$

$$-\frac{\sigma P}{Y}+\frac{P_1}{Y}-\frac{\sigma P_1}{Y} \quad \text{... (24)}$$

and

$$-\frac{\sigma P}{Y}-\frac{\sigma P_1}{Y}+\frac{P_1}{Y} \quad \text{... (25)}$$

respectively. From equations (24) and (25) strain along OY and OZ is zero.

$$\frac{\sigma P}{Y} = \frac{P_1}{Y}(1-\sigma)$$

or

$$\sigma P = P_1\,(1-\sigma)$$

or

$$P_1 = P\frac{\sigma}{1-\sigma} \quad \text{... (26)}$$

Again the strain along OX is given by equation (23)

$$\frac{P}{Y}-\frac{P_1}{Y}2\sigma = \frac{P}{Y}-\frac{P}{Y}\frac{2\sigma^2}{1-\sigma}$$

$$= \frac{P}{Y}\left(1-\frac{2\sigma^2}{1-\sigma}\right) \quad \text{... (27)}$$

$$= \frac{P}{Y_1}$$

where

$$Y_1 = \frac{Y}{\left(1-\frac{2\sigma^2}{1-\sigma}\right)} \quad \text{... (28)}$$

Here the strain is found different from what is obtained in the case of thin rod where transverse contraction or extension does not take place. The value of axial modulus Y_l is given by Eqn. (28).

Hence

$$c_1 = \sqrt{\frac{Y_1}{\rho}} = \sqrt{\frac{Y}{\rho\left(1-\frac{2\sigma^2}{1-\sigma}\right)}}$$

$$= c_{\text{long}}\sqrt{\frac{1}{\left(1-\frac{2\sigma^2}{1-\sigma}\right)}} \text{ as } \sqrt{\frac{Y}{\rho}} = c_{\text{long}}$$

So

$$\frac{c_1}{c_{\text{long}}} = \sqrt{\frac{1-\sigma}{(1-2\sigma)(1+\sigma)}}$$

9.5. VELOCITY OF TRANSVERSE WAVE PROPAGATION ALONG A STRING STRETCHED UNDER TENSION

The string is taken to be uniform in cross-section having constant mass ρ per unit length. It is taken to be perfectly flexible, its stiffness is supposed negligible and it is not subjected to appreciable change in length while in vibration.

A long thin light string stretched under tension T, between two massive and well clamped support. We consider that the undisturbed position of the string as x-axis and the motion is confined in y-axis in the xy plane.

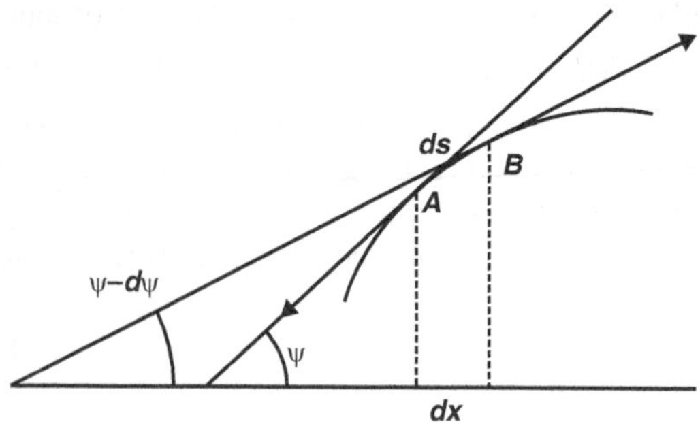

Fig. 6.

An infinitesimally small element of length $AB = ds$ is considered. Its transverse rate change of momentum is

$$\rho dx \frac{d^2 y}{dt^2} \qquad \text{... (29)}$$

and the resultant stretching force of A and B act along the tangent at these points.

The transverse component of the stretching force at A is

$$-T \sin \Psi = -T \frac{dy}{ds} \qquad \text{... (30)}$$

Ψ is the angle which tangent at A makes with the axis of A, Fig. 6.

The transverse component of the stretching force at B is

$$T \sin \Psi_2 = \left\{ T \frac{dy}{ds} - \frac{d}{ds}\left(T \frac{dy}{ds} ds \right) \right\} \qquad \text{... (31)}$$

where $\Psi_2 = \Psi - d\Psi$

Hence the resultant force is from (30) and (31)

$$T \sin \Psi - T \sin (\Psi - d\Psi) = T \frac{d^2 y}{ds^2} ds \qquad \text{... (32)}$$

Since the displacement is small and Ψ and $\Psi - d\Psi$ are small

$$\sin \Psi = \Psi = \tan \Psi = \frac{dy}{dx}$$

$$\sin (\Psi - d\Psi) = \tan (\Psi - d\Psi) = \frac{dy}{dx} - \frac{d^2 y}{dx^2} dx \qquad \text{... (33)}$$

and we may write $\frac{dy}{dx}$ instead of $\frac{dy}{ds}$ in the expression (31). Hence the resultant force on the element AB is written as

$$T \frac{d^2 y}{dx^2} dx \qquad \text{... (34)}$$

Equating the rate of increase of momentum from equations (29) and (34) we have

$$\rho \frac{d^2 y}{dt^2} = T \frac{d^2 y}{dx^2}$$

or

$$\frac{d^2 y}{dt^2} = \frac{T}{\rho} \frac{d^2 y}{dx^2} \qquad \text{... (35)}$$

for the equation of motion of the string.

The general solution of such wave equation

$$y = f(ct - x) + F(ct + x) \qquad \text{... (36)}$$

where c stands for the velocity of propagation of transverse wave along the string.

Substituting the value of $\frac{d^2 y}{dt^2}$ and $\frac{d^2 y}{dx^2}$ from equation (36) in equation (35)

$$c^2 [f''(ct - x) + F''(ct + x)] = \frac{T}{\rho}\left[f''(ct - x) + F''(ct + x)\right]$$

or

$$c = \sqrt{\frac{T}{\rho}} \qquad \text{... (37)}$$

Hence the equation of motion (35) is

$$\frac{d^2 y}{dt^2} = c^2 \frac{d^2 y}{dx^2} \qquad \text{... (38)}$$

9.5. ALTERNATIVE GEOMETRICAL PROOF

Alternative geometrical proof. Let the wave travels with a velocity c along the string from left to right. Imagine the string as a whole is being drawn through a straight smooth tube with the humpy of the wave from right to left with same velocity c, then the hump remains equilibrium in space.

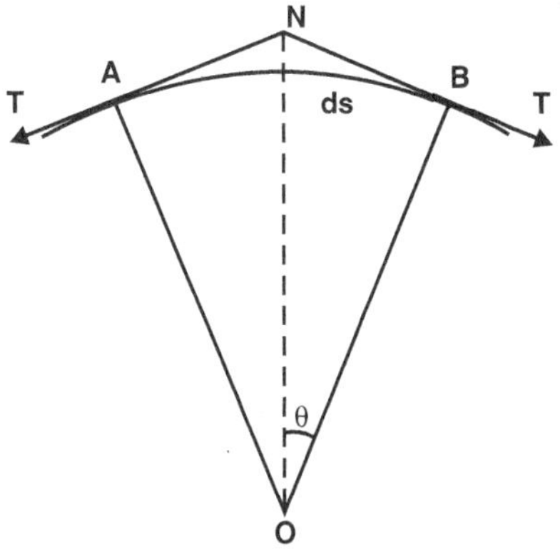

Fig. 7

Let AB be the circular are the tension T at A is along AT and at B is along BT which are tangents at A and B respectively Fig. 7.

Let $\widehat{AB} = ds$

the mass of $\widehat{AB} = \rho ds$

ρ being the mass per unit length of the string.

The centripetal force along this mass is

$$\rho \frac{ds}{r} c^2$$

c is the velocity of the wave, r is the radius of the arc $\widehat{AB}$ and is equal to OA or OB.

The force supplied by the component of the tension T, acting parallel to NO

$$2T \sin\theta = 2T \sin\frac{\angle AOB}{2}$$

$$= 2T \frac{\angle AOB}{2} \quad \text{... (39)}$$

as ∠AOB is very small

$$\therefore \quad 2T \sin\theta = T \frac{\text{Arc } \widehat{AB}}{2} = T\frac{ds}{r} \quad \text{... (40)}$$

Equating (1) and (2)

$$\therefore \quad \rho \frac{ds}{r} \cdot c^2 = T \frac{ds}{r}$$

or

$$c^2 = \frac{T}{\rho}$$

or

$$c = \sqrt{\frac{T}{\rho}} \quad \text{... (41)}$$

SOLVED PROBLEMS

1. If the velocity of sound in air at 0°C is 331.5 metres/sec, calculate its velocity at 30°C.

Velocity of sound at 0°C, $C_0 = 331.5$ metres/sec

The velocity of sound at 30°C is C_t

$$\therefore \quad \frac{C_t}{C_0} = \left(1 + \frac{t}{273}\right)^{1/2}$$

or

$$C_t = C_0 \left(1 + \frac{t}{273}\right)^{1/2}$$

$$= 331.5 \left(1 + \frac{30}{273}\right)^{1/2} \text{ m/sec}$$

$$= 349.24 \text{ metres/sec}$$

2. An increase of pressure of 100 kPa causes a certain volume of water to decrease by 5×10^{-3} per cent of its original volume. Show that the speed of sound in water is about 1414 m/sec. *[C.U. 1991]*

The speed of sound in water is

$$c = \sqrt{\frac{E}{\rho}}$$

where E is the Bulk modulus of elasticity.

$$E = \frac{P}{-\frac{dv}{v}} = \frac{100 \times 10^3}{5 \times 10^{-5}} = 2 \times 10^9 \, Pa$$

Density of water $\rho = 1000$ kg/m^3

$$\therefore \qquad c = \sqrt{\frac{2 \times 10^9}{1000}} \text{ m/sec m/sec}$$

$$= 1414 \text{ m/sec}$$

3. A string is stretched by suspending a load of 5 kg. The mass per unit length of the string is 5 gm/m. The travelling waves are sent through the string by oscillating one of the ends with a frequency of 250 Hz and amplitude 5 mm. Calculate the velocity and the length of the wave. Also write the equation of the travelling wave. Take $g = 10$ m/sec^2

Given $\qquad M = 5$ kg, $g = 10$ m/sec^2

Tension $\qquad T = $ mg

$= 5.10$ N

$= 50$ N

Mass per unit length of the string $\rho = 5$ gm/m

$= 5 \times 10^{-3}$ kg/m

Velocity $\qquad c = \sqrt{\frac{T}{\rho}}$

$$= \sqrt{\frac{50}{5 \times 10^{-3}}} \text{ m/sec}$$

$= 100$ metres/sec

Frequency $\qquad n = 250$ Hz

Wavelength $\qquad \lambda = \frac{c}{n} = \frac{100}{250}$ m $= 0.4$ metre

Amplitude $\qquad a = 5$ mm $= 5 \times 10^{-3}$ metre

The equation of the travelling wave $= a \sin \frac{2\pi}{\lambda}(ct - x)$

$$= 0.005 \sin \frac{2\pi}{0.4}(100t - x)$$

$$= 0.005 \sin 5\pi \, (100t - x)$$

4. A brass rod of length 1 m is damped at its mid point, find the first overtone frequency (γ for brass $= 10 \times 10^{10}$ N/mt^2, density of brass is 8.3×10^3 kg/mt^3.

[Osmania University 2010]

From the problem we have $\frac{3}{2}\lambda_1 = 1$

where λ_1 is the first overtone wavelength

$$\therefore \qquad \lambda_1 = \frac{2}{3}\text{mt}$$

If n_1 is the first overtone frequency then

$$n_1\lambda_1 = \sqrt{\frac{Y}{\rho}}$$

or

$$n_1 = \frac{1}{\lambda_1}\sqrt{\frac{Y}{\rho}}$$

$$= \frac{3}{2}\sqrt{\frac{10 \times 10^{10}}{8.3 \times 10^3}}$$

$$= 5206.5 \text{ Hz}$$

REVIEW QUESTIONS

1. Find an expression for the velocity of sound wave in a gas. How it depends upon pressure temperature, humidity and atmospheric conditions.
2. Deduce the expression for the velocity of longitudinal wave in a solid rod.
3. A flexible string is stretched under tension. How the velocity of transverse wave generated in the string depends upon tension and density of the material of the string?
4. Obtain an expression for the velocity of longitudinal wave in a fluid medium. Write the assumptions you make. [*Pandit Ravisankar University 2010, C.U. 1991*]
5. Discuss the Laplace correction in velocity of sound.
[*Pandit Ravisankar University 2010, Barktullah University 2011*]

PROBLEMS

1. Calculate the velocity of sound in air at 100 °C if the density of air is 0.001293 gm/cm^3 and density of mercury 13.6 gm/cm^3 both at 0°C. Specific heat at constant pressure and constant volume for air are 0.2417 and 0.1715 respectively.

[**Ans.** C_{100} = 388.6 metres/sec]
2. Calculate the velocity of sound in air at S.T.P. (The ratio of the two sp. heats of air = 1.41)

[**Ans.** c = 332 metres/sec]
3. If a wave is represented by $y = a \sin \frac{2\pi}{\lambda}(ct - x)$

prove that $\frac{d^2y}{dt^2} = c^2 \frac{d^2y}{dx^2}$
4. Two wires of steel of the same length are stretched on a sonometer. The tension of the first and second are 8 kg. weight and 2 kg. weight respectively. Find the ratio of the fundamental notes emitted by the two wires when the diameter of the first wire is half that of the second. [*C.U. 1990*]
[**Ans.** Ratio = 4]

10 CHAPTER TRANSVERSE VIBRATION OF STRETCHED STRING AND STRINGED MUSICAL INSTRUMENTS

10.1. FREQUENCY OF A VIBRATING STRING

It is seen that velocity c of the transverse vibration along a string is given by $c = \sqrt{\frac{T}{\rho}}$. T and ρ being the tension and line density of the string respectively. But $c = n\lambda$, λ is the wavelength and n is the frequency of vibration. So

$$n = \frac{c}{\lambda} = \frac{1}{\lambda}\sqrt{\frac{T}{\rho}} \qquad \text{... (1)}$$

when the string vibrates, stationary waves are set up as being reflected from two fixed ends. The fixed ends being always nodes and in between these two nodes there is an antinode when the string vibrates with only one loop, i.e, two fixed ends are two nodes with only one antinode in the middle, l = the length of the string is given by

$$l = \frac{\lambda}{2} \text{ or } n = \frac{1}{2l}\sqrt{\frac{T}{\rho}} \qquad \text{... (2)}$$

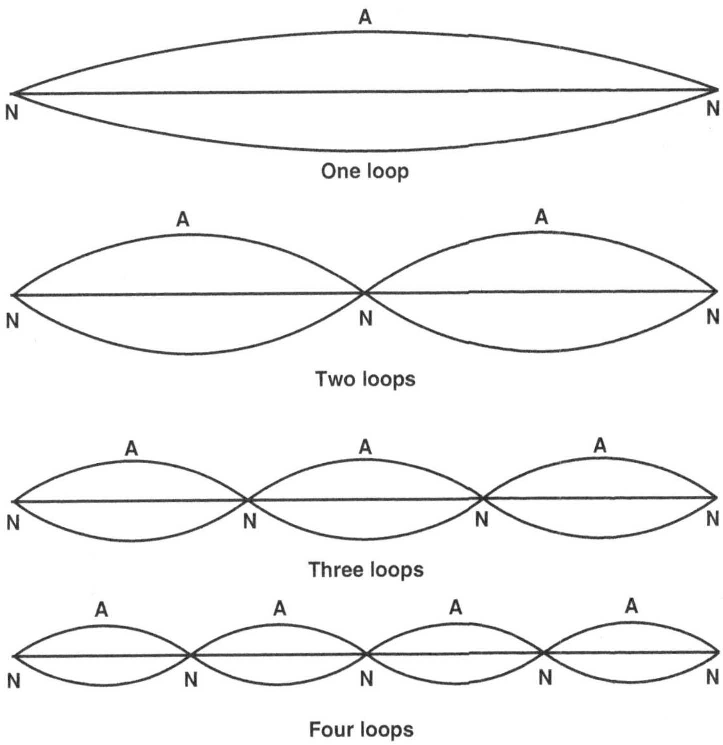

Fig. 1.

The string can vibrate in different modes, within the length of the string, two, three, four etc. number of loops may be formed, keeping always two fixed ends, nodes, as shown in Fig. 1.

So vibration with

One loop $\lambda_1 = 2l, \quad n_1 = \frac{1}{2l}\sqrt{\frac{T}{\rho}}$ ... (3)

Two loops $\lambda_2 = \frac{2l}{2}, \quad n_2 = \frac{2}{2l}\sqrt{\frac{T}{\rho}}$ ... (4)

Three loops $\lambda_3 = \frac{2l}{3}, \quad n_3 = \frac{3}{2l}\sqrt{\frac{T}{\rho}}$ ... (5)

s loops $\lambda_s = \frac{2l}{s}, \quad n_s = \frac{s}{2l}\sqrt{\frac{T}{\rho}}$ ... (6)

Thus it is found the frequency gravest role or fundamental n_1, is (*i*) inversely proportional to length of the vibrating string when tension and line density remain constant (*ii*) directly proportional to the square root of the tension to which the string is subjected when the length and line density remain constant (*iii*) inversely proportional to the square root of the linear density of the string when the length and tension remain constant. If r be the radius of the string and D be density of the material of the string then

$$\pi r^2 D = \rho$$

So the frequency of the note emitted by the vibrating string is inversely proportional to the radius of the string and inversely proportional to the square root of the density of the material of the string when length and tension of the string remain constant.

When the string vibrates with one loop the frequency of the emitted loop (n_1) as in equation (3) is the lowest and the note is the fundamental. When it vibrates with two loops, the frequency (n_2) is doubled the fundamental and the corresponding note is the second harmonic or first overtone as in equation (4). The three loops have a frequency (n_3), thrice that of the fundamental and the frequency (n_3) is the third harmonic or second overtone and so on.

These laws can be verified by means of a sonometer.

The Sonometer. It consists of a sounding box *SB* over which are usually stretched two strings. One string SS is attached to a peg P_1 on one side, while at the other side it passes over a light fraictionless pulley *W* and can be loaded to any desired tension. The second string is kept as a reference wire is stretched between the two pegs P_2P_3 attached to the sounding box *SB*. The peg P_3 can be turned to alter the tension. The string passes through two fixed bridges *FB*. By means of a small separate movable bridge *C* inserted below the wire, any required length of either of the strings can be set in vibration.

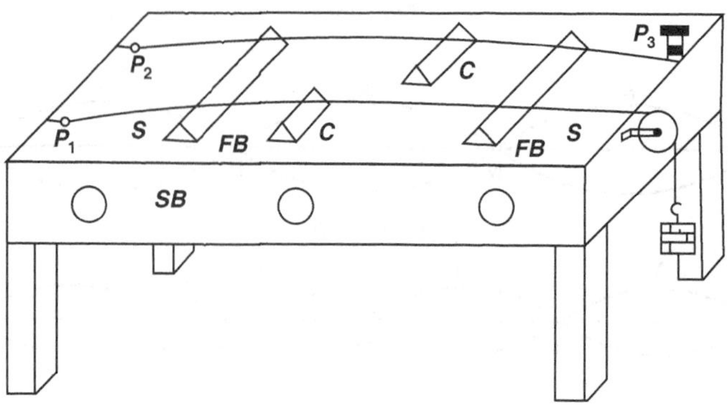

Fig. 1 (*a*).

Verification of laws of transverse vibration of a string

(*i*) **Law of length.** A wire is stretched on a sonometer under a given tension. It is tuned up by adjusting the length by the movable bridge with a tuning fork of known frequency. Let the frequency be n_1 and corresponding length of the wire which vibrate in resonance be l_1. In this way lengths l_1, l_2......l_5 having tuned up with frequencies of the forks $n_1 n_2 ... n_5$, it is found $n_1 l_1 = n_2 l_2 = n_5 l_5$ = const. Thus the law of length is verified.

(*ii*) **Law of tension.** A string of known length and of fixed line density, is tuned up with the frequency of a fork by a adjusting the tension by using known weights and lead shots of known weights. The value of the stretching forces that will bring the wire of given length and linear mass, in unison with each of the forks of known frequencies n_1, $n_2 ... n_5$ are thus determined. Let the corresponding stretching force of tensions are T_1, $T_2 ... T_5$ etc. Then it is found

$$n_1^2/T_1 = n_2^2/T_2 = n_5^2/T_5 = \text{constant}$$

which verifies the law of tension.

(*iii*) **Law of linear mass of the string:** The frequencies of vibrating lengths of the second string on the sonometer of a given linear density and stretched under a given tension is callibrated by using different forks of known frequencies and n~l graph is drawn.

Now the wires of different material and cross section are taken and stretched under a given tension. The frequency of vibration of the string for a given length is determined by the help of the callibrated string. The length of the second string which vibrate in resonance with the, first string *i.e.* in unison is found out and from the graph and corresponding frequency is determined. In this way the different wires of same length and tension but different masses ρ_1, $\rho_2 ... \rho_5$ etc. are set on the sonometer one by one and corresponding vibration frequencies are determined by the help of the second wire. They are n_1, $n_2 ... n_5$ etc. It is found $n_1^2 \rho_1 = n_2^2 \rho_2 = n_5^2 \rho_5$ = const. Thus the law is verified.

10.2. MELDES EXPERIMENT AND VERIFICATION OF THE LAWS OF TRANSVERSE VIBRATION OF A STRING

The string is attached to one of the prong and passes over a smooth pulley. The other end of the string carries the scale pan on which weights are placed. When the fork is excited, the string is set in stationary vibration due to reflection of waves from other end, hence loops are formed which are visible in the string.

(*i*) **Transverse Arrangement**. Fig. 2(*a*). Here the string vibrates up and down with the prong of the fork. As stationary waves are formed due to reflection from the other end fixed, loops with nodes and antinodes are formed. The distance between the ends of a definite number N of loops is located by placing pointed stands at two extreme nodes and the length L is measured, which gives L/N, the length of a single loop between two consecutive nodes. Let l be the average length of a loop then $n = \frac{1}{2l}\sqrt{\frac{T}{\rho}}$ where n is the frequency of vibration, T the tension (which is the weight of the scale pan with standard weights placed on it expressed in dynes), ρ the mass per unit length of the string obtained by weighing a definite length of the same sample of the string.

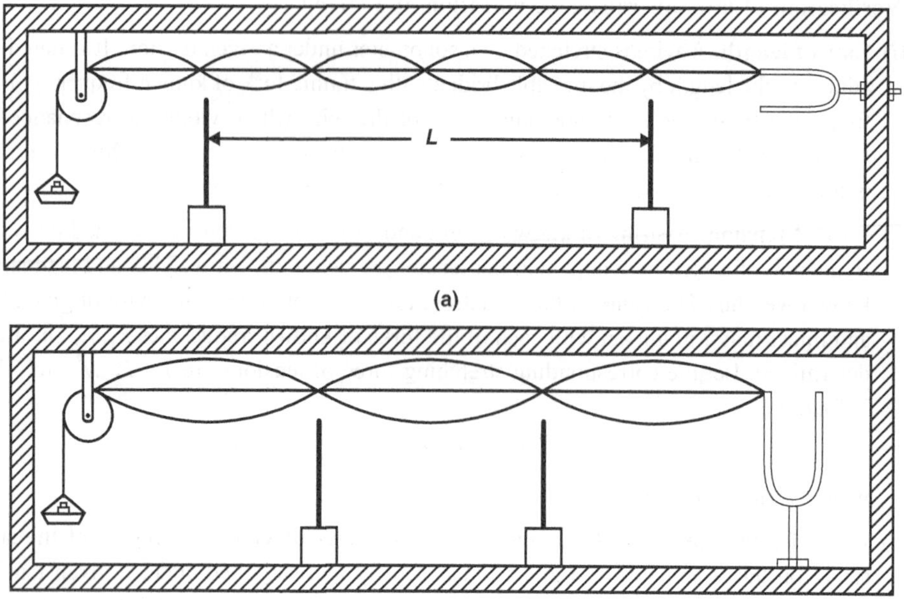

(a)

(b)

Fig. 2.

The nodes, hence the loops are made pronounced by adjusting the tension or the length of the string. In this case, the frequency of the fork is equal to the frequency of the string.

(*ii*) **Longitudinal Arrangement**, Fig. 2 (*b*). In this case, the string is tightened, straight when the prong, during its vibration, is away from the pulley. The string is sufficiently loose when the prong comes nearer to the pulley. Thus the string makes quarter of its vibration during the time the prong makes half its vibration. Thus in this case the frequency of the string is half the frequency of the fork. If the mean length of a loop is l then

$$\frac{n}{2} = \frac{1}{2l}\sqrt{\frac{T}{\rho}} \text{ or } n = \frac{1}{l}\sqrt{\frac{T}{\rho}},$$

10.3. STRINGED MUSICAL INSTRUMENTS

There are many types of stringed musical instruments. They are usually classified into three classes depending on the mode of excitation.

1. Plucked string family of instruments such as Guitar, Harp, Sitar, Tanpura, and Veena etc.
2. Struck string family such as Pianoforte.
3. Bowed string family such as Violin, Essraj.

The vibrations of string produce very insignificant compressions and rarefactions owing to fineness of the string, hence the sound produced is nearly inaudible. So the string is coupled to sound board through the bridge, in order to amplify the feeble sound. The sound board has large area, so produces large compressions and rarefactions in air as forced vibrations which are produced in it by vibrating string. Thus, without creation of energy but only by a process of rapid dissipation of it, the sound board helps to produce louder sounds.

(A) Plucked String Instruments

Guitar. It is one of the popular musical instruments. It has six strings over a flat big pear-shaped hollow sound board B with a sound hole H. The sound board or belly is fitted with a long neck or finger board N which is fitted with frets F at desired intervals of length of the finger board. One end of the strings is fixed in the holes of the bridge T, other end passes over the raised nuts are fixed with tuning keys K on the head. The stretching force is adjusted by proper tuning of the keys. In order to pluck the strings, the finger tops are fitted with finger pecks. A steel cylindrical bar is allowed to move over the frets by left hand in order to give desired effective length to the vibrating string. Fig. 3.

Fig. 3. Guitar

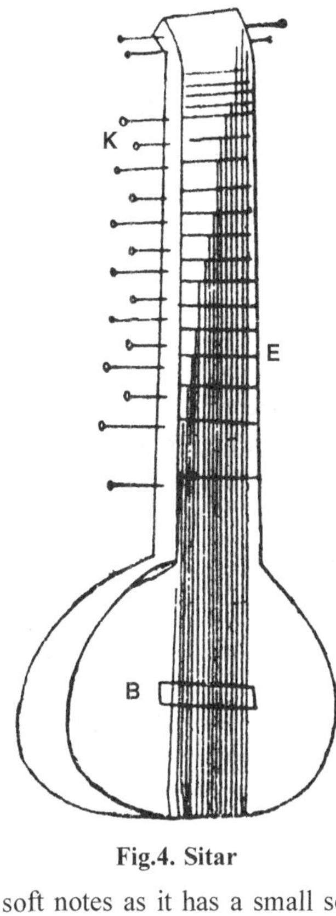

Fig.4. Sitar

Harp. It is a musical instrument which produces soft notes as it has a small sound board. There are number of strings fixed between two bridges each corresponds to a musical note of the musical scale. The length of all the strings can be slightly varied simultaneously in order to change the frequency of each note. The mid-point of the string is touched by palm of the hand in order to produce octave.

Banjo. It produces notes richer in over tones. The emitted notes are weak possess metallic character. There are metallic frets over the finger board, over which the strings are stretched. The plucking of the string is done by finger nails or pointed horn. The effective length of the string is obtained by pressing the string against one of the frets which are so placed so that it is possible to get the notes within the musical scale.

Sitar. It is an Indian musical instrument producing musical notes richer in musical qualities so it is very popular. Fig. 4.

There is a long hollow finger board attached to a hollow hemispherical belly, the upper face of which is a skin membrane. There are eighteen strings all together, seven, of which pass over the ivory bridge placed over the belly and eleven strings pass through the holes of the bridge. One end of these strings are attached to separate pegs fixed on the top and side of the finger board, and the other ends at the bottom of the belly. The tension or stretching force of these strings are adjusted by turning the pegs in order to tune them for particular note. The strings are played by finger nails or finger thimble made of tortoise shell. There are metal frets along the finger board. The player presses the string on the fret by left hand in order to get other notes. The strings below are set in forced vibration and increases the intensity of the musical notes.

Tanpura. There are four strings which when plucked produces somewhat nasal sound due to grazing character of the string with the surface of the bridge. No fret is provided in the instrument.

Veena. It is similar to sitar, only difference is that there are two boards on either side of the finger board.

10.4 GENERAL SOLUTION OF THE EQUATION OF MOTION OF STRING VIBRATING TRANSVERSELY

The differential equation of motion of a vibrating string is (vide Art. 9.5 equation (38).

$$\frac{d^2y}{dt^2} = c^2\frac{d^2y}{dx^2} \quad \text{... (7)}$$

where $$c = \sqrt{\frac{T}{P}}$$

where c is the velocity of transverse wave propagation along the string, T and ρ being the tension under which the string is subjected, and the mass per unit length of the string respectively.

From equation (7) it is found that y the displacement depends both upon the time t and the position of the point x on the string fixed at $x = 0$ and $x = l$

So, $$y = y(t)\,.\,y(x) \quad \text{... (8)}$$

where $y(t)$ is a function of t only and $y(x)$ is only of x.

$$\therefore \quad \frac{\partial y(x,t)}{\partial x} = y(t)\frac{\partial y(x)}{\partial x}$$

and $$\frac{\partial^2 y}{\partial x^2} = y(t)\frac{d^2y(x)}{dx^2} \quad \text{... (9)}$$

similarly $$\frac{\partial^2 y(x,t)}{\partial t^2} = y(x)\frac{d^2y}{dt^2} \quad \text{... (10)}$$

From equations (9) and (10) the equation (7) becomes

$$c^2 y(t)\frac{d^2y(x)}{dx^2} = y(x)\frac{d^2y(t)}{dt^2} \quad \text{... (11)}$$

Dividing both sides by $y(t)\,y(x)$ we get

$$c^2.\frac{1}{y(x)}\frac{d^2y(x)}{dx^2} = \frac{1}{y(t)}\frac{d^2y(t)}{dt^2} = -\omega^2 \quad \text{... (12)}$$

or $$\frac{d^2y(t)}{dt^2} + \omega^2 y(t) = 0$$

This is simple harmonic wave equation. As any point of the string for a particular value of x vibrates simple harmonically, we have

$$y(t) = a \cos \omega t + b \sin \omega t \quad \text{... (13)}$$

where $\omega = 2\pi n$, n being the frequency of vibration.

From equation (10)

$$\frac{d^2y(x,t)}{dt^2} = -y(x)\omega^2 (a \cos \omega t + b \sin \omega t)$$

$$= -\omega^2 y(x)y(t) \quad \text{... (14)}$$

The equation (7) with the help of equations (13) and (14) becomes

$$-\omega^2 y(x)y(t) = c^2 y(t)\frac{d^2y(x)}{dx^2}$$

or $$\frac{d^2y(x)}{dx^2} = -\frac{\omega^2}{c^2} y(x) \quad \text{... (15)}$$

or $$y(x) = \left(a_1 \cos\frac{\omega}{c}x + b_1 \sin\frac{\omega}{c}x\right) \quad \text{... (16)}$$

$\therefore$ $$y(x, t) = y(t).\ y(x)$$

$$= (a\cos \omega t + b\sin \omega t)\left(a_1 \cos\frac{\omega}{c}x + b_1 \sin\frac{\omega}{c}x\right) \quad \text{... (17)}$$

The constants are evaluated from boundary conditions

(*i*) $x = 0$ $\quad y = 0$ for all values of t

(*ii*) $x = l$ $\quad y = 0$ for all values of t ... (17.1)

First condition gives

$$0 = (a\cos \omega t + b\sin \omega t)a_1$$

or $$a_1 = 0$$

Equation(17) becomes

$$y(x, t) = (a\cos\omega t + b\sin\omega t)b_1 \sin\frac{\omega}{c}x$$

Second condition gives

$$0 = b_1 \sin\frac{\omega}{c}l$$

or $$\frac{\omega}{c}l = s\pi$$

where s is a positive integer

or $$\omega = \frac{s\pi c}{l}$$

Equation (17) becomes

$$y(x, t) = \sum_s \sin\frac{s\pi x}{l}\left(A_s \cos\frac{s\pi ct}{l} + B_s \sin\frac{s\pi ct}{l}\right) \quad \text{... (18)}$$

This is the general solution of the differential wave equation of the transversely vibrating string in which stationary waves are set up (vide Art. 8.8 equation (7) compare). s represents the modes of vibrations, $s = 1$ gives the fundamental, $s = 2$ gives octave, $s = 3$ gives third harmonic and so on.

10.5. EIGENVALUES, EIGENFUNCTIONS AND EIGENFREQUENCIES

The equation (16) in Art 10.4 is

$$y(x) = \left(a_1 \cos \frac{\omega}{c} x + b_1 \sin \frac{\omega}{c} x \right) \quad \text{... (16)}$$

satisfies the space part only whatever the value of $\frac{\omega}{c}$ with the boundary conditions in equation (17.1) in Art 10.4 the equation (16) becomes

$$b_1 \sin \frac{\omega}{c} l = 0$$

since $b_1 \neq 0 \quad \therefore \quad \frac{\omega}{c} l = s\pi$

where s is a positive integer

or $\omega_s = \frac{s\pi c}{l}$ with $s = 1, 2, 3, \ldots$ etc.

The restricted values for which the solution exists are called eigenvalues and the corresponding solution are called eigenfunction and the frequencies $\omega_s = \frac{s\pi c}{1}$ are called eigenfrequencies.

The eigenfunctions of a stretched string are orthogonal because they satisfy orthogonality condition

$$\int_0^l \sin \frac{s\pi x}{l} \sin \frac{s'\pi x}{l} dx = \delta_{ss}' \quad \text{... (19)}$$

where $\delta_{ss}' = 0$ when $s \neq s'$

$= \frac{l}{2}$ when $s = s'$

10.6. ENERGY OF THE TRANSVERSELY VIBRATING STRING

The displacement at a given point x at a given time t when it is set in vibration is given by equation (18) of Art 10.4

$$y = \sum_{s=1}^{s} \sin \frac{s\pi x}{l} \left(A_s \cos \frac{s\pi ct}{l} + B_s \sin \frac{s\pi ct}{l} \right) \quad \text{... (18)}$$

$$= \sum_{s=1}^{s} y_s(t) \sin \frac{s\pi x}{l} \quad \text{... (20)}$$

where

$$y_s(t) = \sum_{1}^{s} \left(A_s \cos \frac{s\pi ct}{l} + B_s \sin \frac{s\pi ct}{l} \right) \quad \text{... (21)}$$

$$\therefore \quad \frac{dy}{dt} = \sum_{1}^{s} \sin \frac{s\pi x}{l} \frac{d}{dt} y_s(t)$$

$$= \sum_{1}^{s} \sin \frac{s\pi x}{l} \cdot \frac{s\pi c}{l} \left(-A_s \sin \frac{s\pi ct}{l} + B_s \cos \frac{s\pi ct}{l} \right) \quad \text{... (22)}$$

The kinetic energy of the string of the element $dx = \frac{1}{2} \rho dx \left(\frac{dy}{dt} \right)^2$ ... (23)

Total kinetic energy of the whole length

$$= \frac{\rho}{2}\int_0^l \left(\sum_1^s \sin\frac{s\pi x}{l}\frac{d}{dt}y_s(t)\right)^2 dx \qquad \text{... (23.1)}$$

$$= = \frac{\rho}{2}\left\{\int_0^l \sum_1^s \left(\frac{d}{dt}y_s(t)\right)^2 \sin^2\frac{s\pi x}{l}dx + \int_0^l \sum_1^{m=s}\sum_1^{n=s}\frac{d}{dt}y_m(t).\frac{d}{dt}y_n(t)\sin\frac{m\pi x}{l}\right.$$

$$\sin\frac{n\pi x}{l}dx$$

The integration of the product terms where $m \neq n$ will vanish vide equation (19)

So the total average kinetic energy

$$= \frac{\rho}{2}\int_0^l \sum_1^s \sin^2\frac{s\pi x}{l}dx.\left(\frac{s\pi c}{l}\right)^2\left(A_s^2\sin^2\frac{s\pi ct}{l} + B_s^2\cos^2\frac{s\pi ct}{l} - 2A_sB_s\sin\frac{s\pi ct}{l}\cos\frac{s\pi ct}{l}\right)$$

$$\text{... (24)}$$

$$= \frac{\rho l}{4}\left(\frac{s\pi c}{l}\right)^2\left(A_s^2\sin^2\frac{s\pi ct}{l} + B_s^2\cos^2\frac{s\pi ct}{l} - 2A_sB_s\sin\frac{s\pi ct}{l}\cos\frac{s\pi ct}{l}\right) \qquad \text{... (25)}$$

Potential energy = Work done in changing the configuration = Average force × displacement doing the force

$$\text{Average force} = \frac{1}{2}T\left(\frac{dy}{dx}\right)$$

as at displaced position vertical component of T is $T\frac{dy}{dx}$, initially it was zero.

Potential energy of the element dx is

$$\frac{1}{2}T\frac{dy}{dx}.dy = \frac{1}{2}T\frac{dy}{dx}.\frac{dy}{dx}.dx = \frac{1}{2}T\left(\frac{dy}{dx}\right)^2 dx \qquad \text{... (26)}$$

Total average potential energy is

$$\frac{T}{2}\int_0^l \left(\frac{dy}{dx}\right)^2 dx \qquad \text{... (27)}$$

From equation (20) we get

$$\frac{dy}{dx} = \sum_1^s \frac{s\pi}{l}\cos\frac{s\pi x}{l}y_s(t)$$

$$\text{So total average potential energy} = \frac{T}{2}\sum_1^s\left(\frac{s\pi}{l}\right)^2\left(A_s\cos\frac{s\pi ct}{l} + B_s\sin\frac{s\pi ct}{l}\right)^2\int_0^l \cos^2\frac{s\pi x}{l}dx$$

$$= \frac{\rho l}{4}\left(\frac{s\pi c}{l}\right)^2\left(A_s^2\cos^2\frac{s\pi ct}{l} + B_s^2\sin^2\frac{s\pi ct}{l} + 2A_sB_s\cos\frac{s\pi ct}{l}\sin\frac{s\pi ct}{l}\right) \qquad \text{... (28)}$$

We have used the relation $c = \sqrt{\frac{T}{\rho}}$ or $T = c^2\rho$ in the above equation.

Total average energy is total average kinetic energy + total average potential energy, so the equations (25) and (28) as $\rho l = M$ total mass of the string

$$= M\pi^2\sum_{s=1}^{s}\left(\frac{sc}{2l}\right)^2(A_s^2 + B_s^2)$$

$$= M\pi^2 \sum_{s=1}^{s} \left(\frac{Sc}{2l}\right)^2 (A_s^2 + B_s^2)$$

$$= M\pi^2 \sum_{s=1}^{s} n_s^2 (A_s^2 + B_s^2) \qquad \text{... (29)}$$

as $\qquad \dfrac{sc}{2l} = n_s$ the frequency of the sth mode.

Thus the total average energy depends upon the total mass of the string M and square of the amplitude $(A^2_s + B^2_s)$and square of the frequency of a particular harmonic considered.

10.7. MOTION OF STRETCHED STRING EXERTED BY PLUCKING

The differential equation of motion of string vibrating transversely is given by

$$\frac{d^2y}{dt^2} = c^2 \frac{d^2y}{dx^2} \qquad \text{... (30)}$$

The solution of the above equation (30) is given by equation (18) as

$$y = \sum_{s=1}^{s} \sin\frac{s\pi x}{l}\left(A_s \cos\frac{s\pi ct}{l} + B_s \sin\frac{s\pi ct}{l}\right) \qquad \text{... (18)}$$

In the case, the string excited by plucking initial displacement is produced by plucking aside at any point $x = a$ and suddenly releasing it. The ordinate of the plucked point P is h at $t = 0$.

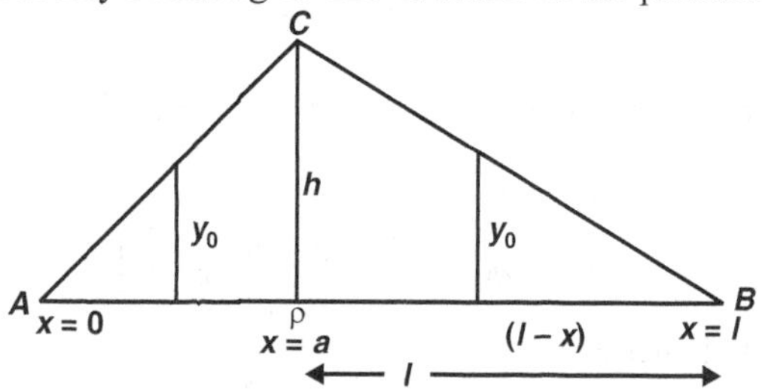

Fig. 5.

Between the points $x = 0$ and $x = a$, the displacement at any point x is

$$y_0 = \frac{h}{a}x \qquad \text{... (31)}$$

and between the points $x = a$ to $x = b$ the displacement

$$y_0 = \frac{h}{l-a}(l-x) \qquad \text{... (32)}$$

Initial conditions for the plucked string is at $t = 0$ $y = y_0$ ... (33)

and $$\left(\frac{dy}{dt}\right)_{t=0} = 0 \qquad \text{... (34)}$$

From equation (34) the equation (18) becomes for sth mode

$$B_s \frac{s\pi c}{l} \sin\frac{s\pi x}{l} = 0 = 0$$

or $$B_s = 0 \qquad \text{... (35)}$$

From the condition (33) the equation (18) becomes

$$y_0 = A_s \sin\frac{s\pi x}{l} \qquad \text{... (36)}$$

Multiplying both sides by $\sin\dfrac{s\pi x}{l}dx$ and integrating

$$\int_0^l y_0 \sin\frac{s\pi x}{l}\,dx = \int_0^l A_s \sin^2\frac{s\pi x}{l}\,dx = A_s\frac{l}{2}$$

or

$$A_s = \frac{2}{l}\int_0^l y_0 \sin\frac{s\pi x}{l}\,dx \qquad \text{... (37)}$$

Substituting the values of y_0 from equations (31) and (32) equation (37) becomes

$$A_s = \frac{2}{l}\left[\int_0^a \frac{h}{a}x\sin\frac{s\pi x}{l}\,dx + \int_a^l \frac{h}{(l-a)}(l-x)\sin\frac{s\pi x}{l}\,dx\right] \qquad \text{... (38)}$$

As there is discontinuity at $x = a$ the integration 0 to l is made two folds, 0 to a and a to l

or

$$A_s = \frac{2h}{l}\left[\frac{1}{a}\int_0^a x\sin\frac{s\pi x}{l}\,dx + \frac{1}{(l-a)}\int_a^l (l-x)\sin\frac{s\pi x}{l}\,dx\right]$$

$$= \frac{2h}{l}\left[\frac{1}{a}\left\{-\left(\frac{1}{s\pi}\right)x\cos\frac{s\pi x}{l} + \left(\frac{l}{s\pi}\right)^2\sin\frac{s\pi x}{l}\right\}_0^a\right.$$

$$\left.+\frac{1}{(l-a)}\left\{-\left(\frac{1}{s\pi}\right)(l-x)\cos\frac{s\pi x}{l} - \left(\frac{l}{s\pi}\right)^2\sin\frac{s\pi x}{l}\right\}_a^l\right]$$

or

$$A_s = \frac{2h}{l}\left[\left(\frac{l}{s\pi}\right)\cos\frac{s\pi a}{l} + \left(\frac{l}{s\pi}\right)^2\frac{1}{a}\sin\frac{s\pi a}{l}\right.$$

$$\left.-\left(\frac{l}{s\pi}\right)\cos\frac{s\pi a}{l} + \left(\frac{l}{s\pi}\right)^2\frac{1}{(l-a)}\sin\frac{s\pi a}{l}\right]$$

$$= \frac{2h}{\pi^2}\frac{l^2}{a(l-a)}\frac{1}{s^2}\sin\frac{s\pi a}{l} \qquad \text{... (39)}$$

Substituting the values of A_s from equation (39) and B_s from equation (35) in equation (18)

$$y = \frac{2hl^2}{\pi^2 a(l-a)}\sum_1^s \frac{1}{s^2}\sin\frac{s\pi a}{l}\sin\frac{s\pi x}{l}\cos\frac{s\pi ct}{l} \qquad \text{... (40)}$$

Hence the amplitude of the sth harmonic

$$a_s = \frac{2hl^2}{\pi^2 a(l-a)}\cdot\frac{1}{s^2}\sin\frac{s\pi a}{l}\sin\frac{s\pi x}{l} \qquad \text{... (41)}$$

As $a_s \propto \frac{1}{s^2}$ so the amplitude of the higher harmonics decreases very rapidly with the order of harmonics.

Further if the string is plucked say at $\frac{l}{5}$ *i.e.*, $a = \frac{l}{5}$

$$\sin\frac{s\pi a}{l} = \sin\frac{s\pi}{5}$$

which is equal to zero if $s = 5, 10, 15$ etc. Consequently a_5, a_{10}, a_{15} are zeros. Thus harmonics which have a node at the plucked point are all absent. This is Young-Helmholtz law or Young's law.

But in the class of plucked string family instrument Young-Helmholtz's law is apparently violated *i.e.*, $a_s \neq 0$.

10.8. CONFIGURATION OF THE STRING AT SUBSEQUENT TIME

The string is fixed at A , $x = 0$ and at B, $x = l$ initially. It is plucked at C, $x = a$ into a shape ACB. When the string is let go the bend up point C is levelled out, comes to its original mean position of the string giving at two discontinuities travelling down CA and CB respectively. Initially $y = \frac{h}{a}x$ between $x = 0$ to $x = a$ and $\frac{h}{(l-a)}(l-x)$ between $x = a$ to $x = l$. The subsequent motion is represented by superposition of discontinuities travelling with velocity c. These are given by the equations stated above moving with half the amplitudes of the above. There will be change of phase π when these continuities will be reflected from $x = 0$ and $x = l$. The subsequent resultant configurations are given by superimposing [Fig. 6(a)] these discontinuities [Fig. 6(b)]. It is found that at any instant the string is three straight pieces. Two end pieces are parallel to portions of the string in the original position and are at rest. The middle portion is always, either extending or withdrawing at each end.

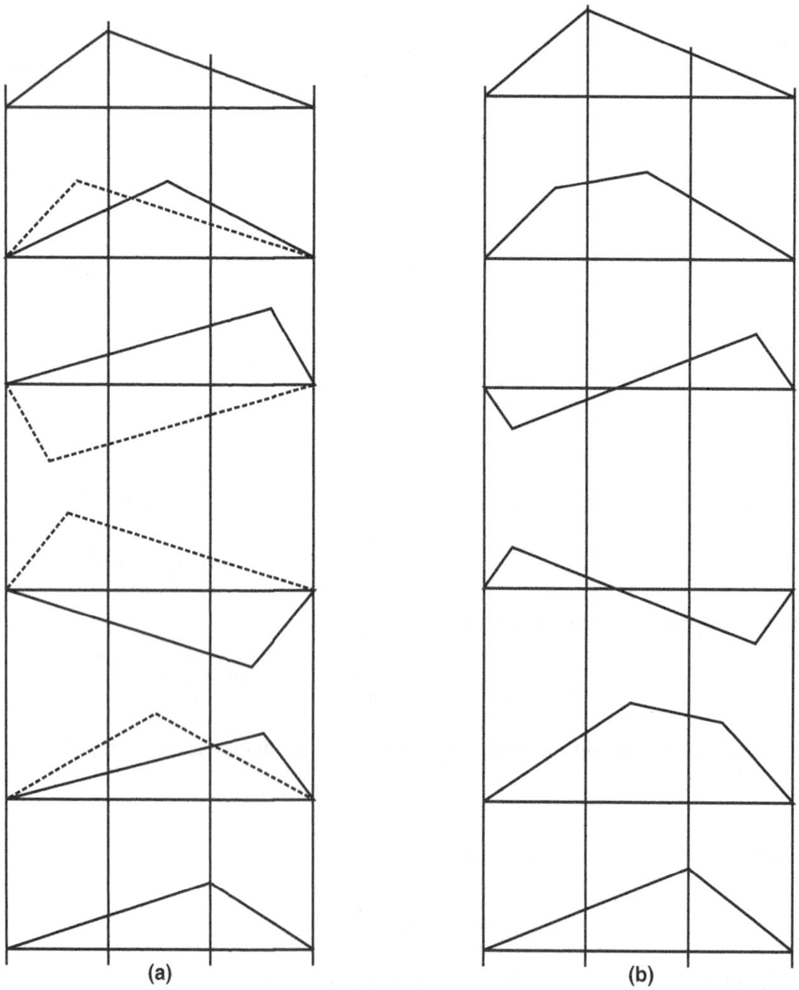

Fig. 6.

10.9. EFFECT OF FINITE LENGTH OF PLUCKING

In Art. 10.7 it is assumed that the string is plucked at a point $x = a$ by a sharp object, by finger nail or by a pointed piece of horn known as 'plectrum' where the generation of harmonics in greater number takes place, when plucked, very near the end, and tone becomes brilliant.

Let h_1 and h_2 be the initial displacement at $x = a$ and at $x = b$ at $t = 0$. So the displacement at any point initially is given by

$$(y_0)_x = \frac{h_1}{a} x \text{ between } x = 0 \text{ to } x = a \quad \ldots (42)$$

$$= h_2 + \frac{h_1 - h_2}{b - a}(b - x) \text{ between } x = a \text{ to } x = b \quad \ldots (43)$$

$$= \frac{h_2}{l - b}(l - x) \text{ between } x = b \text{ to } x = l \quad \ldots (44)$$

As in eqn. (38) Art. 10.7 we have

$$A_s = \frac{2}{l}\left[h_1 \int_0^a x \sin\frac{s\pi x}{l} dx + \int_a^b \left\{h_2 + \frac{h_2 - h_1}{b - a}(b - x)\right\} \sin\frac{s\pi x}{l} dx + \frac{h_2}{l - b}\int_b^l (l - x)\sin\frac{s\pi x}{l} dx\right]$$

or

$$A_s = \frac{2l}{\pi^2 s^2}\left\{\frac{h_1}{a}\sin\frac{s\pi a}{l} + \frac{h_2}{l - b}\sin\frac{s\pi b}{l}\right\} \quad \ldots (45)$$

If $h_1 = h_2 = h$ say

$$A_s = \frac{2hl}{\pi^2 s^2}\left\{\frac{1}{a}\sin\frac{s\pi a}{l} + \frac{1}{l - b}\sin\frac{s\pi b}{l}\right\} \quad \ldots (46)$$

when $b = a + da$ *i.e.,* the length of contact during contact is made small

$$A_s = \frac{2h \cdot l^2}{\pi^2 a(l - a)} \frac{1}{s^2} \sin\frac{s\pi a}{l}$$

which is same as eqn. (39) of Art. 10.7

Thus the small length of plucking has no effect on the quality of the note.

10.10. RIGIDITY OF STRING AND ITS EFFECT

So far we have assumed the vibrating string fixed at $x = 0$ and $x = l$ is flexible. The potential energy is given by the expression given in Art. 10.6 Eqn. (27). But when the string is stiff the potential energy as given by the equation mentioned above requires correction by taking into consideration the potential energy of bending like a rigid rod. This modified potential energy leads to produce the change of frequency of overtones generated. These frequencies do not bear simple ratio 1 : 2 : 3 etc. as are found in the case of flexible string. Hence the notes emitted by such a string is not musical.

10.11. YIELDING OF THE BRIDGE AND ITS EFFECT

It has been assumed so far that, there is no vibration in the string beyond $x = 0$ and $x = l$ where the string is fixed. In practice these two points are obtained by allowing the string to rest over two sharp bridges. But these bridges yield, and the effect produced on the vibration of the string depends upon the effective mass of the bridges and the force of restitution per unit displacement of the bridges. This modifies the kinetic energy of the string given in Eqn. (23.1) of Art. 10.6 due to yielding of the bridges. It is found that when the mass of the bridge tends to zero and the co-efficient of restitution per unit displacement of the bridge tends to be infinite, the string is fixed at $x = 0$ and

$x = l$. When the mass of the bridge is small and co-efficient of restitution of the bridge is large, the frequencies of the overtones are as 1, 2, 3 etc. Here the effective length extends beyond l and the result is that the frequencies of the partials are lowered. When the mass becomes infinite and co-efficient of restitution is zero the notes become unmusical and frequencies do not bear harmonic ratios. When mass is zero and co-efficient of restitution is zero there are antinodes at both ends which is not possible in practice. Detail calculation is beyond the scope of this *book*[1].

10.12. STRUCK STRING

The pianoforte is one of the popular musical instrument whose string is excited by blow of a felt hammer which rebounds after impact. This is a key board instrument. On pressing the keys the hammer by the action of a lever strikes the string. A second felt, called the damper, comes into contact with the string in order to stop the vibration as soon as the finger is removed from the key. By pressing a pedal the damper can be put out of action and the vibration of the string is allowed to die away of its own rates. The string is generally struck at one-seventh of the length of the string in order to weaken the seventh harmonic and its multiple which is not desirable for good music.

There are several theories in order to explain the different observations. The rigorous proof is difficult. These theories are classed mainly in two ways. In one class of theories, definite laws of force exerted by impinging hammer are assumed. *Helmoltz*[2] assumed that the force is of the form $F \sin pt$ where he remarked that p depends upon the mass of the hammer and its elastic property. The magnitude of p increases as the value of elastic properties increases and mass decreases. *Delemer*[3] assumed that the force exerted by the hammer during impact is a constant quantity. *Lamb*[4] assumed that the force exerted by the hammer on the string becomes appreciable, after a certain instant, when it comes in contact, and leaves the string long after the pressure falls into a small value. The other methods are based on functional solution without assuming any law of force. Following this method *Kaufmann*[5] worked out two cases for impinging hard hammer, striking at the mid point of the string and very near one end, where he assumed, the shorter segment of the string vibrates like a rigid rod. Bhargava and Ghosh tried to extend the same for elastic hammer. *Das*[6] considered the case of sami-infinite string *Kar-Ghosh*[7] tried to solve the problem on the assumption that the string behave like a loaded string so long the hammer is in contact with the string. These theories failed to explain the observed facts.

Rigorous dynamical theory is given by *M. Ghosh*[8] applying Heavyside Operational method. All previous theories come as special cases from this theory, and is able to explain the observed facts made by previous workers including Davis and others.

This simple classical theory of the struck string is given below:

The equation of motion of the vibrating string is given by

$$\frac{d^2y}{dt^2} = c^2 \frac{d^2y}{dx^2} \qquad \text{... (1)}$$

[1]Sound Vol. I — Rayleigh A . 135

Introduction to Theoretical Physics — K.C. Kar II art. 5 Page 48.

[2]Helmholtz — Sensation of lane translated by Ellis.

[3]Delemer Amr. Sci. de Brussel (1905 - 1906).

[4]Lamb — Dynamical Theory of Sound.

[5]Kaufmann — Ann. der Physik (1895)

[6]Das — Proc. Ind. Assr. (1926)

[7]Kar-Ghosh — Phil. Mag (1930)

[8]M. Ghosh — Ind. J. Phy. (1938, 1938-1941)

The solution of the above equation is given by (vide Art. 10.4 Eqn. (18).

$$y = \sum_{s=1}^{\infty} \sin \frac{s\pi x}{l}\left(A_s \cos \frac{s\pi ct}{l} + B_s \sin \frac{s\pi ct}{l}\right) \quad \text{... (18)}$$

The initial conditions are

(*i*) $t = 0,\quad y = 0$ for all points along the string ... (47)

(*ii*) $t = 0,\quad \frac{dy}{dt} = v_0$ at struck point $x = a$ ... (48)

but $\frac{dy}{dt} = 0$ at all other points along the string

From equation (47) $\quad A_s = 0$ gives

$$y = \sum_{1}^{\infty} \sin \frac{s\pi x}{l} B_s . \sin \frac{s\pi ct}{l} \quad \text{... (49)}$$

From equation (48)

$$\frac{dy}{dt} = \sum_{1}^{s} \frac{s\pi c}{l} B_s \sin \frac{s\pi x}{l} \cos \frac{s\pi ct}{l} \quad \text{... (50)}$$

at $\quad t = 0, \quad \left(\frac{dy}{dt}\right)_{t=0} = v_0 = \Sigma \frac{s\pi c}{l} B_s \sin \frac{s\pi x}{l}$ ... (51)

or multiplying both sides by $\sin \frac{s\pi x}{l} dx$ and integrating

$$\int_0^l v_0 \sin \frac{s\pi x}{l} dx = \int_0^l \sum_1^s \frac{s\pi c}{l} B_s \sin^2 \frac{s\pi x}{l} dx$$

$$= \sum_1^s \frac{s\pi c}{l} B_s \frac{l}{2} \quad \text{... (52)}$$

From equation (52) $\quad B_s = \sum_1^s \frac{2}{s\pi c} \int_0^l v_0 \sin \frac{s\pi x}{l} dx$

$$= \sum_1^s \frac{2}{s\pi c}\left[\int_0^a v_0 \sin \frac{s\pi x}{l} dx + \int_a^{a+da} v_0 \sin \frac{s\pi x}{l} dx + \int_{a+da}^{l} v_0 \sin \frac{s\pi x}{l} dx\right] \quad \text{.... (53)}$$

The hammer strikes at $x = a$ extending up $a + da$ where da is the length of contact. As $v_0=0$ at all points except between a and $a + da$ where $\left(\frac{dy}{dt}\right)_{t=0} = v_0$ so all other integrations except between a and $a + da$ will vanish, or

$$B_s = \sum_{s=1}^{s} \frac{2}{s\pi c} \int_a^{a+da} dx\, v_0 \sin \frac{s\pi x}{l} = \frac{2v_0 da}{s\pi c} \sin \frac{s\pi a}{l} \quad \text{...(54)}$$

Hence y is given by

$$y = \frac{2v_0 da}{\pi c} \sum \frac{1}{s} \sin \frac{s\pi a}{l} \sin \frac{s\pi x}{l} \sin \frac{s\pi ct}{l} \quad \text{... (55)}$$

The amplitude is found to be inversely proportional to the order of harmonics. Hence the intensity of the tones though fall off with the increase of the order of tones, but less rapidly than the plucked string instruments.

$$\text{If} \qquad \frac{l}{a} = s \text{ that is } l = sa, \ \sin\frac{s\pi a}{l} = 0$$

that is the harmonics which have a node at $x = a$, in other words at the point of striking, will be absent.

It is observed that the force imparted by the hammer to the string is not instantaneous but there is a finite duration of contact, depending upon the mass, elasticity and the position of impact of the hammer on the string. M. Ghosh observed and measured the duration of impact of hard and elastic hammers by photographic method. In the case of felt hammer, duration of impact depends upon the velocity of impact. It is further observed that pressure within the time of impact is fluctuating, without being steady during the growth and decay. The prediction as given in the rigorous theory of Ghosh was experimentally corroborated by *Devis and others*[1] by oscillographic method who used a hammer of piazo elastic crystal with platinised head.

10.13. BOWED STRING — VIOLIN FAMILY OF MUSICAL INSTRUMENT

This family of instruments which are in common use are Violin, Viola, Viono-cello, Esraj and Dilruba.

The mechanism of excitement of strings is same in all cases, only difference are their shape of the resonating box.

There are four strings, wire or catgut in each of the instruments, *Violin, Viola, Viono-cells* mounted on resonating box. The strings pass over a bridge *B* which communicate the vibration to the wood of the resonating box. The tension in the string is adjusted by turning the pegs *K*. The resonating box or belly of the instrument is hollow and has two holes *H H*. The notes other than given by the vibrating strings are made by pressing the string on finger board *F* at correct point which alters the vibrating length. There is no fret or other guides so that the player will be guided so as to where the finger is to be placed.

Fig. 8. Violin

The relative intensities of different harmonics depend upon several factors mainly upon the bowing pressure, width of the bow, point of bowing and on the degree of reinforcement of these harmonic over-tones by resonator and other vibrating parts of the instrument.

Esraj and Dilruba. Both instruments are of same type but they differ in shape of the resonating box. In Dilruba, the resonating box is boat shaped and in Esraj is spherical with two semi-circular cut on two sides. This box is covered with flat skin membrane over which the bridge is placed. There

[1]N. Devis, J. Mc. Caig., M. Normal Phil. Mag (1939).

is a long partially hollow finger board *E*, over which semi-curved metal frets *F* across the length of the finger board *E*. The finger board is joined with the resonating box or belly. Four steel wires are passing over the ivory bridge *B*. They are stretched over the finger board *F* and tied to the pegs *K*, below, there are 15 or 17 wires passing through the holes of the bridge. They also are stretched over the finger board and are tied to the side pegs *P*. These pegs enable to adjust the stretching force for tuning. These instruments are played with the help of a bow. The underneath wires, the air chamber and the membrane help much for large intensity of sound, produced by the instruments.

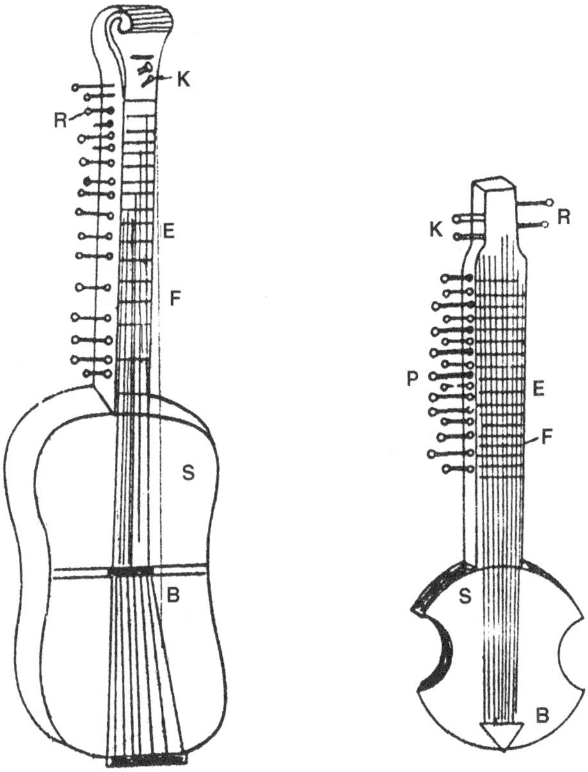

(*a*) **Dilruba** (*b*) **Sitar**

Fig. 8.

The motion of a bowed string is a special case of maintained vibration where the forcing system has no independent frequency of its own. The frequency of the forcing system is automatically adjusted to the natural frequency of the maintained system. Vibration of any string may be maintained by a single bow. The maintained vibration is a special case of forced vibration. The average rate of supply of energy by the bow is equal to the average rate of dissipation of energy through the bridges over which the strings are stretched, to the side of the resonating box on which the bridges are fixed and ultimately to the atmosphere as sound waves. During maintenance of vibration the supply of energy should not change discontinuously during forward and backward motion of the string. It should be rather constant.

10.14. MODULUS OF OPERATION OF THE BOW

Motion of the bow maintains the vibration of the string. A well resinated bow moves along the stretched string, it drags the string with it, because of static friction between the bow and the string. Two segments of the strings on two sides of the bow become more and more inclined to the bow. So, the restoring force acting on the stretched string gradually increases. Soon the magnitude of the restoring force exceeds the magnitude of the frictional force and the string slips towards the original position which overshoots due to inertia, finally the string comes to rest. It is again dragged by moving bow which is kept in contact with the string all along. These two operations namely the dragging of the string by the bow and slipping of the same are repeated rapidly so long the bow is moving while it is in contact with the string. The static friction during forward motion of the string is greater than the dynamic friction operative during slip. The maintenance of vibration of the string is due to difference of these two, as more work is done by the static friction on the string.

The time displacement curve of a given point of such a vibrating string is two steps zig-zag. To explain the formation of time displacement curve Raman assumed: Two velocity waves are travelling along the string in opposite directions. The velocity waves parallel straight lines passing through two ends $x = 0$ and $x = l$ with an intervening discontinuity. From the resultant of these two velocity diagram, it is possible to explain the form of time-displacement curve (Raman Bulletin 15 page 24-25 Indian Asso. Cult. Science 1918).

10.15. HELMHOLTZ LAWS OF BOWED STRING

First Law. The forward velocity of the bowed point of the string is same as velocity of the bow.

Raman's[1] *Explanation.* Since it is the case of maintained vibration, the rate of supply of energy by the bow during forward motion should be the same during the backward slip. There will be dynamical friction if the forward velocity of the string be less than that of the bow. As the velocity of the string suddenly changes with the direction from forward to backward, the frictional force must change which causes the change in the supply of energy suddenly. This will violate the fundamental law of maintained vibration unless during forward motion there is statical friction not dynamical friction. The limiting value of the statical friction is always greater than the dynamical friction, the average rate of supply of energy during forward motion when statical friction is called into play is equal to the average rate of supply of energy during backward slip when dynamical friction is called into play. Hence Helmholtz law is supported by Raman's arguments.

Kar[2] pointed out that this law cannot be rigorously correct as the idea of *bowed point* is an ideal one. There is *bowed region* has the bow as some width. Certainly there is one point within the bowed region which has same forward velocity of the string. Experiments show that the bowed point lies within the bowed region. Any other point within the bowed region has a velocity other than the forward velocity of the string. There the velocity is less than the velocity of the bow. Still

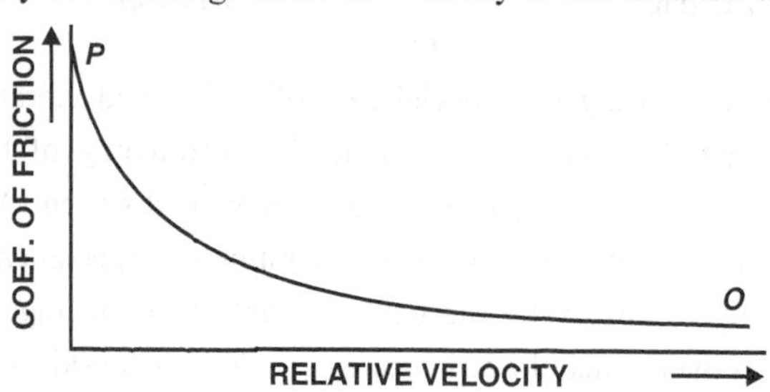

Fig. 9.

the constancy of energy supply is not violated. Fig. 9. represents the variation of co-efficient of friction with the relative velocity. *OP* represents the limiting value of statical friction. Dynamical friction decreases with increase of relative velocity and attains a constant value. When any point within the bowed region is moving with small relative velocity in the forward direction, the

[1]C.V. Raman — Bull. Indian Asso. Culv. Sci. 1912-15.

[2]Kar. K.C. Theoretical Physics, Vol. II

dynamical co-efficient of friction is greater than the asymptotic small constant value during the backward slip when the magnitude of the relative velocity is very large. Hence the frictional force which is given by the product of dynamical co-efficient of friction and the relative velocity is same in the forward and backward directions. Thus the rate of supply of energy by the bow is uniform throughout.

Second Law. When a stretched string is bowed near one end, the velocities of any *point* in the string in forward and backward directions are separately constant and their ratio is equal to the ratio of the segments into which the *point* divides the string and Time-displacement curve is two step zig-zag.

Explanation. If the observed point is at a distance $x = \frac{l}{4}$ from one end, l being the total length of the string where forward and backward velocities are v_1 and v_2 then according to the statement in the law

$$\frac{v_1}{v_2} = \frac{l/4}{l - l/4} = \frac{1}{3} \qquad \text{... (56)}$$

As $BD = 2\gamma$ where γ is the amplitude $2\gamma = v_1\tau_1 = v_2\,\tau_2$ where τ_1 and τ_2 are the time during which the observed point moves with velocities v_1 and v_2.

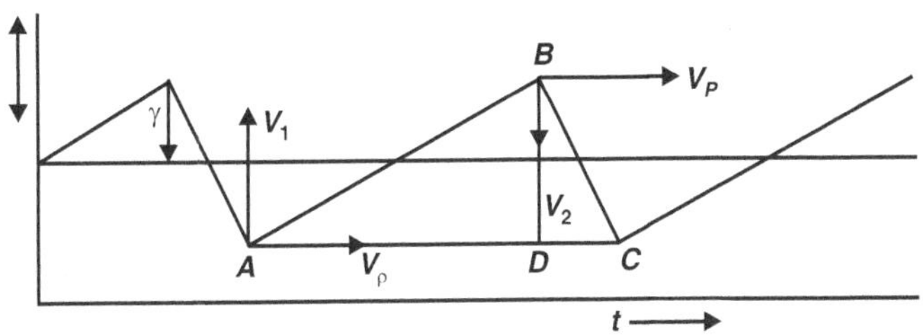

Fig. 10.

So $$\frac{v_1}{v_2} = \frac{\tau_2}{\tau_1} = \frac{DC}{AD}$$

Suppose the observed point is at a distance $x = a$ from the fixed end $x = 0$ ($x = \frac{l}{4}$ in the case considered above). Within the time period of vibration of the string $T\left(=\frac{2l}{c}\right)$ the transverse velocity changes discontinuously twice. The discontinuity of velocity produced at $x = a$ travels towards the end $x = l$ traversing a distance $(l - a)$ and is reflected from the fixed end. This continues to travel in the opposite direction and reaches the point $x = a$ again when the second discontinuous change of velocity is observed at $x = a$. This discontinuity further moves to the end $x = 0$ and is reflected, third discontinuity is observed when it comes to $x = 0$ again after transversing the distance $2a$. The interval between two successive discontinuities observed are $\frac{2(l-a)}{c}$ and $\frac{2a}{c}$ respectively, so

$$\frac{\tau_2}{\tau_1} = \frac{2a/c}{2(l-a)/c} = \frac{a}{l-a} = \frac{1}{3} \qquad \text{... (57)}$$

where $$a = \frac{l}{4}$$

Thus the law is explained.

Third Law. The sum of the forward and backward velocities at all points of the string under the same condition of bowing is a constant quantity *i.e.*, $v_1 + v_2$ = constant and is independent of x.

If y_0 is the amplitude and v_1 and v_2 are the forward and backward velocities at the observed point, then

$$v_1 + v_2 = \frac{2\gamma_a}{\tau_1} + \frac{2\gamma_a}{\tau_2}$$

$$= 2\gamma_a \frac{\tau_1 + \tau_2}{\tau_1 \tau_2}$$

$$= \frac{2\gamma_a T}{\tau_1 \tau_2} = \frac{2\gamma_a \dfrac{2l}{c}}{\dfrac{2a}{c} \cdot \dfrac{2(l-a)}{c}}$$

$$= \frac{\gamma_a lc}{a(l-a)} = \frac{2\gamma_a l^2}{Ta(l-a)} \qquad \text{... (58)}$$

which is independent of x.

At mid-point $\qquad \gamma_a = \gamma_0$

when $\qquad a = \dfrac{l}{2}$

$$v_1 + v_2 = \frac{8\gamma_0}{\tau} \qquad \text{... (59)}$$

From (58) and (59)

$$\frac{2\gamma_a l^2}{Ta(-a)} = \frac{8\gamma_0}{T}$$

whence

$$\frac{\gamma_0}{a(l-a)} = \frac{4\gamma_0}{l^2} \qquad \text{... (60)}$$

This is based upon the third law and can easily be verified by measuring the amplitude at different places of the string under same conditions.

10.16. EXPERIMENTAL VERIFICATION OF THE LAWS

The laws are verified by recording the time displacement curve of the vibrating bowed string. Photographic method is adopted and the shadow-graph of the particular point of the vibrating string is obtained on a photographic paper moving on a carrier behind the slit duly illuminated by electric arc lamp. The slit can be placed at any desired position to expose any point of the string. It is generally placed at an *aliquot* part of the length of the string. The instrument known as Vibrograph consists of a light tight wooden box, blacken inside, with a fine slit S in the front. A string particularly a fine catgut is fixed between two points A and B. At A there is a screw arrangement to adjust the tension of the string B is fixed with a hook. Behind the slit and within the box there are two horizontal grooves through which a slide carrying a photographic plate can move parallel to the string, under the action of gravity. One chord fixed at one end of the carrier passes through a small hole in the side of the box and runs down a smooth pulley fixed on the box at the point of the holes, the end carries the weights. The chord tied to the other end of the carrier passes through another hole on the other side of the box and kept tied to a peg, and is released when the carrier will move down the groove under the action of the fall of weight due to gravity.

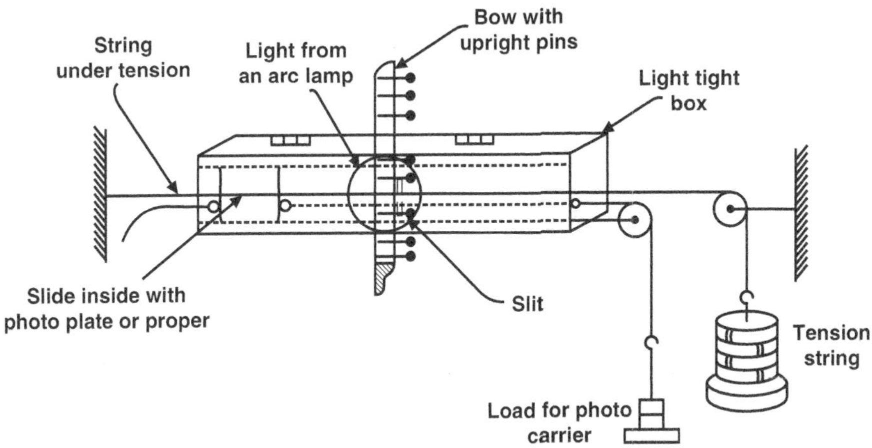

Fig. 11.

A well resined very narrow bow, with paper-pins fixed at one side of the wooden arm of the bow is used. These points are fixed parallel to each other and perpendicular to the arm of the bow. The slit is placed at aliquot part of the string. The point is illuminated by arc lamp. The bow is placed very close to the slit so that one edge of the hairs of the bow coincides the edge of the slit without obstructing it, and keeping the pins parallel to the string. The bow is drawn smoothly in one direction and simultaneously the carrier is released to move along the groove, while moving, the sensitised paper or plate will record the time-displacement curve of the point exposed with some parallel lines due to pins moving with the bow. The paper is developed and washed. It is found, the images of the pins are parallel in most cases to the forward motion of the string, proving the velocity of the bow at bowed point is equal to the velocity of the string in the forward direction as in Fig. 12.

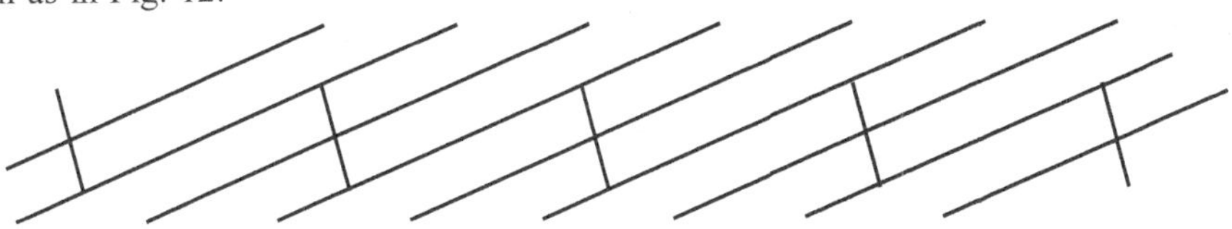

Fig. 12.

To verify the second and third laws the bow without pins is allowed to play at $\frac{l}{15}$ of the string so as to elicit the fundamental strongly. The slit is placed at different *aliquot* parts of the string $\frac{l}{7}\frac{l}{6}\frac{l}{5}\frac{l}{4}\frac{l}{3}\frac{l}{2}$. The time displacement curves are photographed as before in Fig. 13. By means of a travelling microscope $\frac{DC}{AD}$ (Fig. 13) for each curve along with the amplitude is measured separately, and in each case $\frac{DC}{AD}$ is equal to the corresponding.

$$\frac{a}{l-a}$$

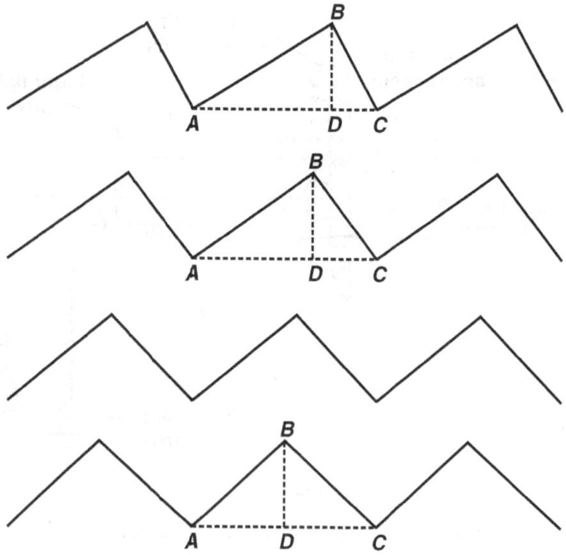

Fig. 13.

It is also found $\frac{\gamma_a}{a(l - a)}$ = constant which is equal to the value for midpoint $\frac{4\gamma_0}{l^2}$.

10.17. WOLF NOTE

In the bowed string instrument of violin family, a particular type of howling note is emitted some times without eliciting a soft pure tone. When the bow is moved with certain pressure and velocity, the belly of the instrument and air inside it which ordinarily acts as a system producing flat response, act as a resonator. An undesirable effect in the form of a howling noise known as *Wolf Note* is emitted when pitch of the string coincides with the pitch of the certain harmonics particularly, of upper one of the wooden belly. G.W. White concluded from the photograph obtained in this experiment that the pitch of the wolf note coincides with the pitch of the best resonance of the belly of the instrument.

C.V. Raman gave a further explanation from the experimental observation. In order to produce steady musical note to elicit the fundamental strongly a minimum critical bowing pressure is to be given by the bow for a given bowing distance and velocity. If velocity increases this value of the minimum critical pressure required, will increase proportionately, to the increase of the velocity of the bow. The body of the instrument with air inside, ordinarily acts as a system producing flat resonance. But for the large yielding of the bridge the belly of the instrument is strongly excited by sympathetic resonance. The dissipation of energy increases. The fundamental which requires greater energy ceases to be excited or becomes very feeble and octave is excited instead. At this stage, belly cannot vibrate in resonance and fundamental tone reappears. Thus this cyclic change of fundamental tone and octave becomes finally responsible to produce howling sound. This explanation corroborated by experimental observation, made, by simultaneously photographing the vibration of the belly and the string.

10.18. HELMHOLTZ KINEMATIC THEORY OF BOWED STRING

The string of length l fixed at $x = 0$, and $x = l$, stretched under tension $T = c^2\rho$ is set in vibration by the application of the bow, c being the velocity of transverse wave propagation along the string of linear density ρ.

Equation of motion of the vibrating string is as before

$$\frac{d^2y}{dt^2} = c^2\frac{d^2y}{dx^2} \qquad \text{... (1)}$$

where the general solution is (vide Art. 10.4)

$$y = \sum_1^s \sin\frac{s\pi x}{l}\left(A_s\cos\frac{s\pi ct}{l} + B_s\sin\frac{s\pi ct}{l}\right) \qquad \text{... (18)}$$

So the expression for displacement of bowed point $x = a$ is

$$y_a = \sum_1^s \sin\frac{s\pi a}{l}\left(A_s\cos\frac{s\pi ct}{l} + B_s\sin\frac{s\pi ct}{l}\right) \qquad \text{... (61)}$$

The velocity at $x = a$ is obtained by differentiating the above expression.

$$\frac{dy_a}{dt} = \dot{y}_a = \sum_1^s \sin\frac{s\pi a}{l}\cdot\frac{s\pi c}{l}\left(-A_s\sin\frac{s\pi ct}{l} + B_s\cos\frac{s\pi ct}{l}\right) \qquad \text{...(62)}$$

Multiplying both sides by $\sin\frac{s\pi ct}{l}\cdot dt$ and integrating over the time period T of the fundamental

$$\sum_1^s\int_0^T \dot{y}_a\sin\frac{s\pi ct}{l}dt = \sum_1^s\int_0^T \sin\frac{s\pi a}{l}\cdot\frac{s\pi c}{l}\left(-A_s\sin^2\frac{s\pi ct}{l} + B_s\cos\frac{s\pi ct}{l}\sin\frac{s\pi ct}{l}\right)dt$$

$$= \frac{s\pi c}{l}\sin\frac{s\pi a}{l}\left(-A_s\frac{T}{2}\right)$$

or

$$A_s = -\frac{2l}{s\pi cT\sin\frac{s\pi a}{l}}\int_0^T \dot{y}_a\sin\frac{s\pi ct}{l}dt \qquad \text{... (63)}$$

Multiplying by $\cos\frac{s\pi ct}{L}$ and integrating over T we have as before,

$$B_s = \frac{2l}{s\pi cT\sin\frac{s\pi a}{l}}\int_0^T \dot{y}_a\cos\frac{s\pi\, ct}{l}dt \qquad \text{... (64)}$$

The string moves with forward velocity v_1 for a time $t = 0$ to $t = \tau_1$ and with a backward velocity v_2 for a time $t = \tau_1$ to $t = \tau_2 = T - \tau_1$, so Eqs. (63) and (64) become

$$A_s = \frac{2l}{s\pi cT\sin\frac{s\pi a}{l}}\left[v_1\int_0^{\tau_1}\sin\frac{s\pi ct}{l}dt\right] - v_2\int_{\tau_1}^T\sin\frac{s\pi ct}{l}dt$$

$$B_s = \frac{2l}{s\pi cT \sin\frac{s\pi a}{l}}\left[v_1\int_T^{\tau_1}\cos\frac{s\pi ct}{l}\,dt - v_2\int_{\tau_1}^{T}\cos\frac{s\pi ct}{l}\,dt\right] \quad \text{... (65)}$$

On integration

$$A_s = -\frac{4l^2(v_1+v_2)}{(s\pi c)^2 T\sin\frac{s\pi a}{l}}\left(1 - \cos\frac{s\pi c\tau_1}{l}\right)$$

$$= -\frac{T}{\pi^2}\frac{(v_1+v_2)}{\sin\frac{s\pi a}{l}}\cdot\frac{1}{s^2}\sin^2\frac{s\pi c\tau_1}{2l} \quad \text{... (66)}$$

as $\quad 2l = cT$

and

$$B_s = \frac{4l^2(v_1+v_2)}{(s\pi c)^2 T\sin\frac{s\pi a}{l}}\cos\frac{s\pi c\tau_1}{2l}\sin\frac{s\pi c\tau_1}{2l}$$

$$= \frac{T}{\pi^2}\frac{(v_1+v_2)}{\sin\frac{s\pi a}{l}}\frac{1}{s^2}\cos\frac{s\pi c\tau_1}{2l}\cdot\sin\frac{s\pi c\tau_1}{2l} \quad \text{... (67)}$$

Eq. (61) with the help of (66) and (67) becomes

$$y_a = \sum_1^s \frac{T}{\pi^2}\frac{(v_1+v_2)^2}{\sin\frac{s\pi a}{l}}\cdot\frac{1}{s^2}\sin\frac{s\pi a}{l}\sin\frac{s\pi c\tau_1}{2l}$$

$$\times\left(-\sin\frac{s\pi c\tau_1}{2l}\cos\frac{s\pi ct}{l} + \cos\frac{s\pi c\tau_1}{2l}\cdot\sin\frac{s\pi ct}{l}\right)$$

$$= \frac{T}{\pi^2}(v_1+v_2)\sum_1^s\frac{1}{s^2}\sin\frac{s\pi c\tau_1}{2l}\times\sin\frac{s\pi c}{l}\left(t-\frac{\tau_1}{2}\right) \quad \text{... (68)}$$

As τ_1 and τ_2 are interchangeable it does not matter if the forward motion takes place before the backward motion, or backward motion takes place before the forward motion so we can write down

$$y_a = \sum_1^s B_s\sin\frac{s\pi c\tau_2}{2l}\sin\frac{s\pi c}{l}\left(t-\frac{\tau_2}{2}\right) \quad \text{... (69)}$$

where B_s stands for $\frac{T(v_1+v_2)}{\pi^2}\cdot\frac{1}{s^2}$ which is the amplitude of sth harmonic.

This equation suggests that the general equation for the displacement at any point of the bowed string can be written as

$$y = A\sum\frac{1}{s^2}\sin\frac{s\pi x}{l}\sin\frac{s\pi c}{l}\left(t-\frac{\tau_2}{2}\right) \quad \text{... (70)}$$

or

$$y = A\sum\frac{1}{s^2}\sin\frac{s\pi x}{l}\sin\frac{s\pi ct}{l} \quad \text{... (71)}$$

Shifting the origin of time by $\frac{\tau_2}{2}$

where $$A = \frac{T(v_1 + v_2)}{\pi^2} = s^2 B \text{ and as } \frac{\tau_2}{T} = \frac{a}{l}, \frac{\tau_2}{\tau_1} = \frac{a}{a - l}$$

from second law which gives

$$\sin \frac{s\pi c\tau_2}{2l} = \sin \frac{s\pi\tau_2}{T} = \sin \frac{s\pi a}{l}$$

and writing x for a the observed points

$$\sin \frac{s\pi c\tau_2}{2l} = \sin \frac{s\pi x}{l}$$

and from third law $(v_1 + v_2)$ is independent of x and is given by vide eq.(58), Art. 10.15

$$v_1 + v_2 = \frac{2\gamma_a l^2}{Ta(l - a)}$$

and A in eq. (71) becomes

$$A = \frac{2\gamma_a l^2}{\pi^2 a(l - a)} \qquad \text{... (72)}$$

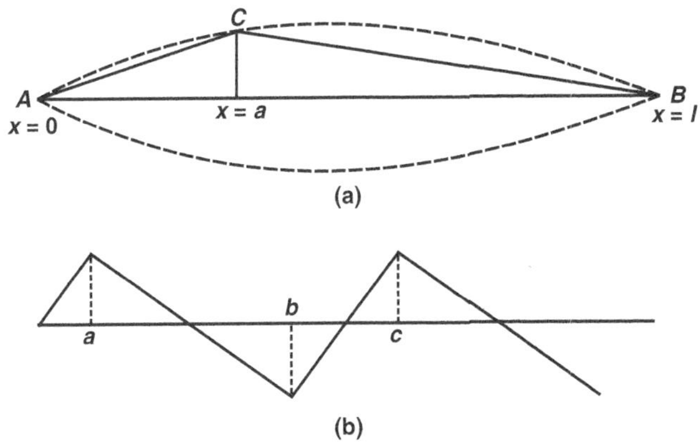

Fig. 14.

Let AB, Fig. 14 represent the equilibrium position of the string. Its form will be similar to ACB during vibration, its form is obtained by compounding two straight lines AC and BC intersecting at C. This point of intersection moves with a constant velocity along two flat parabolic arc in one or the other side, in which equilibrium position of the string is the common chord AB. In Eq. (72) the change of observed point a produces the change of γ_a the amplitude. So Eq. (72) represents pair of straight lines, the co-ordinates a, and γ_a of the point of intersection satisfy the Eq. (72). This gives the motion of the whole string. The motion of a single point may be deduced from Eq. (72). The two parts ab and bc of time displacement curve Fig. 14 of the time of every vibration is proportional to two parts of the string which are separated by the observed point.

SOLVED PROBLEMS

1. What will be the frequency of the note emitted by a wire of 0.3 m in length when stretched by a weight 25 kg if 2 metres of wire are found to weigh 4.79×10^{-3} kg.

   We know that the frequency of the fundamental note is given by

$$n = \frac{1}{2l}\sqrt{\frac{T}{\rho}}$$

Here $l = 0.3$ m

$T = 25$ kg wt $= 25 \times 9.8$ N

$\rho =$ Mass per unit length $= \dfrac{4.79 \times 10^{-3}}{2}$ kg/m

$= 0.002395$ kg/m

$\therefore$ $$n = \frac{1}{2 \times 0.03}\sqrt{\frac{25 \times 9.8}{0.002395}} \text{ Hz}$$

$= 533.3$ Hz

2. A stretched string of length l fixed at its ends, is plucked by a distance 'b' at a point distance 'a' from one of its ends. Find the energy of the sth harmonic.

The displacement at a point x on the string at time t for the sth harmonic is

$$y(x, t) = \frac{2bl^2}{\pi^2 a(l-a)}\frac{1}{s^2}\sin\frac{s\pi a}{l}\sin\frac{s\pi x}{l}\cos\frac{s\pi ct}{l}$$

$$= A_s \sin\frac{s\pi x}{l}\cos\frac{s\pi ct}{l}$$

where $$A_s = \frac{2bl^2}{\pi^2 a(l-a)}\frac{1}{s^2}\sin\frac{s\pi a}{l}$$

The kinetic energy of the element dx

$$= \frac{1}{2}\rho dx\left(\frac{dy}{dt}\right)^2$$

where ρ is the mass per unit length of the string.

or $$\text{Kinetic energy} = \frac{1}{2}\rho\frac{s^2\pi^2c^2}{l^2}A_s^2\sin^2\frac{s\pi x}{l}\sin^2\frac{s\pi ct}{l}dx$$

$$\text{The maximum kinetic energy} = \frac{1}{2}\rho\frac{s^2\pi^2c^2}{l^2}A_s^2\cdot\sin^2\frac{s\pi x}{l}dx$$

= Total energy of the plucked string of the element dx

$\therefore$ Total energy of the entire string

$$= \frac{1}{2}\rho\frac{s^2\pi^2c^2}{l^2}A_s^2\int_0^l \sin^2\frac{s\pi x}{l}dx$$

$$= \frac{1}{4}\rho\frac{s^2\pi^2c^2}{l^2}A_s^2 l$$

$$= \rho l\frac{s^2l^2c^2}{4l^2}A_s^2$$

If $M = \rho l$ is the mass of the string and $n_s = \dfrac{sc}{2l}$ the frequency of the sth mode, then total energy is $\pi^2 M n_s^2 A_s^2$

3. A uniform string of length l is stretched between its fixed ends $x = 0$ and $x = l$. Obtain an expression for the transverse displacement $y(x, t)$ of the string when it is struck at the centre so that the velocity varies linearly from zero at the ends to v_0 at the centre.

[Burd. U. 1995)]

With initial conditions we have the displacement equation as

$$y(x, t) = \sum_{s=1}^{s} B_s \sin\frac{s\pi x}{l} \sin\frac{s\pi ct}{l}$$

The velocity at any point

$$y'(x, t) = \frac{dy}{dt} = \frac{\pi c}{l}\sum_{1}^{s} s\, B_s \sin\frac{s\pi x}{l}\cos\frac{s\pi ct}{l}$$

The initial condition at $t = 0$

$$y'_0\, y'(x, 0) = \frac{\pi c}{l}\sum_{s=1}^{s} s\, B_s \sin\frac{s\pi x}{l}$$

Multiplying both sides by $\sin\dfrac{s\pi x}{l}\,dx$ and integrating we get

$$\frac{\pi sc}{l} B_s = \int_0^l y'_0 \sin\frac{s\pi x}{l}\,dx$$

But the initial conditions are given by

$$y'_0 = \frac{v_0}{l/2}x \text{ for } 0 < x < \frac{l}{2}$$

$$= \frac{v_0(l-x)}{l/2} \text{ for } \frac{l}{2} < x < l$$

So
$$\frac{\pi cs\, B_s}{2} = \int_0^{l/2} \frac{v_0}{l/2} x\sin\frac{s\pi x}{l}\,dx + \int_{l/2}^{l} \frac{v_0(l-x)}{l/2}\sin\frac{s\pi x}{l}\,dx$$

or
$$\frac{\pi csB_s}{2} = \frac{2v_0}{l}\left[\int_0^{l/2} x\sin\frac{s\pi x}{l}\,dx + \int_{l/2}^{l}(l-x)\sin\frac{s\pi x}{l}\,dx\right]$$

Integrating and simplifying the above equation we get

$$\frac{\pi csB_s}{2} = \frac{4v_0 l}{s^2\pi^2}\sin\frac{s\pi}{2}$$

or
$$B_s = \frac{8v_0 l}{s^3\pi^3 c}\sin\frac{s\pi}{2}$$

$\therefore$
$$y(x, t) = \frac{8v_0 l}{\pi^3 c}\sum_{1}^{s}\frac{1}{s^3}\sin\frac{s\pi}{2}\sin\frac{s\pi x}{l}.\sin\frac{s\pi ct}{l}$$

REVIEW QUESTIONS

1. How stationary wave is set up in vibrating string? Upon what factors the frequency of vibration will depend? How are they related to frequency?
2. Describe Melde's experiment and state the use of the same. How far can the laws of vibration of a string be verified in this experiment?

3. (*a*) Deduce the differential wave equation of transversely vibrating string. Find the expression giving the displacement at any time.
   (*b*) Find an expression for velocity of sound in a uniformly stretched string. [*Barkatullah University 2011*]
4. Deduce the expression for the energy of the string vibrating transversely.
5. Briefly discuss the formation of stationary waves for the transversely vibration of a string under tension and fixed at the ends. *[C.U. 1997]*
6. A string light and flexible stretched under tension over two bridges at a distance l is plucked at a distance a from one of the bridges. Find the displacement at a given point x at subsequent time. How does the harmonics present depend upon the point of plucking?
7. Give a general survey of development of the theory of vibration of a string excited by striking.
8. State the laws of Bowed string. How was the first law explained by Raman and Kar? Explain the action of Bow on the string? How are the laws of Bowed string verified experimentally?
9. Set up the wave equation for a string under tension is form of a Fourier series when two ends are fixed. *[C.U. 1971]*
10. If x and t denote the position and time co-ordinate respectively and c is a velocity show that any function of $x - ct$ or $x + ct$ represents a plane progressive wave. Derive the differential equation of such waves. Write down the solution to represent plane harmonic waves and identify the constants of solution. *[C.U. 1970]*
11. Analyse the motion of a plucked string and show that its vibration obey Young's rule.
12. Analyse the motion of a struck string. When are some harmonics expected to be absent?
13. Discuss the motion of bowed string.
14. Write a short note on stringed musical instruments. How is the energy of a vibrating string of an instrument communicated in the air?

PROBLEMS

1. In the transverse arrangement of Melde's experiment, the string vibrates in 4 loops when the tension is 0.40 kg. wt. Determine the tension when the string vibrates in 2 loops in the longitudinal arrangement. [**Ans.** 0.40 kg. wt]
2. A string of length l is tightly stretched with a tension T between its two fixed ends $x = 0$ and $x = l$. A hammer blow is given to a small part of length b of the string at a distance a from the end $x = 0$; as a result of which the part b of the string has an initial velocity v_0, the rest of the string is initially undisturbed. Show that the subsequent vibration of the string is given by

$$y(x,t) = \frac{4v_0 t}{\pi^2 v}\sum_{s=1}^{\infty}\frac{1}{s^2}\sin\left(\frac{s\pi a}{l}\right)\sin\left(\frac{s\pi b}{2l}\right).\sin\left(\frac{s\pi x}{l}\right)\sin\left(\frac{s\pi ct}{l}\right)$$

   where $v = \sqrt{\frac{T}{\rho}}$; ρ being the linear density of the string. Assume tension remains constant.
3. A stretched string of length L supported at its ends is pulled perpendicular to its length by a distance H at its midpoint and then released. Find the amplitude of the fundamental at a distance $\frac{L}{4}$ from one end. $\left[\textbf{Ans.}\ \text{Amplitude} = \frac{4\sqrt{2}H}{\pi^2}\right]$
4. A string of length l fixed at both ends has a fundamental frequency f_0. The string is plucked at a point (l/4) from one end and immediately afterwards an observer touches the string lightly at its midpoint. What is the lowest frequency excited in this case? What will be frequency of the next harmonic? [C.U. 1988]
   [**Ans.** $2f_0$, $6f_0$]

11 CHAPTER

VIBRATION OF A BAR

11.1. LONGITUDINAL VIBRATION

In determining the expression of velocity of propagation of longitudinal wave in a thin bar, we obtained the equation of motion (vide Art. 9.3)

$$\frac{d^2\xi}{dt^2} = c^2 \frac{d^2\xi}{dx^2} \qquad \text{... (1)}$$

where ξ represents the displacement due to longitudinal wave at any point x at time t $c = \sqrt{\frac{Y}{\rho}}$ represents the velocity of propagation, Y is the Young's modulus and ρ is the density of the material of the rod, respectively.

In case of stretched string under tension, two ends are always fixed but in the case of the thin rod vibrating longitudinally, it is possible to keep the rod with

(*i*) Two ends clamped.

(*ii*) Clamped at the midpoint, two ends free.

(*iii*) Two ends free,

(*iv*) Clamped at one end other end being free.

The solution of equation (1) as obtained in Art. 10.4 eqn. (17)

$$\xi = (a_1 \cos \omega t + b_1 \sin \omega t)\left(a_2 \cos \frac{\omega}{c} x + b_2 \sin \frac{\omega}{c} x\right) \qquad \text{... (2)}$$

where $\omega = 2\pi n$, n being the frequency of vibration.

(I) When both ends are fixed the boundary conditions are

(*i*) $x = 0$ $\xi = 0$ for all values of t

(*ii*) $x = l$ $\xi = 0$ for all values of t

when there be no permanent tension, and

(*iii*) $x = l$ $\xi = \alpha$ a small constant if there be permanent tension.

From condition (*i*) $a_2 = 0$ so the equation (2) becomes, adjusting the constants

$$\xi = \xi = \sin \frac{\omega}{c} x(A \cos \omega t + B \sin \omega t) \qquad \text{... (3)}$$

From condition (*ii*)

$$\sin \frac{\omega}{c} l = 0 \text{ gives } \omega = \frac{s\pi c}{1} \qquad \text{... (4)}$$

where s is a whole number or equation (3) can be written

$$\xi = \sum_s \sin \frac{s\pi x}{l}\left(A_s \cos \frac{s\pi ct}{l} + B_s \sin \frac{s\pi ct}{l}\right) \qquad \text{... (5)}$$

From condition (*iii*) for zero value of ω

$$\xi = A_0 \sin \frac{\omega}{c} x$$

or
$$\alpha = A_0 \sin \frac{\omega}{c} l$$

whence
$$\xi = \alpha \frac{\sin \frac{\omega}{c} x}{\sin \frac{\omega}{c} l} = \alpha \frac{x}{l} \qquad \text{... (6)}$$

retaining only the first term.

Hence the complete value of

$$\xi = \alpha \frac{x}{l} + \sum_s \sin \frac{s\pi x}{l} \left(A_s \cos \frac{s\pi ct}{l} + B_s \sin \frac{s\pi ct}{l} \right) \qquad \text{... (7)}$$

The series of tones form a complete harmonic scale. The period of gravest mode is $\frac{2l}{c}$ and of the sth mode is $\frac{2l}{sc}$.

(*ii*) When the midpoint of the rod is clamped

(*a*) at $x = \frac{l}{2}$, $\xi = 0$

and (*b*) at $x = 0$ and $x = l$, $\frac{d\xi}{dx} = 0$ for all values of t.

$$\frac{d\xi}{dx} = 0 \text{ at } x = 0 \text{ gives from Eqn. (2)}$$

$$b_2 = 0 \qquad \text{... (8)}$$

Eqn. (2) becomes

$$\xi = \cos \frac{\omega}{c} x (A \cos \omega t + B \sin \omega t) \qquad \text{... (9)}$$

when
$$\frac{d\xi}{dx} = 0 \text{ at } x = l \text{ gives}$$

$$\omega = \frac{s\pi c}{l} \qquad \text{... (10)}$$

and
$$\xi = 0 \text{ at } x = l/2 \text{ gives}$$
$$s = (2k + 1) \qquad \text{... (11)}$$

where k is a positive integer hence given by 0, 1, 2, 3,

$$\xi = \sum_0^k \cos(2k+1) \frac{\pi x}{l} \left(A_{(2k+1)} \cos \frac{(2k+1)ct}{l} + B_{2k+1} \sin \frac{(2k+1)ct}{l} \right) \qquad \text{... (12)}$$

Here only odd harmonics are produced. The period of the gravest mode $T = \frac{2l}{c}$ and any other harmonic

$$T_{(2k+1)} = \frac{2l}{(2k+1)c} \qquad \text{... (13)}$$

The rod vibrates with antinodes at two ends and nodes at the middle.

(*iii*) Rod two ends free gives

$$\frac{d\xi}{dx} = 0 \quad \text{at } x = 0 \quad \text{and } x = l$$

As in Eqns. (9), and (10) without applying the condition (11) we have

$$\delta = \sum_{1}^{s} \cos\frac{s\pi x}{l}\left(A_s \cos\frac{s\pi ct}{l} + B_s \sin\frac{s\pi ct}{l}\right) \qquad \text{... (14)}$$

Here all the modes are excited with antinode at two ends.

(*iv*) Rod fixed at one end.

In this case fixed end $x = 0$ is a node *i.e.* at $x = 0$, $\xi = 0$ and free end $\frac{d\xi}{dx} = 0$ at $x = l$.

The condition $x = 0$, $\xi = 0$ when applied in Eqn. (2) gives $a_2 = 0$

The condition $\frac{d\xi}{dx} = 0$ at $x = l$ gives

$$\cos\frac{\omega}{c}l = 0$$

or

$$\omega = (2k + 1)\frac{\pi}{2}.\frac{c}{l} \ .k = 0, 1, 2, \text{etc.}$$

or

$$\xi = \sum_{0}^{k} \sin(2k+1)\frac{\pi x}{2l}\left(A_{2k+1}\cos(2k+1)\frac{\pi ct}{2l} + B_{2k+1}\sin(2k+1)\frac{\pi ct}{2l}\right) \quad \text{... (15)}$$

Here only odd harmonics are developed. The period of the gravest mode is given by $\frac{4l}{c}$.

A rod clamped at the middle, excited to longitudinal vibration is of special interest as it is used in Kundt's tube (vide Chap. 15.4) experiment where rod is excited by drawing a wet cloth or resinated leather. The dynamics of exciting vibration by such a method has not yet been developed. But the dynamics of a rod clamped at one end and excited by striking by hard and elastic load has been developed by M. Ghosh and S.K. Ghosh[1]. (I. J.P. 1951). It is beyond the scope of this book so it is not discussed.

11.2. TRANSVERSE VIBRATION OF A BAR

A bar or a rod vibrating transversely, differs from the vibration of a string. The strings are stretched under tension and fixed at two ends, harmonic waves travel along the length of the string with a velocity independent of wavelength. But in the case of a bar non-harmonic waves are set up with no definite velocity. This motion is due to the sum of separate harmonic components, each moves with a velocity appropriate to its own wavelength.

In case of the string the restoring force is the tension. In case of the bar stiffness is called into play. The stiffness becomes more and more significant as the ratio of the diameter and the length increases and tension becomes less and less significant. The transition stage between a wire affected by tension and a bar affected by stiffness is sufficiently complicated. So a simple case of a bar is considered, unaffected by tension, with following simplifying assumption.

(*i*) The vibrating bar is straight,

(*ii*) It is of uniform cross-section and density,

(*iii*) It is not subjected to tension or compression,

(*iv*) The amplitude of vibration is small and curvature of bending is also small and may be represented by d^2y/dx^2.

(*v*) The rotatory effect during bending may be neglected for further simplification.

A straight bar is placed along the axis of X. It produces the transverse displacement along Y-axis. When the bar is bent as shown in Fig.1 the upper part of the bar is elongated and lower part is contracted and in between these two parts, there is neutral surface represented by $N N_1$ which remains unaltered in length due to bending.

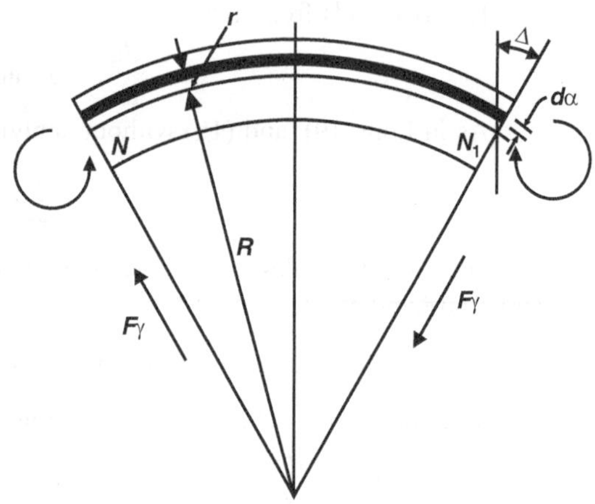

Fig. 1.

At the distance r from the neutral surface a portion of the bar of area $d\alpha$ and of length dx is considered, $\alpha = \Sigma\, d\alpha$ being the total area of cross-section of the bar. Let Δ be the extension produced by dx, relative to neutral axis, on bending, so a longitudinal force dF_x is acting on the area $d\alpha$ having Young's modulus of Y the material, is given by

$$dF_x = d\alpha Y \frac{\Delta}{dx} \qquad \text{... (16)}$$

From geometrical consideration

$$\frac{dx + \Delta}{R + r} = \frac{\Delta}{r} = \frac{dx}{R}$$

so

$$\frac{\Delta}{dx} = \frac{r}{R} \qquad \text{... (17)}$$

where R is the radius of curvature of the neutral axis given by

$$\frac{1}{R} = \frac{d^2y}{dx^2} \qquad \text{... (18)}$$

By the help of eqns. (17) and (18) eqns. (16) becomes

$$dF_x = d\alpha Y r . \frac{d^2y}{dx^2} \qquad \text{... (19)}$$

The total longitudinal force over the element dx and area of cross-section α, will vanish as, above the neutral surface *i.e.*, on convex side, the force produces extension of length, while on the lower portion of the neutral surface *i.e.,* on the concave portion there is contraction of length by a force acting in opposite direction. Thus the net effect of the force is zero.

or

$$F_x = \int dF_x = 0$$

But these two equal and opposite forces will constitute a couple, the moment of the couple or Bending moment M is given by

$$M = \int r dF_x$$

and by eqn. (19)

$$M = \int \frac{Y}{R} r^2 d\alpha$$

$$= \frac{Y}{R} \int r^2 d\alpha$$

$$= \frac{Y}{R} \alpha\, k^2$$

$$= Y\alpha k^2 . \frac{d^2 y}{dx^2} \quad \text{... (20)}$$

where k is the radius of gyration of the area α

and $$\frac{1}{R} = \frac{d^2 y}{dx^2}$$

The bending moment M varies with x along the length of the rod. They are M at x and $M + \frac{dM}{dx} dx$ at $x + dx$. The transverse force $(F_y)_x$ at x and $(F_y)_{x+dx}$ at $x + dx$ applied at two ends of the element dx which is responsible for bending produces, is equivalent to a shear having a moment $F_y dx$. The resultant force gives the element and acceleration $\frac{d^2 y}{dt^2}$. As the force $(F_y)_x$ and $(F_y)_{x+dx} = \left(F_y + \frac{dF_y}{dx} dx \right)$ are unequal the resultant force

$$\frac{dF_y}{dx} dx = \rho\alpha \frac{d^2 y}{dt^2} dx$$

where $\rho\ \alpha\ dx$ is the mass of the element dx.

or $$\frac{dF_y}{dx} = \rho\alpha \frac{d^2 y}{dt^2} \quad \text{... (21)}$$

The shear moment $F_y\ dx$ and the bending moment

$$\left(M + \frac{dM}{dx} dx \right) - M = \left(\frac{dM}{dx} \right) dx \quad \text{... (22)}$$

will cause the transverse slice of thickness dx and of cross-section α to rotate through a small angle

$$\theta = \frac{dy}{dx}$$

during bending from equilibrium position.

or $$I \frac{d^2\theta}{dt^2} = \frac{dM}{dx} dx + Fy \quad dx \quad \text{... (23)}$$

or $$\rho \alpha k^2 \frac{d^2}{dt^2} \left(\frac{dy}{dx} \right) = \frac{dM}{dx} + Fy \quad \text{... (24)}$$

as $$I = \rho dx\ \alpha\ k^2 \quad \text{... (25)}$$

On differentiating with respect to x

$$\rho\alpha k^2 \frac{d^4 y}{dt^2 dx^2} = \frac{d^2 M}{dx^2} + \frac{dF_y}{dx} \quad \text{... (26)}$$

From eqns. (20) and (21) as

$$\frac{d^2 M}{dx^2} = Y \alpha k^2 \frac{d^4 y}{dx^4}$$

and $$\frac{dF_y}{dx} = \rho\alpha \frac{d^2 y}{dt^2}$$

Eqn. (26) becomes

$$\rho\alpha k^2 \frac{d^4 y}{dt^2 dx^2} = Y\alpha k^2 \frac{d^4 y}{dx^4} + \rho\alpha \frac{d^2 y}{dt^2} \quad \text{... (27)}$$

Hence the equation of motion of transversely vibrating bar

$$\frac{d^2 y}{dt^2} - k^2 \frac{d^2 y}{dt^2 dx^2} + c_1^2 k^2 \frac{d^4 y}{dx^4} = 0 \quad \text{... (28)}$$

where $c_1 = \sqrt{\dfrac{Y}{\rho}}$

the velocity of longitudinal wave. The second term represents the rotatory motion of the bar which is small and hence can be neglected and eqn. (28) becomes

$$\frac{d^2 y}{dt^2} + c_1^2 k^2 \frac{d^4 y}{dx^4} = 0 \quad \text{... (29)}$$

11.3. DEPENDENCE OF VELOCITY OF TRANSVERSE WAVE IN A BAR ON FREQUENCY

If we represent the solution of Eqn. (29) as a plane progressive wave the displacement can be represented by

$$y = Ae^{i\omega\left(t - \frac{x}{c}\right)} \quad \text{... (30)}$$

whence $\dfrac{d^2 y}{dt^2} = -\omega^2 y$

and $\dfrac{d^4 y}{dx^4} = \dfrac{\omega^4}{c^4} y$ ... (31)

So Eqn. (29) can be written as with the help of Eqn. (31)

$$c_1^2 k^2 \frac{\omega^4}{c^4} y = -\omega^2 y \quad \text{... (32)}$$

or $c^4 = \omega^2 c_1{}^2 k^2$ as $\omega = 2\pi n$

n being the frequency of vibration so

$$c = \sqrt{2\pi c_1 k} \; n^{\frac{1}{2}} \quad \text{... (33)}$$

Thus it is found that the velocity of transverse wave in a bar depends upon the frequency.

11.4. GENERAL SOLUTION OF THE WAVE EQUATION (29)

Let us assume the bar executes undamped simple harmonic motion so the solution may be of the type

$$y = y_x \cdot y_t$$
$$= y_x . a \sin(\omega t + \phi)$$

Putting this value of y in Eq. (29), we get

$$\frac{d^4 y_x}{dx^4} = \frac{\omega^2}{c_1^2 k^2} y_x \quad \text{... (34)}$$

or $$\frac{d^4 y_x}{dx^4} = v^4 y_x \quad \text{... (35)}$$

where $$v^4 = \frac{\omega^2}{c_1^2 k^2}$$

Let the general solution be

$$y_x = A_1 e^{\beta x} \quad \text{... (36)}$$

by the help of Eq. (36), Eqn. (35) gives

$$\beta^4 = v^4$$

or $$\beta = \pm v, \pm iv$$

or $$y = y_x y_t = \sin(\omega t + \phi)\,\{A \cosh vx + B \sinh vx + C \cos vx + D \sin vx\} \quad \text{... (37)}$$

where A, B, C, and D are constants to be determined from boundary conditions.

The boundary conditions are for

(*a*) Free End

(*i*) The curvature or bending moment

$$\frac{d^2 y}{dx^2} = 0$$

(*ii*) Transverse force, neglecting rotation

$$\frac{d^3 y}{dx^3} = 0$$

(*b*) Clamped End (rigid)

(*i*) Displacement $y = 0$

(*ii*) Slope $\frac{dy}{dx} = 0$

(*c*) Supported End (rigid knife edge support)

(*i*) $y = 0$

(*ii*) $\frac{d^2 y}{dx^2} = 0$

11.5. A. CLAMPED FREE BAR

Clamped at $x = 0$

From Eq. (37) with the help of condition (b)

(*i*) $A + C = 0$

or $C = -A$

(*ii*) $B + D = 0$

$D = -B$

or Eq. (37) becomes

$$y = \sin(\omega t + \phi)\{A(\cosh vx - \cos vx) + B(\sinh vx - \sin vx)\}$$

From condition (a) at $x = l$

(*i*) $A(\cosh vl - \cos vl) = -B(\sinh vl + \sin vx)$

(*ii*) $A(\sinh vl - \sin vl) = -B(\cosh vl + \cos vl)$

Multiplying crosswise and simplifying

$$\sec vl = -\cosh vl$$

The solution is obtained graphically for v by plotting

$$y = \sec vl$$

and $$y = -\cosh vl$$

and finding the point of intersection of two curves which are given by

$$vl = 1.88, 4.69, 7.85, 11.00, 14.14$$

For gravest mode $vl = 1.88$ gives,

$$n = \frac{\omega}{2\pi}$$

$$= \frac{v^2 c_1 k}{2\pi}$$

$$= \frac{c_1 k}{2\pi} \cdot \frac{(1.88)^2}{l^2}$$

$$= \frac{k}{2\pi}\sqrt{\frac{Y}{\rho}} \cdot \frac{(1.88)^2}{l^2} \qquad \text{... (39)}$$

Hence the frequency of different overtones are given by

$$n_1 : n_2 : n_3 :: (1.88)^2 : (4.69)^2 : (7.85)^2$$

Unlike the modes of vibrations of transversely vibrating string or longitudinally vibrating bar, transversely vibrating bar produces no harmonic overtone. Dynamical theory of such a bar excited by impact has been developed by M. Ghosh, K. D. Roy and B. B. Banerjee and experimental verification of the theory was made by B. B. Banerjee.

B. Free-Free Bar – *both ends are free*

Applying the conditions given in (a)

$$\frac{d^2y}{dx^2} = \frac{d^3y}{dx^3} = 0 \text{ at } x = 0 \text{ and } x = l$$

We get from Eq. (37)

$$\frac{d^2y}{dx^2} = v^2 \sin(\omega t + \varphi)(A \cosh vx + B \sinh vx - C \cos vx - D \sin vx)$$

$$\frac{d^2y}{dx^2} = v^3 \sin(\omega t + \varphi)(A \sinh vx + B \cosh vx + C \sin vx - D \sin vx)$$

At $x = 0, A = C, B = D$

At $x = l,$

$$A(\cosh vl - \cos vl) = -B(\sinh vl - \sin vl)$$

and $$A(\sinh vl + \sin vl) = -B(\cosh vl - \cos vl)$$

This on solving gives

$$\cosh vl = \sec vl$$

As before, we get

$$vl = 4.73, 7.85, 10.99, 14.14 \text{ etc.}$$

The frequency of the gravest mode is given by

$$n_1 = \frac{k}{2\pi}\sqrt{\frac{Y}{\rho}} \frac{(4.73)^2}{l^2} \qquad \text{... (40)}$$

The frequency of the notes emitted are in the ratios

$$n_1 : n_2 : n_3 :: (4.73)^2 : (7.85)^2 : (10.99)^2 \ldots\ldots\ldots$$

So they are not in harmonic ratio

C. Bar Supported at Both Ends

Applying the conditions (B) in Eq. (37)

At $x = 0$

(*i*) $A + C = 0$ and

(*ii*) $-A + C = 0$

This gives $A = C = 0,$

So $y = \sin(\omega t + \varphi)(B \sinh vx + D \sin vx)$

At $x = l$

(*i*) $B \sinh vl + D \sin vl = 0$

(*ii*) $B \sinh vl - D \sin vl = 0$

gives $\sin vl = 0$

or $vl = s\pi$ where s is a whole number

Hence $$n_s = \frac{k\pi}{2l^2}\sqrt{\frac{Y}{\rho}}.s^2$$

so the frequency of different modes are in ratio

$$n_1 : n_2 : n_3 :: 1 : 4 : 9.$$

11.6. TUNING FORK

Chladni regarded the tuning fork as a free bar bent in the form of *U* elongated. The nodes gradually approach each other, as a bar is bent gradually at the midpoint Fig. 2. The vibration of

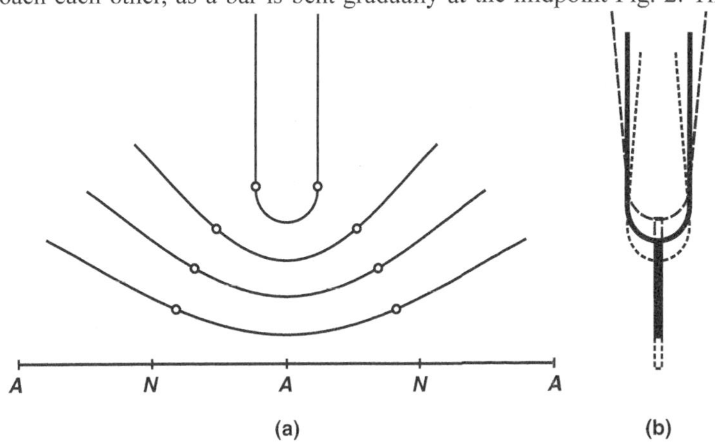

Fig. 2.

antinode, in between the two nodes, is small compared to that of the free ends. The addition of stem at the midpoint causes the nodes to approach closer. It is equivalent to, adding a mass at antinode and to increase the stiffness of the fork between two nodes causing the reduction of vibration of the midpoint. This small vibration of the stem Fig. 2(*b*) fluctuates the sounding board or the resonators when the stem is pressed into contact with them. The energy of vibration of the fork is rapidly transmitted to them through the stem.

Rayleigh regarded the tuning fork as two clamped free bar mounted parallel to each other on a heavy rigid block of metal. The frequency of the gravest note is given by as before. Art. 11.5. Eq. (39) on approximation cited before

$$n_1 = \frac{k}{2\pi}\sqrt{\frac{Y}{\rho}}\frac{(1.88)^2}{l^2}$$

The radius of gyration for a rectangular bar of thickness t about the neutral surface is $\frac{t}{\sqrt{12}}$

and $c_1 = \sqrt{\frac{Y}{\rho}} = 5.24 \times 10^5$ cm/sec for steel

so $$n_1 = 8.46 \times 10^4 . \frac{t}{l^2}$$

The frequency of a bar clamped at one end and free at the other, may be lowered either by loading the free end or reducing the area of cross-section of the bar near the fixed end. Similarly the frequency may be increased by reducing the area of cross-section of the bar at the free end or by shortening the length of the bar. The frequency of bar may be adjusted by using a sliding collar and rigidly fixing at any desired position. The nearer the sliding collar (weight) to the free end, lower is the frequency. These principles are generally adapted in practice in the construction and tuning of a tuning fork and in harmonium reeds.

11.7. TEMPERATURE DEPENDENCE OF A TUNING FORK

The change of temperature is of the serious cause of the change of frequency of a tuning fork. This is mainly due to the change of elastic property of the material of the fork.

The value of Young's Modulus Y, the value of the density ρ of the material of the fork and the length l of the prong of the tuning fork change with temperature as given by

$$Y_t = Y_0(1 - \in t)$$
$$\rho_t = \rho_0(1 - 3\,\alpha t)$$
$$l_t = l_0(1 + \alpha t)$$

where the quantities with suffix t and 0 represent the respective values at t°C and 0°C, $\in$ is the temperature co-efficient of Young's Modulus, α the linear temperature co-efficient of length,

whence it is found $\frac{t}{2\pi}\sqrt{\frac{Y_t}{\rho_t}}\frac{(1.88)^2}{l_t^2}$

$$= \frac{k}{2\pi}\sqrt{\frac{Y_0}{\rho_0}}\frac{(1.88)}{l_0^2}\frac{\left(1 - \frac{\in t}{2}\right)}{\left(1 - \frac{3\alpha}{t}t\right)(1 + 2\,\alpha t)}$$

$$= n_0\left(1 - \frac{\in - \alpha}{2}t\right)$$

It is generally found that the change of frequency of the fork depends more upon the change of elasticity than upon the change of length and density. Guillaume discovered Elinver in which has very small temperature co-efficient of elasticity. It is one of the new nickel steel alloy. This is now used for the construction of tuning forks, as it is of invariable elasticity like 'inver' which has very small expansion co-efficient and is used for the construction of best watches and chronometers.

11.8 ELECTRICALLY MAINTAINED TUNING FORKS

P, P are the prongs of the tuning fork, rigidly fixed on metal stand T by nut and screw. M is an electromagnet, C the platinum points. By adjusting the screw S tipped with platinum wire, it is made to contact the spring top C with a small platinum sheet fixed on the flat surface of the spring. This platinum contact saves from damage of the contact-joint from sparking. When the electrical circuit is closed, current flows through the metal stand, the body of the prong, the attached spring, *CS* the contact point, through the coil *M* of the electromagnet. The prongs are attracted. The contact point *CS* is separated. The electromagnet is demagnetised as current ceases to flow. Evidently the prongs are released and again comes in contact and the operation is repeated. In this way, the prongs set in vibration. During the motion from normal position towards the electromagnet

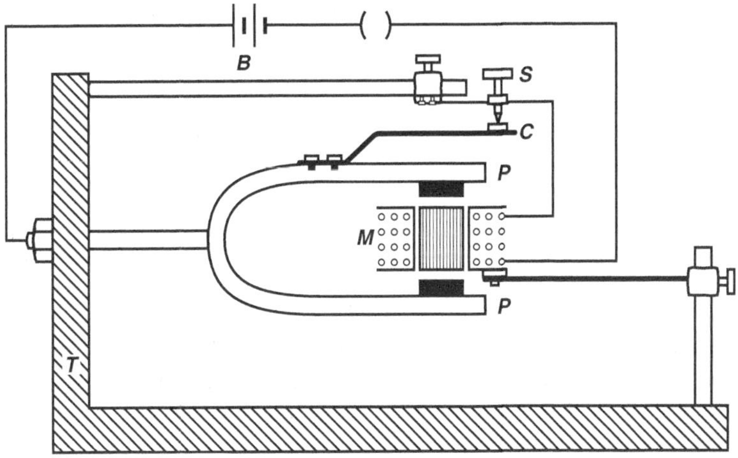

Fig. 3.

in $\frac{T}{4}$ and back in time $\frac{T}{4}$, when there is no current in the electromagnet. As soon as it comes in contact, the current sets in the prong goes further from the normal position in time $\frac{T}{4}$. At this stage, the current which has already been set up in electromagnet opposes the motion of the prongs and no useful work is done, the prongs come back from extreme position to its normal position in time $\frac{T}{4}$, where T is the time period of vibration of the fork. During this motion, the current assists the motion, and useful work is done. The net impulse is given at this interval $\frac{T}{2}$ to the fork in maintaining the vibration of the fork during the other interval $\frac{T}{2}$. In the case of non-inductive circuit, the current remains steady from the start, maintenance of vibration becomes difficult as the net impulse is zero. But, as electromagnet includes sufficient inductance in the circuit, the current grows exponentially, hence the total charge passing, during the first half is less than that during the second half, so the resultant impulse helps to maintain the vibration. The large inductance in the circuit produces large e.m.f. during break which will causes *A* the arcing at the contact point *C*. A suitable condenser is generally placed parallel to the contact point in order to stop the spark.

11.9. VALVE MAINTAINED TUNING FORK

Up to a frequency 100 c.p.s. a tuning fork can be efficiently maintained electrically as described above. The vibration of a tuning fork for higher frequencies are maintained by thermo-ionic valves. Two electromagnets M_1M_2, one in the grid circuit and other in plate circuit of the valve are used. The coils of electromagnets facing the prongs are wound in the opposite directions. Variable condensers are used across the electromagnets to effect tuning in the circuit grid bias batteries are introduced.

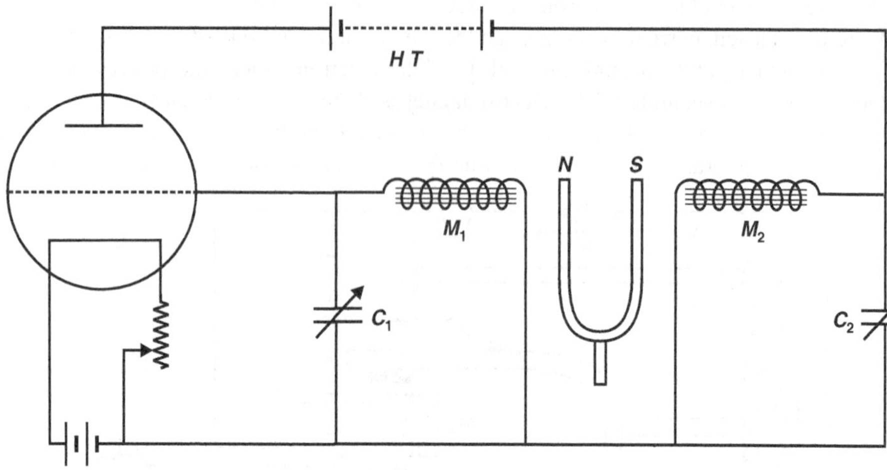

Fig. 4.

A vibration of the prong N of the fork generates some induced e.m.f. in the grid coil of the magnet so that amplified fluctuating current is obtained in the plate circuit passing through the coil of the electromagnet M_2 so that it attracts S prong with south polarity. The method was further developed by Dye as a precision time standard.

SOLVED PROBLEMS

1. The longitudinal vibration of a bar free at both ends are initiated by striking the free end $x = l$ so that at time $t = 0$ the velocity is zero except over a small length dl at $x = l$. Find the expression for the displacement.

The displacement of the bar free at both ends is given by equation (14) as

$$\xi = \sum_{1}^{s} \cos\frac{s\pi x}{l}\left(A_s \cos\frac{s\pi ct}{l} + B_s \sin\frac{s\pi ct}{l}\right)$$

Since at $t = 0$, $\xi = 0$ for all values of x

So $$\sum_{s=1}^{s} \cos\frac{s\pi x}{l} A_s = 0$$

or $$A_s = 0$$

Hence $$\xi = \sum_{1}^{s} B_s \cos\frac{s\pi x}{l} \sin\frac{s\pi ct}{l}$$

The velocity $$\dot{\xi}_0 = \left.\frac{d\xi}{dt}\right/_{t=0} = \sum_{1}^{s} B_s \frac{s\pi c}{l} \cos\frac{s\pi x}{l}$$

Multiplying both sides by $\cos\dfrac{s\pi x}{l}\,dx$ and integrating

$$\int_0^l \dot{\xi}_0 \cos\frac{s\pi x}{l}\,dx = B_s \frac{s\pi c}{l}\int_0^l \cos^2\frac{s\pi x}{l}\,dx = B_s \frac{s\pi c}{l}\cdot\frac{l}{2}$$

or $$B_s = \frac{2}{s\pi c}\int_{l-dl}^{l} \dot{\xi}_0 \cos\frac{s\pi x}{l}\,dx = \frac{2(-1)^s}{s\pi c}\int_{l-dl}^{l} \dot{\xi}_0\,dx$$

$\therefore$ $$\xi = \frac{2}{\pi c}\left(\int_{l-dl}^{l} \dot{\xi}_0\,dx\right)\sum_{1}^{s} \frac{(-1)^s}{s} \cos\frac{s\pi x}{l} \sin\frac{s\pi ct}{l}$$

REVIEW QUESTIONS

1. Investigate the longitudinal vibration of a rod

   (*i*) Clamped at both ends (*ii*) clamped at the middle. [C.U. 1964]

2. Develop the theory of transverse vibration of bar. Discuss the general characteristics with reference to velocity, frequency, etc.
3. Why the transverse vibration of a rod is not utilised in the construction of any musical instrument?
4. What are the factors determining the frequency of a tuning fork? Why the vibration of a fork is used to produce pure tone?
5. Explain and describe giving the practical details for maintaining the vibration of a tuning fork electrically. Why this is used as frequency standard?

PROBLEMS

1. Given that

$$y = \cos \omega t \left[A \cosh \frac{\omega x}{v} + B \sinh \frac{\omega x}{v} + C \cos \frac{\omega x}{v} + D \sin \frac{\omega x}{v} \right]$$

where $\omega = (1.88v/l)$ is the fundamental frequency of transverse vibration of a bar clamped at one end and free at the other. If the amplitude at the free end $x = l$ is 'a'; determine the constants A, B, C and D in terms of 'a'

12 CHAPTER
VIBRATION OF MEMBRANES AND PLATES

12.1. (i) MEMBRANE

Let a membrane of uniform mass per unit area α is stretched under constant tension T in x-y plane. A very small displacement ω takes place along xz axis while sets in vibration, the tension T remains constant.

The membrane is displaced by an excess of pressure p on the concave side. Initial area before displacement is $a \times b$, a and b are the two sides of the rectangular piece. After displacement $d\omega$ the final area is $(a + da) \times (b + db)$

The increase in area is

$$(adb + bda) = \left(\frac{db}{b} + \frac{da}{a}\right) \times ab \quad \dots (1)$$

The work done by the excess of pressure when displacement is $d\omega$ along z-axis is

$$p.\ ab\ .\ d\omega \quad \dots (2)$$

This is balanced by the increase of potential energy due to increase in area

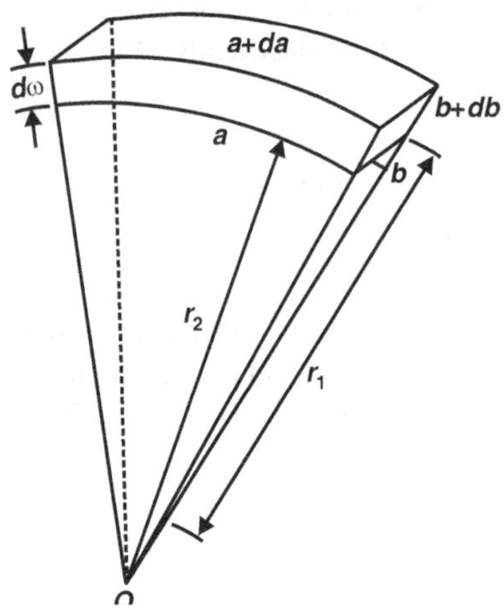

Fig. 1.

$$T.ab\left(\frac{da}{a} + \frac{db}{b}\right) \quad \dots (3)$$

As $$\frac{d\omega}{r_2} = \frac{da}{a}, \quad \frac{d\omega}{r_1} = \frac{db}{b}$$

where r_1 and r_2 are the radii of curvature of the arc a and b Fig. 1. They are given by

$$\frac{1}{r_1} = \frac{d^2\omega}{dx^2}, \quad \frac{1}{r_2} = \frac{d^2\omega}{dy^2}$$

So $$p.\ ab.\ d\omega = T\left(\frac{da}{a} + \frac{db}{b}\right) ab$$

$$= Tab.d\omega\left(\frac{1}{r_1} + \frac{1}{r_2}\right) \quad \dots (4)$$

$$= Tab.d\omega\left(\frac{d^2\omega}{dx^2} + \frac{d^2\omega}{dy^2}\right) \quad \dots (5)$$

or $$p = T\left(\frac{d^2\omega}{dx^2} + \frac{d^2\omega}{dy^2}\right) \quad \dots (6)$$

As p is the force per unit area is given by

$$p = \sigma \frac{d^2y}{dt^2} \qquad \text{... (7)}$$

where σ is the maso per unit area.

Hence
$$\frac{d^2y}{dt^2} = \frac{T}{\sigma}\left(\frac{d^2\omega}{dx^2} + \frac{d^2\omega}{dy^2}\right) \qquad \text{... (8)}$$

where
$$\frac{T}{\sigma} = c^2$$

which is the equation of motion of a vibrating membrane.

(*i*) Rectangular membrane of boundary $x = 0$ to $x = a$, and $y = 0$ to $y = b$ of two sides.

The membrane is taken of uniform thickness and areal mass density σ and vibrates with small amplitude while stretched under constant tension T. The boundary conditions are

$$\omega = 0, \text{ at } x = 0 \text{ and } x = a$$

also at
$$y = 0 \text{ and } y = b.$$

Following the method as adopted in the case of transverse vibration of the string, the general solution may be written as

$$\omega = \sum\sum \sin\frac{m\pi x}{a}\sin\frac{n\pi y}{b} \times (A_{mn}\cos p_{mn}t + B_{mn}\sin p_{mn}t) \qquad \text{... (9)}$$

where
$$p_{mn}{}^2 = c^2\pi^2\left(\frac{m^2}{a^2} + \frac{n^2}{b^2}\right) \qquad \text{...(10)}$$

or frequency of natural vibration.
$$v_{nm} = \frac{p}{2\pi} = \frac{1}{2}\sqrt{\frac{T}{\sigma}\left(\frac{m^2}{a^2} + \frac{n^2}{b^2}\right)} \qquad \text{... (11)}$$

The constants A_{mn} and B_{mn} are determined from boundary conditions.

$$\omega = \omega_0 \text{ and } \frac{d\omega}{dt} = v_0 \text{ at } t = 0$$

Hence we have

$$\omega_0 = \sum_{m=1}^{\infty}\sum_{n=1}^{\infty} \sin\frac{m\pi x}{a}.\sin\frac{n\pi x}{b} A_{mn}$$

or
$$A_{mn} = \sum_{m=1}^{\infty}\sum_{n=1}^{\infty} \frac{4}{ab}\int_0^a\int_0^b \omega_0 \sin\frac{m\pi x}{a}\sin\frac{n\pi x}{b}\,dxdy$$

Similarly

$$B_{mn} = \sum\sum \frac{4}{abp_{mn}}\int_0^a\int_0^b v_0 \sin\frac{m\pi x}{a}\sin\frac{n\pi x}{b}\,dxdy$$

The lowest mode of vibration corresponds to $m = 1$, $n = 1$ for which nodal lines are on the boundary where $\omega = 0$.

The fundamental frequency v_{11} Eqn. (11) depending on T and σ is analogous to the case of the vibration of a string. Among the allowed frequencies which are harmonics to the fundamental are $v_{22} = 2v_{11}$, $v_{33} = 3v_{11}$ etc. There are other frequencies which are not harmonics. When a and

b are comparable and nearly equal, two extra frequencies v_{12} and v_{21} come in between first and second harmonics. Six extra frequencies $v_{13}\, v_{31}\, v_{23}\, v_{32}$ and $v_{14}\, v_{41}$ come between the second and third. Similarly 10 extra frequencies come between third and fourth and so on.

From Eqn. (9) number and position of nodal lines ($\omega = 0$ for all values of t) parallel to the axes of x and y for different values of m and n are given by

$$y = \frac{b}{n}, \frac{2b}{n} \ldots \frac{n-1}{n} b \text{ parallel to } x\text{-axis}$$

$$x = \frac{a}{m}, \frac{2a}{m} \ldots \frac{m-1}{m} b \text{ parallel to } y\text{-axis}$$

Their numbers are $n + 1$ and $m + 1$ including those at boundary.

In the case of square membrane $a = b$, so frequency

$$v_{mn} = \frac{1}{2a}\sqrt{\frac{T}{\sigma}(m^2 + n^2)}$$

Gravest mode $m = 1\; n = 1$

$$v_{11} = \frac{c}{2a}\sqrt{2}$$

higher mode $m = 1, n = 2 \;\; v_{12} = \frac{c}{2a}\sqrt{5}$

and $m = 2, n = 2$ and so on $v_{22} = \frac{c}{2a}\sqrt{8}$

where $c = \sqrt{\frac{T}{\sigma}}$

Thus all the modes of vibrations do not form harmonic series so the resultant note is a noise

(*ii*) **Circular membrane.** The first step to be adopted in the case of a circular membrane of fixed boundary, the general differential equation of motion as given in eqn. (8) is reduced to plane polar co-ordinates which becomes

$$\frac{d^2\omega}{dt^2} = c^2\left(\frac{d^2\omega}{dr^2} + \frac{1}{r}\frac{d\omega}{dr} + \frac{1}{r^2}\frac{d^2\omega}{d\theta^2}\right)$$

The subsidiary condition to be satisfied at the boundary is that

$$\omega = 0 \text{ at } r = a$$

The exact procedure is beyond the scope of this treatise. The nodal lines as obtained from the analysis are concentric circles and diameters. The relative frequencies of the tones emitted including fundamental do not form harmonic series as in kettle drum.

There are some Indian musical instruments with circular stretched membrane under tension backed by air tight vessel such as **Mridanga Pakhwaj** and **Tabla**. The membrane is loaded at the central region producing variation of density along the radius. A few harmonic overtones are generated by skilled players.

12.2. VIBRATION OF PLATES

The experimental study of the vibration of plates, begins as early as 1787 by Chladne in a very simple and novel method. He demonstrated the positions of nodal lines by sprinkling fine sand, over the surface of the plate sets in vibration. Variety of patterns were obtained by fixing the plate at the centre of symmetry by bowing and touching certain points on the edge. These variety of pattern obtained, give evidence about the complexity of the problem in mathematical treatment. Chladni gave 52 figures with square plate, 43 with circular, 30 hexagonal, 52 rectangular, 26 elliptical, 15 semi-circular and 25 triangular plates.

The vibrating plates are used as percussion instruments in music in different forms. The complex problem of flat plate is made further complicated when the plate is curved such as gongs etc.

REVIEW QUESTIONS

1. Deduce the differential equation of vibration of a rectangular membrane fixed on four sides under tension. All the modes of vibration do not form harmonic series as a result of which resultant note is noise. Explain the statement "how the note can be made musical".
2. Write a short note on the vibration of membrane and plates.

13 CHAPTER VIBRATION OF AIR COLUMN

13.1. VIBRATION OF AIR COLUMN

The differential equation of motion of longitudinal wave travelling in a gaseous medium vide Art. 9.1 eqn. 3 is given by

$$\frac{d^2\xi}{dt^2} = c^2 \frac{d^2\xi}{dx^2} \quad ...(1)$$

where $$c^2 = \frac{E}{\rho}$$

c is the velocity of wave propagation, E is the bulk modulus, ρ is the density of the medium at rest.

As in Art. 10.4 eq. (17) the solution of the eqn. (1) is given by

$$\rho = (a\cos\omega t + b\sin\omega t)\left(a_1 \cos\frac{\omega}{c}x + b_1 \sin\frac{\omega}{c}x\right) \quad ...(2)$$

(1) *For pipe open at both ends – open pipe*

$$\frac{d\xi}{dx} = 0 \text{ at } x = 0 \text{ and } x = l \text{ for all values of } t.$$

From Eq. (2)

$$\frac{d\xi}{dx} = \frac{\omega}{c}\left(-a_1 \sin\frac{\omega x}{c} + b_1 \cos\frac{\omega}{c}x\right)(a\cos\omega t + b\sin\omega t)$$

gives $$\frac{d\xi}{dx} = 0 \text{ at } x = 0,$$

or $$-a_1 \cdot 0 + b_1 \times 1 = 0$$

gives $b_1 = 0$ and at $x = l$, $\frac{d\xi}{dx} = a\sin\omega l = 0$

gives $$\frac{\omega l}{c} = s\pi$$

or $$\omega = \frac{s\pi c}{l}$$

where s is a positive integer $s = 1, 2, 3, 4$ ... etc.

So $$\xi = \sum_s b_1 \cos\frac{s\pi x}{l}\left(a\cos\frac{s\pi ct}{l} + b\sin\frac{s\pi ct}{l}\right)$$

can be written in the form

$$\xi = \sum_s \cos\frac{s\pi x}{l}\left(A_s \cos\frac{s\pi ct}{l} + B_s \sin\frac{s\pi ct}{l}\right)$$

where A_s and B_s are constants and are different for different values of s.

As $$\rho = 2\pi n = \frac{s\pi c}{l}$$

so $$n = \frac{s}{2l} \cdot c$$

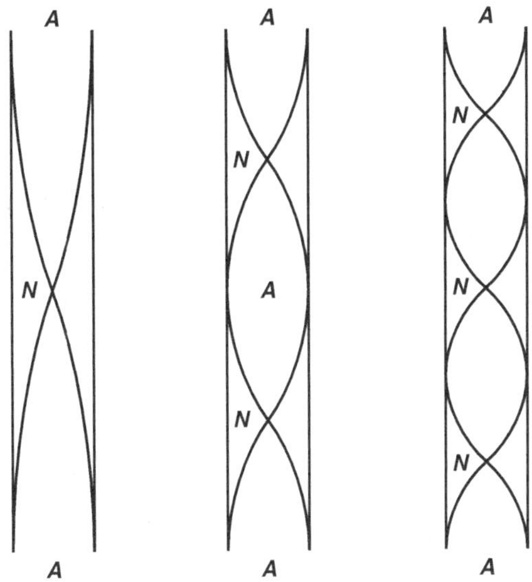

Fig. 1.

Here a pulse of compression starting from the mouth of the pipe is reflected from the opposite open end as rarefaction *i.e.,* with reversal of phase, travels along the length of the pipe and reaches the mouth, where another reflection takes place with reversal of phase *i.e.,* reflected pulse becomes compression like the initial pulse Fig. 1. Thus initial state of the pulse is restored after two complete traverse of the length of the pipe. Here all the harmonics are present.

(*ii*) *Pipe closed at one end and open at the other.*

Here (*i*) at open end

$$\frac{d\xi}{dx} = 0$$

at $x = 0$ and (*ii*) at closed end. $\xi = 0$ *a*; $x = l$

for all values of *t*,

At (*i*) open end gives

$$b_1 = 0$$

(*ii*) closed end gives

$$a_1 \cos\frac{\omega}{c}l = 0$$

or $$\frac{\omega}{c}l = \frac{\pi}{2}, \frac{3\pi}{2}, \frac{5\pi}{2}, \ldots \frac{2k+1}{2}\pi$$

or $$\omega = (2k+1)\frac{\pi c}{2l}$$

where $2k + 1$ is an odd integer, *so, only odd harmonics* are present.

So
$$\xi = \cos\frac{2k+1}{2l}\pi x\left\{A_{(2k+1)}\cos\frac{(2k+1)}{2l}\pi ct + B_{(2k+1)}\sin\frac{2k+1}{2l}\pi ct\right\}$$

writing
$$2k + 1 = r, \text{ where } r = 1, 3, 5, \ldots \text{ etc.}$$
$$\xi = \cos\frac{r\pi x}{2l}\left(A_r\cos\frac{r\pi c}{2l}t + B_r\sin\frac{r\pi ct}{2l}\right)$$

The frequency n of the pipe is given by
$$n = \frac{\omega}{2\pi} = \frac{rc}{4l}$$

The initial pulse of compression travels along the length of the pipe and reaches the closed (*i*) end where reflection takes place without change of phase *i.e.*, as compression, and travels again in the opposite direction along the length, and reaches the open end where (*ii*) reflection takes place with change of phase *i.e.*, compression as rarefaction. This rarefaction travels again along the length of the pipe and (*iii*) is reflected from the closed end as rarefaction and on (*iv*) reaching the open end it becomes compression. Thus in order to repeat the cycle the pulse is required to travel four times the length of the pipe.

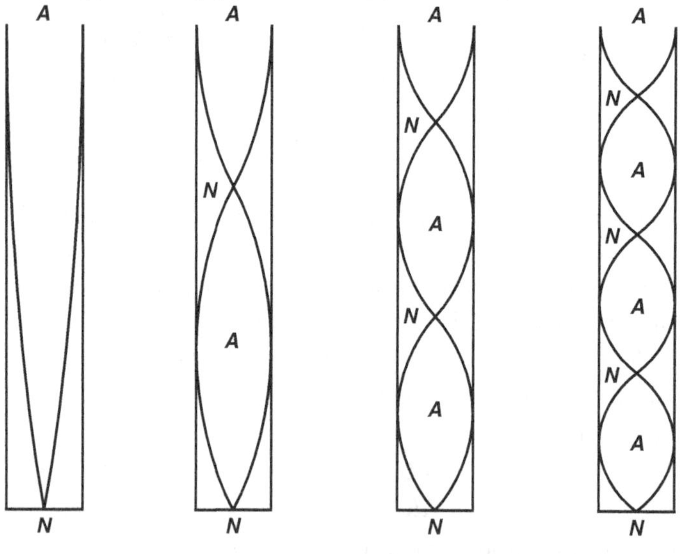

Fig. 2.

The Fig. 1 and Fig. 2 indicate a few possible modes of vibration of closed and open pipe. For a given frequency, the length of the open pipe is twice the length of the closed pipe.

13.2. VORTEX SOUND

It is very commonly observed that when a peg is fixed at the midway of a stream of water, the stream of water swirls down the peg.

Vortices or eddies are formed behind the obstacle when placed in stream line flow of a fluid. These eddies arise from the viscous drag between the surface of the obstacle and the fluid.

Similar phenomenon takes place when a cylindrical obstacle such as a rod or a wire is placed in a stream of air. Double sets of eddies are set up down the obstacle. These vortices are developed on opposite sides of the obstacle. They revolve in opposite directions and they are detached and are carried along the stream. The stream of air passed the wire—the obstacle, is thus sets into transverse vibration, as right handed and left handed eddies are formed and above a particular velocity of the stream, they are detached. This effect produces alternating thrust which causes the wire to vibrate.

The frequency N of the alternating thrust which depends upon the diameter d of the wire and the speed of rotation 'v' but does not depend upon the tension and the length of the wires is given by

$$N = 0.185\frac{v}{d}$$

The natural frequency of the wire for sth mode is given by

$$n_s = \frac{s}{2l}\sqrt{\frac{T}{\rho}}$$

l, T and ρ are the length, tension and linear density of the wire, and s is the order of the modes. With change of the speed of rotation 'v' resonance occur with a particular mode of vibration of the string where $N = n_s$, the sound is intensified.

The sound so produced is called **Aeolian Tone**.

Several wires with different diameters are fixed in a frame. Their tensions are so adjusted as to make their natural frequencies of vibration of the fundamental mode are the same. Wind passes through them may excite some of the wires and produce appropriate over-tones. This instrument is known as **Aeolian Harp**.

There are many examples of Aeolian tone. The singing of the telegraphic wire, whistling of wind through trees etc., are the examples of the phenomenon.

13.3. (a) JET TONE

Two rows of vortices rotating in opposite directions are formed on the two sides of the central stream lines, Fig. 3 which takes the sinous course when a jet of air issues from a narrow slit. A weak and uncertain note called jet note is heard. The frequency N of the note emitted is given by $N = 0.045\frac{v}{a}$

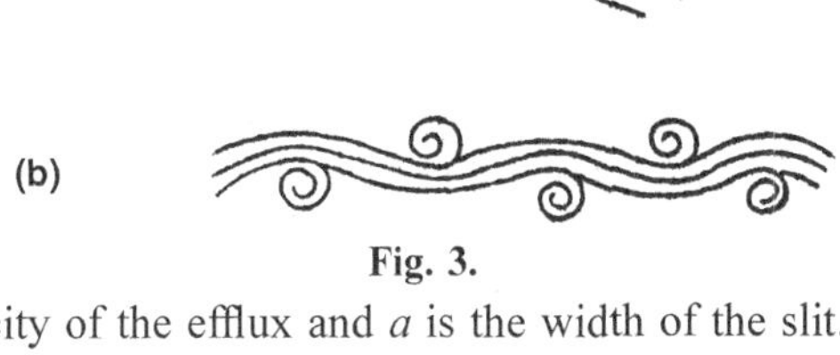

Fig. 3.

where v is the velocity of the efflux and a is the width of the slit.

(*b*) **Edge Tone.** This is developed when a blade-shaped slit of air or any gas, streams out from the slit and allowed to impinge a sharp wedge Fig. 4. The distance of the wedge is adjusted; when the distance is very small no sound is developed. As the distance increases, a clear note is heard at a critical distance d_0. This distance depends upon the velocity V of the impinging gas. On further increase of the distance, the frequency of the emitted note first falls off continuously, until there is abrupt change of frequency of octave higher. With the further increase of the distance, the frequency again falls off till the second rise of frequency is heard. With further increase of distance of the note ceases.

If the velocity of efflux increases when the distance is kept constant, the frequency increases and jumps to octave above.

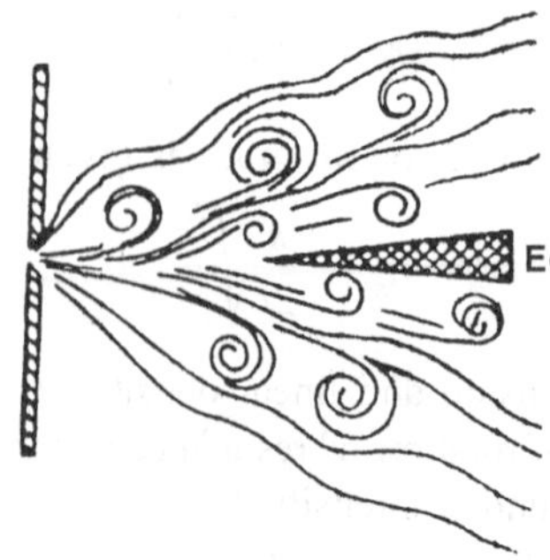

Fig. 4.

In the case of edge tone, the issuing lamina of air has a tendency to form vortex filaments on either edge of the stream as it enters the undisturbed air on the other side of the slit. When the air strikes the edge, the space between the slit and the edge acts as a form of resonator so that the distance d to the edge becomes equal to, or multiple of the wavelength of the system of the vortices, which is equal to the distance between the successive vortices in the same row. Suppose a vortex leaving the outer wall of the slit, moves onward and strikes the edge when next vortex leaves the surface of the slit. The similar process takes place for the system of vertices in the other row, revolving in the opposite directions. The secondary vortices are probably formed on the boundary layer along side of the edge of the wedge.

The frequency of the edge tone is given by the frequency with which eddies strike the edge.

If v be the velocity of the eddies with which they move towards the edge and strike it, and l be the distance between the vortices in the same row then frequency n is given by

$$n = \frac{v}{l}$$

As $l = d_0$, the minimum distance of the wedge from the slit because one vortex strikes the edge while the next is generated at the surface of the slit, and $v = aV$ where 'a' is constant V is the velocity of efflux of the stream, as v the velocity of vortex system relative to the stationary fluid though less than 'V' but is proportional to it, then

$$\frac{V}{nd_0} = \text{constant},$$

As vortex passes by the wedge, it sucks in air to one side or the other according to the eddies. The arrival of anticlockwise eddy near the wedge produces the compression on the same side and clockwise produces rarefaction. Now a train of compressions and rarefactions moves forward and produces a note known as edge tone. The period of edge tone as already shown is the interval between two successive eddies of the same type strike the edge.

13.4. (i) ORGAN PIPE

There are two types of organ pipe. (i) Flue pipe Fig.5(a). This is generally known as *flue organ pipe* or *fixed lip* type organ pipe. Air is blown through the mouth piece N at a suitable pressure. A blade like stream of air directed by the languet L issues from the slit of flue C, impinges upon the edge E. It produces edge tone. The air column and the eddy system are coupled together, *i.e.*, the edge tone at the mouth, is coupled to the natural frequency of the column of air in the tube. But slight departure from resonance between the pipe tone and generated edge tone does not alter the time period.

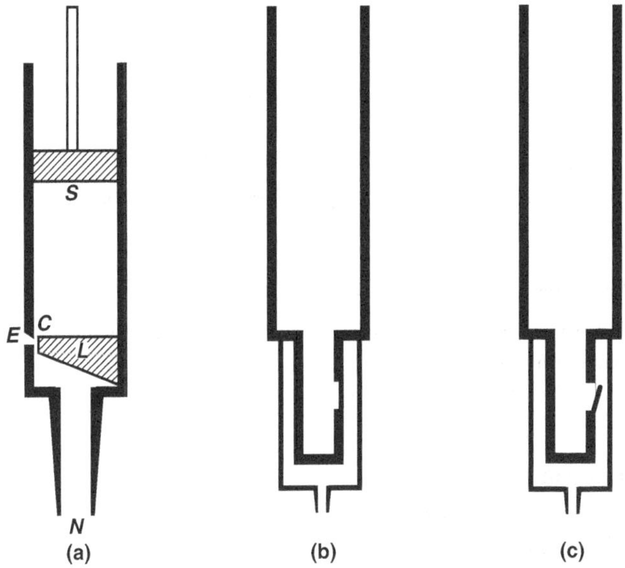

Fig. 5.

(*ii*) **Reed Pipe.** A tongue like reed vibrates in the front of an opening, Fig. 5(*b*) and Fig. 5(*c*), or a chink. A series of rapid puffs of air enter the pipe which are responsible for the vibration of the enclosed air column. The reeds are generally of two types (*i*) *Floating reed* moves both inside and outside when the chink is slightly bigger than the reed (*ii*) the *beating reed* moving only outward where the chink is slightly smaller than the reed as in the case of clarionet.

The organ pipes may be of square cross-section where it is generally made of wood, or of metal of circular cross-sections. An adjustable stop S closes the organ pipe at or near about the end so that the speaking length is from the opening C to S.

A large end correction is necessary at the blown end, so vibrating length is greater than the geometric length. When the tube is narrow, the end correction will be small. The sound formed in a narrow tube is likely to be rich in harmonics. For wide tube, the end correction given by 0.6 R at the open end, and 3.8 R at the mouth, where R is the radius of the tube, are to be made.

13.5. OVER BLOWING AND UNDER BLOWING OF FLUE TYPE ORGAN PIPE

The organ pipe should be designed so that there must be all conditions necessary for production of edge tone while the tube itself is capable of giving a series of partial tones. Both together, form a coupled system.

It is already stated that the increase of the flow of velocity *i.e.,* the pressure of blowing, results the formation of overtones. The formation of overtone is known as *over blowing the pipe.* Now the system—the tube tone and the jet tone are still coupled when the blowing pressure is increased. The pipe enters to impose one on the other of its own tone upon the vortex production at the mouth. The distance between the consecutive vortices in the same row will be altered in such a way that the natural vibration of the next over tone of the pipe is maintained, as the frequency jumps from a lower value to a higher value.

Under Blowing

When the velocity of efflux of air is gradually decreased from the normal velocity, the frequency of the edge tone will be far below the pipe tone and *will gradually become feeble*. On the other hand the pipe tone which is still coupled with edge tone, will try to impose one or the other of its over-tones upon the formation of vortex and the fundamental is excited feebler.

In the case of overblown pipe, the tones are louder than the normal but in the case of under blown pipe are feebler.

13.6. (i) PHENOMENON INSIDE THE PIPE—EXPERIMENTAL OBSERVATION

(*i*) A paper membrane formed on a ring and covered with fine sand is suspended like a scale pan and is lowered inside a sounding pipe. The sand particles will be agitated most, at antinodes and minimum at nodes. Thus the position of nodes and antinodes in a sounding pipe is *located* by this simple experiment.

(ii) Konig's Manometric Flame

A chamber C is divided into two compartments by an elastic diaphragm say a thin sheet of rubber membrane. The portion m is open to outside through a speaking tube T. The front chamber C is supplied with coal gas which passes through an exit tube with a fine pin hole burner. The tube T is sealed by glue or wax to any point of an organ pipe on which a hole is made. The pressure fluctuations in sound wave is admitted through T, set the membrane in vibration due to pressure fluctuation on the gas within the front chamber m. Consequently the flame is caused to jump up and down with a frequency equal to that of the incident wave. Since the vibration is very rapid, they cannot be observed when the flame is looked at directly. To see the changes in the size of the flame, a rotating mirror is used in front of it. As the mirror rotates in the vertical axis it throws the image of the flame in the successive intervals. The flame can also be photographed by a rapidly moving photographic plate.

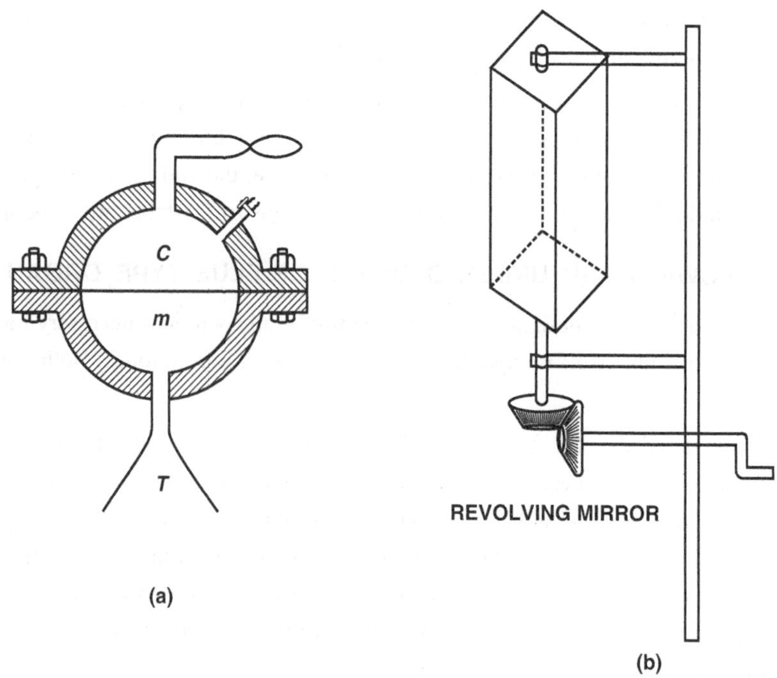

Fig. 6.

From the movement of the flame as observed by the help of the revolving mirror or from the photographs, the position of nodes and antinodes in the pipe is easily observed.

(iii) Richardson's Manometric Capsule

It is used to explore the pressure variation in a pipe.

The capsule widens out conically from a short cylindrical piece, to a flange which supports a tightly stretched rubber membrane *R* having a high natural frequency compared with the range of frequencies of the pipe under investigation. A small mirror about one *mm* square is attached by rubber solution at any point mid way between the centre and the edge of the membrane. With a lamp and a scale placed a metre away the instrument has a sensibility of 1 cm for a pressure equal to one thousandth of an atmosphere. The callibration of the instrument is carried out by statically applying various small known pressures and noting the corresponding deflectives of the image of the slit of light on the scale. The cylindrical portion of this capsule is scaled with a hole made on a sounding pipe by soft wax to observe the pressure amplitude at the point in the sounding tube.

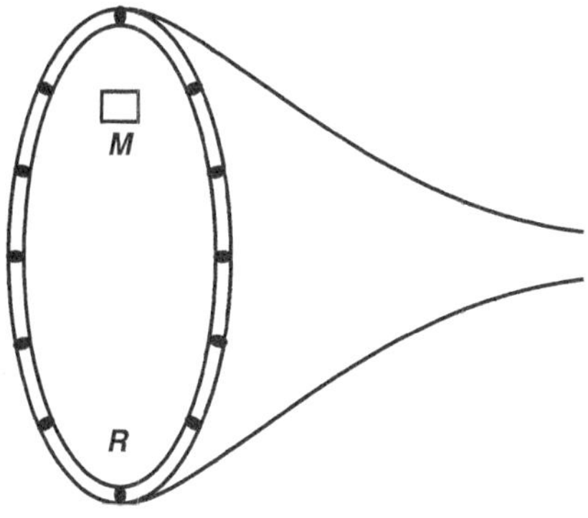

Fig. 7.

13.7. GALTON'S WHISTLE

It is a small sized organ pipe *C* the length which can be varied by means of a screwed piston attached to micrometer *M*. Air is blown under suitable pressure through the nozzle *N* fitted with a screw *S* which helps to vary the distance between the edge of the pipe *C* and the annular slit. The pressure of air blust is adjusted so as to set the small pipe with resonant vibration at natural frequency. The frequency of the note is determined from the graduation in the micrometer head *M*. The note of very high frequency even above the audible limit (ultrasonic vibration) is produced by this whistle.

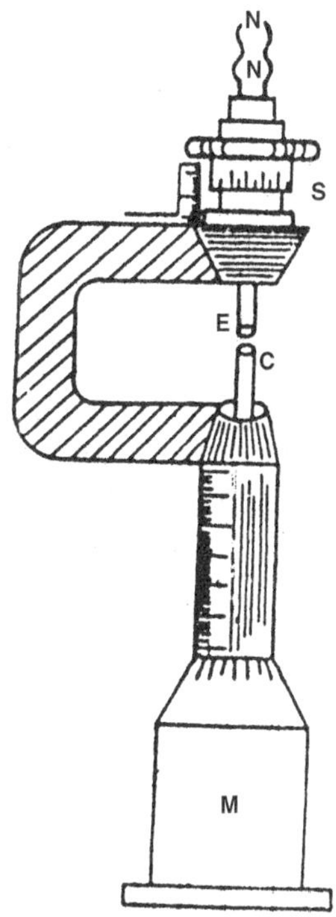

Fig. 8.

SOLVED PROBLEMS

1. Two cylindrical pipes of the same length but one closed and the other open are sounded together. The frequency of the second overtone in the closed pipe is 300 Hz higher than that of the first overtone of the open pipe. Calculate the fundamental frequency of the closed pipe.

   The angular frequency for pipe closed at one end and open at the other is given by

$$\omega_k = (2k+1)\frac{\pi c}{2l}$$

or

$$2\pi n_k = (2k+1)\frac{\pi c}{2l}$$

or
$$n_k = (2k+1)\frac{c}{4l}$$

The frequency of the second overtone ($k = 3$) of the closed pipe is $n_3 = \frac{7c}{4l}$

and the frequency of the first overtone ($k = 2$) of the pipe open at both ends

$$n_2 = \frac{c}{l} = \frac{4c}{4l}$$

From the given condition we have

$$n_3 - n_2 = 300$$

or
$$\frac{7c}{4l} - \frac{4c}{4l} = 300$$

or
$$\frac{c}{4l} = 100 \text{ Hz}$$

The frequency of the fundamental ($n = 1$) of the pipe closed at one end is

$$n = \frac{c}{4l} = 100 \text{ Hz}$$

REVIEW QUESTIONS

1. Investigate the modes of vibration of an air column in a pipe (1) closed at one end and (2) open at both ends. Do they produce musical note?
2. What is jet tone and edge tone? How they are produced?
3. How flue organ pipe is excited? Give the modes how the vibration is set up in such a pipe. What is over blowing and under blowing of a pipe?
4. Write a note on (*i*) Richardson's manometric capsule (*ii*) Konig manometric flame. How they are used to explore the character of vibration inside the organ pipe?
5. Describe Galton's whistle. What is the range of frequency of the note it can emit?

PROBLEMS

1. Two open pipes of length 1 metre and 1.05 metres produce 5 beats in 6 seconds when each is sounding its fundamental note. Calculate the frequencies of the two notes.

[**Ans.** 175 Hz and 166.7 Hz]

14
CHAPTER

HEAT MAINTAINED SOUND

14.1. SINGING FLAME

The sound emitted by a jet of hydrogen burning in a pipe open at both ends were observed as early as 1771. The tube is found to 'speak' in a favourable condition. Thereafter it was found that other combustible gases such as carbon monoxide, coal gas are also competent to produce the same effect. It was further observed that the note emitted is due to intermittent heating of air near the jet and the frequency of the note emitted, is due to intermittent heating of air near the jet and

the frequency of the note is related to the length of the tube excited. Rayleigh tried to give an explanation about the maintenance of the vibration.

In Fig. 1, S is the supply gas tube from the reservoir R with a pinhole at the upper end of the tube in which the gas burns in a flame. The flame is introduced in a wide tube T of glass ordinarily of length 30 to 40 cm. and of diameter 1.5 to 2 cm. When the flame is at a distance 8 to 10 cm from the lower end of the wider tube, it begins to sing spontaneously. The size of the flame is gradually reduced by regulating the gas supply which will pursue the singing of the flame. The frequency of the note excited equal to the natural frequency of the tube $\frac{c}{2l}$, where c is the velocity of propagation of sound wave and l is the length of the tube. This suggests that stationary wave is set up.

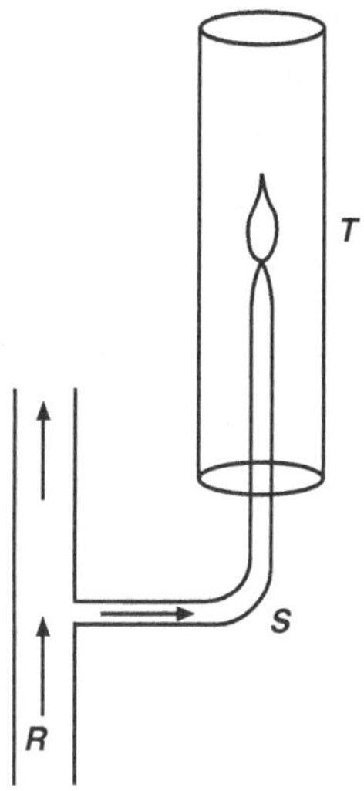

Fig. 1.

Sondauss showed that singing was due to intermittent heating of air. A satisfactory explanation was first given by Rayleigh. He pointed out that the fundamental factor of importance is the phase of heat supply relative to the vibration of the gas in the singing tube. The heat must be supplied at the instant of greatest condensation and heat must be removed at the instant of greatest rarefaction. The driving force which is the heat supply here must be in phase with the condensation in order to maintain the vibration. The gas tube and the air tube require to have stationary wave.

E.G. Richardson showed experimentally by the help of Konig manometric flame and rotating mirror that the phase of heat supply is same as predicted by Rayleigh.

It is found for powerful singing, stationary wave in the tube is set up and in the gas tube also. There is always a node at the jet and antinode at the *reserviour* of the gas. According to Rayleigh, the gas tube should be of length $\frac{\lambda}{4}, \frac{3\lambda}{4}, \frac{5\lambda}{4}$ where λ is the wavelength of the note emitted in the tube. But when the supply tube is of length equal to $\frac{\lambda}{2}$ or any other multiple of $\frac{\lambda}{2}$ the tube will not be excited.

14.2. GAUGE TONE OR RIJKE TUBE (1869)

Rijke placed a piece of metal wire gauge in the lower half of the vertical tube and heated it red hot by a flame. The air in the tube emit strong note as soon as the flame is withdrawn. The sound persists as long as the difference in the temperature of air and gauge persists.

Bosscha and Riess (1859-1860) observed a complimentary phenomenon. If a current of hot air impinge upon a cold gauge, sound is produced but here gauge must be placed at upper half of the tube and not as before in the lower part, in order to obtain principal note of the tube. The upper part of the brass tube is kept cool by water circulation. This also kept the gauge comparatively cool.

A continuous sound is produced when a metal gauge stretched across the lower part of a vertical tube is heated by passing a continuous electric current.

The vibration is set up at first by convection current. The cold air is convected through the hot wire gauge in the upward direction, takes up the maximum amount of heat, as a result there is maximum temperature variation which increases the velocity amplitude. A compression travels in the upward direction. The compression is reflected from the open end as rarefaction. This has a temperature nearly equal to heated gauge. So, on reaching the gauge no heat or a very small amount of heat is received from the gauge. So heat supply becomes intermittent. This requires investigation about the mechanism of maintenance.

In this connection, Rayleigh remarked that variable transfer of heat depends upon the motion of vibration, while the effect of transfer depends upon the variation of pressure. The gauge must therefore be placed where both effects are sensible *i.e.*, neither near a node nor near a loop. About a quarter of the length of the tube from the lower or upper end as the case may be, appears to be most favourable position.

14.3. TRAVELYAN ROCKER

This is also an example for maintenance of vibration by heat in a solid block of metal.

It consists of a narrow long copper prism P, from one of whose long edges, a semi-cylindrical groove is cut off, thus making two parallel edges R or ridges in place of one. The end of this prism is connected to a metal rod ending with a sphere of the same metal, Fig. 2(a). A long block of lead L which has a large thermal co-efficient of expansion and heat conductivity, which is less than that of the material of the rocker. The lead block has slight convexity on the upper surface and is placed on the table.

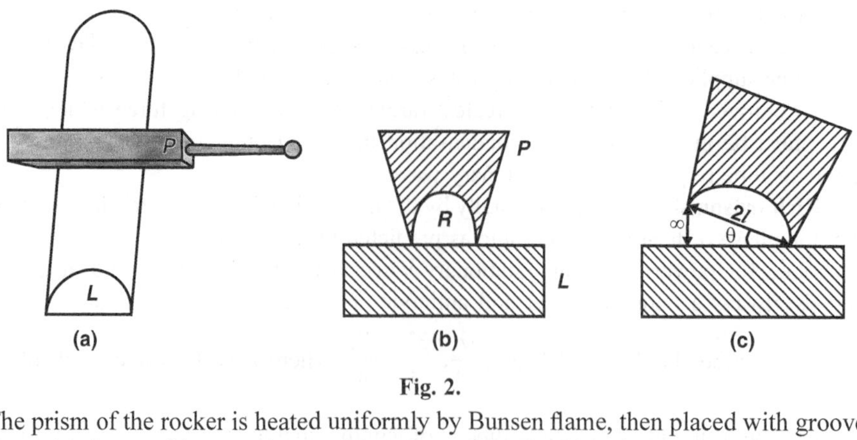

Fig. 2.

The prism of the rocker is heated uniformly by Bunsen flame, then placed with groove downwards in a slanting position on clean cold surface of the lead block. The handle ending with the bob rests on the table. The vibration of the rocker starts on the ridges on the side of the groove, alternately tilts from one side to the other.

When the rocker is placed on the lead block one of the ridges touches the lead block earlier or with slightly greater pressure than the other. The heat communicated to the lead block along this line, expands the lead slightly and tilts the rocker on that edge. The rocker, therefore, alternately rises up and comes down under the action of gravity on two edges, thus produces rocking motion and results in producing sound which is maintained as long as there is rocking motion *i.e.*, as long as the rocker is hot. The sound at the beginning is rough, and later, gives place to ones markedly musical in character.

Assuming the motion of the rocker about two edges is under its own weight the equation of motion is given by

$$mk^2 \frac{d^2\theta}{dt^2} = -mgl \qquad \text{... (1)}$$

where m is the mass of the rocker, k is the spin radius (radius of gyration) θ is the angular derivation which the plane containing the edges makes with the horizontal surface of the block; $2l$ is the distance between the ridges. On integrating

$$k^2 \int_0^{T/4} \frac{d^2\theta}{dt^2} dt = -gl \int_0^{T/4} dt$$

or

$$k^2 \left[\frac{d\theta}{dt}\right]_0^{T/4} = -gl[t]_0^{T/4} \qquad \text{... (2)}$$

or

$$0 - \omega = -\frac{gl}{k^2}\frac{T}{4}$$

or

$$\omega = \frac{gl}{k^2}\frac{T}{4} \qquad \text{... (3)}$$

If a be the amplitude, maximum displacement

$$\theta_{max} = \frac{a}{2l} \qquad \text{... (4)}$$

During the interval $\frac{T}{4}\omega$, changes from 0 to ω. Hence the mean value during the interval is $\frac{\omega}{2}$

or

$$\frac{\omega}{2} = \frac{4\theta_m}{T} = \frac{2a}{lT}$$

or

$$\omega = \frac{4a}{lT} \qquad \text{... (5)}$$

From Eqs. (3) and (5)

$$g\frac{l}{k^2}T = \frac{4a}{T} \qquad \text{... (6)}$$

or

$$T^2 = \frac{4a}{g}\frac{k^2}{l^2} \qquad \text{...(7)}$$

So

$$n^2 = \frac{gl^2}{4ak^2} \qquad \text{... (8)}$$

n being the frequency, which gives

$$n^2a = \frac{gl^2}{4k^2}$$

$= d$ constant for a particular Rocker.

Richardson and others carried out the experiment, and supported the gravity theory, and have shown that n^2a; is fairly constant, the length of the handle does not affect the value of n^2a; the amplitude a increases with the rise of temperature.

B.N Chakravarty carried out an extensive experiment and concluded that the rocker when gives musical sound is far from being simple rocking movement. The elastic vibration of the rocker and the handle play an important part. At earlier stage simple rocking occurs.

REVIEW QUESTIONS

1. What explanation can you give for

   (*i*) singing flame

   (*ii*) singing of tube by heated wire gauge?

2. Give theory of Travelian Rocker. Does the theory enable to explain the observation of different workers in this direction?

CHAPTER 15

Velocity of Sound and its Measurements

15.1. INTRODUCTION.

The expression for the velocity of sound in gases and effect of pressure, temperature, humidity, wind velocity, and the temperature gradient with height in the atmosphere have been considered in Chapter 9. Here different methods of measurement are discussed.

The earliest method of determination of velocity of sound in open air was to measure the time interval between the seeing of flash of light and hearing the report of firing a gun by an observer at a distant point. But this time interval t is the time taken by the sound wave to travel the distance d so that the velocity c of sound wave is obtained and is given by d/t, assuming the velocity of light is too large (3×10^8 metres per second). To eliminate the effect of blowing of the wind, similar measurements were made in the opposite direction and mean value of the velocity of sound wave obtained. The method involves the personal error, that is the time interval between the reception of stimulus of seeing and hearing is different and corresponding response for different persons. This was minimised by increasing the distance.

Regnault's Electrical Method. In order to eliminate the personal error, Regnault applied electrical method for registering the time interval of seeing the flash and hearing the roar of gun firing.

The arrangement is shown in the diagram Fig. 1 (a). The gun G is placed in front of a fine wire W forming a part of the electrical circuit in transmitting station *T. S.* at *a* distance. R is a receiver of sound, with a diaphragm d in receiving station *RS*. Behind d there is a contact point C. The diaphragm is pushed back by an impulse of sound; the circuit *R dc EMB* is completed *EM* is an electromagnet in the circuit in front of which there is a soft iron S attached to a style *St.* The style touches a rotating drum D with constant speed, wrapped with a shooted paper.

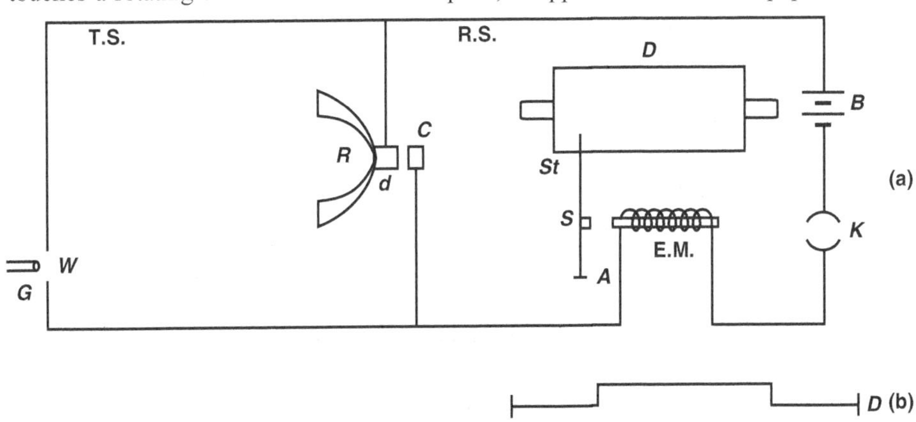

Fig. 1.

The gun G is fired at transmitting station, this blows the wire W off. The current in the circuit is broken off. The style which was tracing a straight line Fig. 1(*b*) on the shooted paper is displaced towards left as *EM* is demagnetised, owing to breaking off the current. The style will continue to trace the line on the drum parallel to the original line. When sound reaches *R*, the diaphragm '*d*' is pushed backward and touches *C* and circuit is momentarily completed, the electromagnet *EM* attracts the style towards right. Thus two marks are left on the line traced upon the drum, once when the gun is fired and again when the sound wave reaches *R*. From the known speed of rotation of the drum, the time interval between the two kinks on the line is determined. The distance between *W* and *R* which is about thousand metres being known velocity of sound is calculated.

To eliminate the effect of wind velocity a reciprocal arrangement is made.

15.2. HEBB'S METHOD

This method depends upon the determination of the wavelength of sound λ for known frequency n of the wave, whence the velocity $c = n\lambda$ is obtained.

Two parabolic mirrors M_1 and M_2 of Plaster of Paris about 5 feet (152.4 cm) diameter and 15 inches (38.1 cm) focal length are placed face to face coaxially at a considerable distance. At the focus of each is placed a carbon microphone F_1 and F_2 each connected in series with a battery B_1 or B_2 and a primary winding of a transformer P_1 or P_2. A whistle with a large value of frequency and which can be adjusted for different frequencies is placed at the focus of M_1. The sound emitted at the focus of M_1, after reflection from two mirrors will converge at the focus of M_2. The resultant sound is received by the telephone T connected to the secondary coil S of the transformer is the algebraic sum of two sounds received by F_1 and F_2. Now M_2 and F_2 are removed gradually along the axis. At certain distance of M_2 F_2 from M_1 F_1, sound will be reinforced and at a intermediate position it will be annulled depending on the difference of phase of two sounds received by F_1 and F_2. The distance between two successive maxima is found to be equal to the distance between two successive minima each of which is equal to the wavelength of the emitted sound whose frequency is known. The velocity of sound wave in open air is calculated.

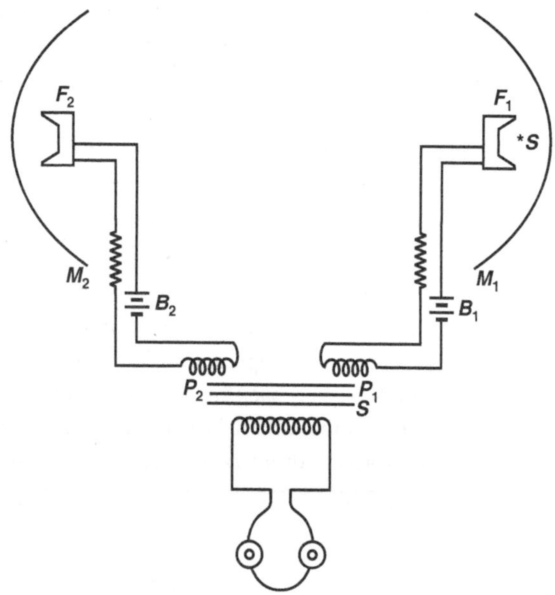

Fig. 2.

15.3. VELOCITY OF SOUND WAVE WITHIN THE LIMIT OF THE SMALL LABORATORY

1. *Resonance of Air Column.* A vibrating tuning fork of given frequency held upon the mouth of a tube may excite the air column *AB,* in a tube *AC* to resonance, when the length of the air column is adjusted by raising or lowering the reservoir *R* of water. The reservoir *R* can be placed at a suitable position by means of a clamp in the stand *S*.

Let l_1 be the smallest length of air column, λ the wavelength then

$$l_1 = \frac{\lambda}{4}$$

as the open end is an antinode, and closed end is the consecutive node vide Art. 13.1 Fig. 2 The open end may not be just the position of antinode but is at a point little above the open end. A correction is made by Rayleigh

$$(l_1 + 0.6r) = \frac{\lambda}{4}$$

where r is the radius of the tube. Hence velocity

$$c = 4n(l_1 + 0.6r)$$

The end error may be eliminated by taking a second observation when the tube will respond for third harmonics. Assuming x to be the end error and for smallest length l_1,

$$c = 4n(l_1 + x)$$

If l_2 be the length of the tube giving third harmonics, i.e., second resonance

$$c = \frac{4}{3}n(l_2 + x)$$

From these two equations

$$c = 2n(l_2 - l_1)$$

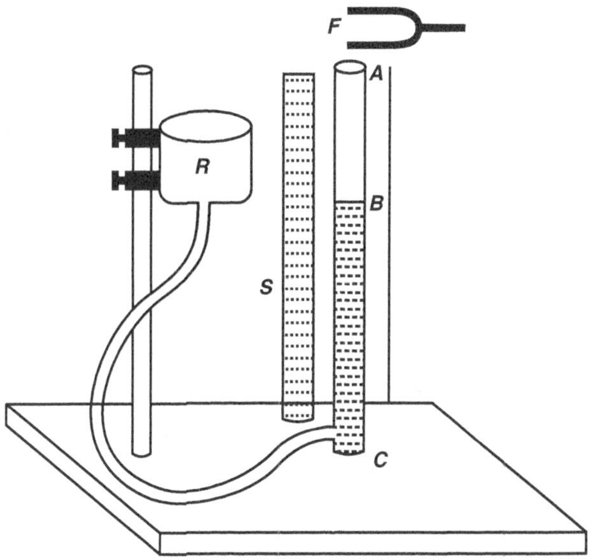

Fig. 3.

The velocity of sound wave is obtained by exciting the column of air to resonance by a vibrating tuning fork for different harmonics.

Owing to large damping of the vibration of air the resonance is not very sharp, hence it is difficult to obtain accurate result. Again the viscous force and heat loss near the wall of the tube will tend to lower the value of the velocity in the case of narrow tube.

15.4. KUNDT'S TUBE METHOD

It is one of the simplest method by which (*i*) stationary wave in a gas can be produced or demonstrated (*ii*) velocity of sound in a gas even when it is supplied in small quantities can be found out, (*iii*) The velocity of sound in a solid rod or its Young's modulus can be determined. (*iv*) The ratio of the specific heats of a gas can be found out.

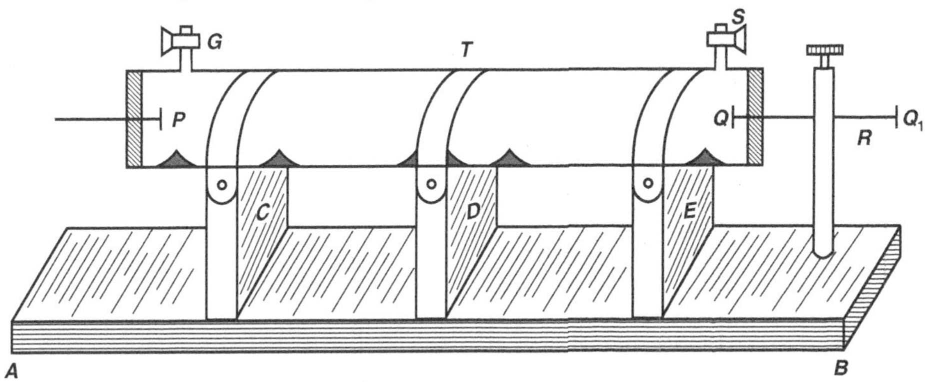

Fig. 4.

A long tube of glass T about 150 cms long and 4 cms in diameter is fixed horizontally over a wooden table AB with the help of three wooden blocks C, D, E fitted with clamps. In some cases, the tube is fitted with two side tubes S, G with stopper at two ends of the glass tube T. One end of the tube is closed by a rubber cork through which passes a movable piston. The other end is also closed by a rubber cork through which passes a rod of metal or glass, ending with a disc Q of nearly the same cross-section as that of the tube. Other end of the rod carries a similar disc Q_1. The rod R is firmly clamped at the middle point. The tube is dried thoroughly, and lycopodium powder is spread over, within the tube throughout the length between two discs P and Q at two ends. The rod R is set into longitudinal vibration by drawing a resinated leather (wet cloth in case of glass rod) and position of the piston P is adjusted by drawing in or out. The resonance will occur between air column inside the tube and the rod. This is indicated by the formation of striation and heaping up of the lycopodium powder at regular intervals of length.

Due to vibration of the rod R, the piston Q attached to it is set into vibration. The compressional waves get reflected at the piston P at the other end and two trains of waves moving along the same line of propagations in opposite direction form stationary wave in the tube.

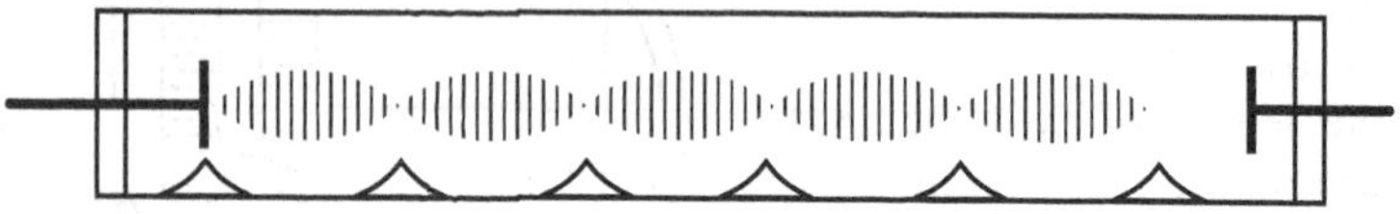

Fig. 5.

Lycopodium powder is heaped at nodes and striations occur at loops, the longest ridge being at antinode. The mean distance l between two successive heaps except the last two at ends is equal to $\lambda/2$. But for the precision of measurement it is better to measure the distance between the two consecutive longest ridges the mean value of which is $\lambda/2$.

15.5. VELOCITY OF SOUND IN AIR OR OTHER GASES

Let l_a be the mean length of the distance between two antinodes obtained by measuring the distance between two longest ridges in striations which gives

$$l_a = \frac{\lambda_a}{2}$$

λ_a being the wavelength of sound in air and c_a is the velocity in air. Let l_s be the length of the solid rod clamped at the middle. The two ends of the rods are two consecutive antinodes, so

$$l_s = \frac{\lambda_s}{2}$$

where λ_s is the wavelength of the longitudinal wave in the solid rod. c_s is the velocity of the longitudinal wave in the solid given by

$$c_s = \sqrt{\frac{Y}{\rho_s}}$$

Y being the Young's modulus and ρ_s the density of the material of the rod. So, as resonance is set up

$$n = \frac{c_a}{\lambda_a} = \frac{c_s}{\lambda_s}$$

or

$$\frac{c_a}{c_s} = \frac{\lambda_a}{\lambda_s} = \frac{l_a}{l_s}$$

or

$$c_a = c_s \frac{l_a}{l_s} = \sqrt{\frac{Y}{\rho_s}} \frac{l_a}{l_s} \qquad \dots (1)$$

again $$c_a = \sqrt{\frac{\gamma P}{\rho_a}} \text{ and } \frac{P}{\rho_a} = \frac{c^2}{3} \text{ (from kinetic theory of gases)}$$

where c^2 is the square mean of the molecular velocity

$$c_s = \sqrt{\frac{\gamma P}{\rho_a}} \frac{l_s}{l_a} \quad \text{... (2)}$$

$$= \sqrt{\frac{\gamma c^2}{3}} \frac{l_s}{l_a} \quad \text{... (3)}$$

from eq. (1) c_a the velocity of sound in the gas is found if the velocity of longitudinal wave in solid is known and (2) velocity of longitudinal wave in solid c_s is known in terms of c_a or from eq. (3), the ratio of specific heats of the gas γ is known in terms of pressure and density of the gas from eq. (4) Young's modulus of the material of the solid can be found out. From eq. (1) root mean square value of the molecular velocity of the gas can be obtained.

To find the velocity of any other gas other than air, the gas introduced by the side tube *G* and displaced air comes out through *S*. Here Kundt's tube as modified by Andrade (vide art. 15.6) is generally used.

The effect of pressure and temperature on the velocity of sound of gases can also be studied by this experiment. Gas under different pressure is introduced into the tube and corresponding velocities are determined. In order to study the variation of the velocity with temperature, the tube is surrounded by a jacket through which steam is allowed to flow. For higher temperature, the tube is surrounded with heating coil through which electric current is passed.

15.6. EXPLANATION OF THE FORMATION OF STRIATION IN KUNDT'S TUBE

In addition to the collections of dust particles in heaps at nodal points, the dust tube exhibits a regular series of fine rib like striations at the bottom of the tube. While the vibrations are in progress the fine dust particles were found to be disposed in thin laminae, transverse to the tube extending upwards to a certain distance from the bottom. The effect is pronounced at the antinodal region and disappears in the neighbourhood of nodes, where heaps are collected. When vibration stops the laminae necessarily fall and in so doing lose much of their sharpness but remain visible as transverse rib like lines.

Konig tried to explain the above phenomenon by calculating the forces operative in the case of two spherical dust particles which are not close to each other. A single spherical obstacle experience no force from an alternating current. But presence of a neighbouring particle alters the condition. For simplicity, when two spheres are considered at a moderate distance *d* apart and are so situated that their centres are parallel to the stream, they will repel each other with a force of magnitude $6\pi\rho_a r_1^3 r_2^3 \frac{v^3}{d^4}$ and when they are placed perpendicular to the stream, they will attract each other with a force of magnitude $3\pi\rho_a r_1^3 r_2^3 \frac{v^2}{d^4}$ where r_1 and r_2 are the radii of the particles, v^2 is the square mean of the velocity of the stream and ρ_a is the density of air. The result of these forces between the neighbouring particles produces a tendency to aggregate in laminae. These laminae are finally responsible for the formation of striations.

Andrade's Explanation. Andrade devised a Kundt's tube which was excited, not by longitudinal vibration of a rod, but by employing a telephone diaphragm and a loud speaker unit connected to a valve oscillator of controllable frequency and excited the dust tube into resonance.

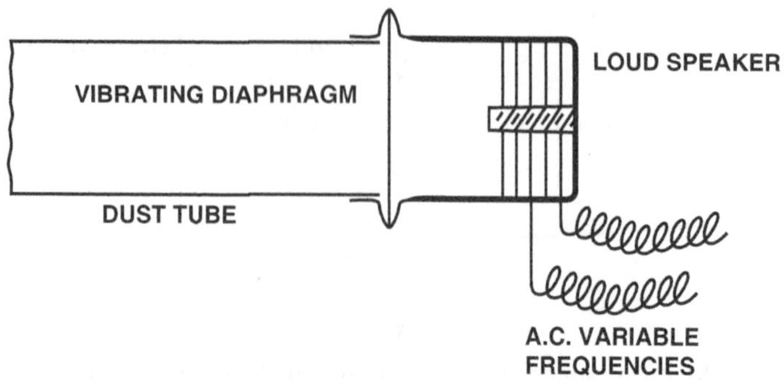

Fig. 6.

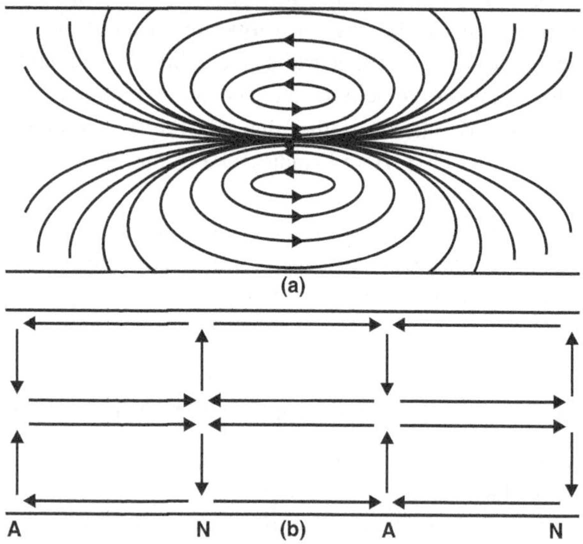

Fig. 7.

In this method, striations are readily obtained which increase towards antinodes, over and above a clearly defined disc of dust particles of surprising sharpness. These discs extend across the tube like a barrier. In order to investigate the motion of air inside the tube, tobacco smoke was introduced. The smoke particles were used as tracer of motion of air. The motion of the smoke particles exhibited the general circulation of air between nodes and antinodes and vortex motions were seen round every particle. As the intensity of sound increases, the smoke particles exhibit the circulation in air in a very striking manner. The circulation takes place along the wall of the tube from antinode to node Fig. 7(*a*) and returning up the axis of the tube as was predicted by Rayleigh Fig. 7(*b*). In vibrating air column, it is found experimentally that all forces between the spheres which was supposed irrational are vortex motions. So Konig's theory is proved to be wrong.

REVIEW QUESTIONS

1. Discuss the different methods of determining the velocity of sound in atmospheric air.
2. How do you determine the velocity of sound in Kundt's tube? Describe how the ratio of two specific heats of a gas is worked out. For an accurate measurement in Kundt's tube will you measure nodal distance or antinodal distance? Give the reasons for your answer.
3. Describe Kundt's tube method of determining the velocity of sound in a gas mentioning in detail the precautions necessary and explain the formation of striations.
4. State and explain what further light has been thrown on the nature of the movement of the gas inside the tube by the works of Andrade and others.

16 CHAPTER

FREQUENCY AND ITS MEASUREMENT

16.1. INTRODUCTION.

There are various methods of measurement of frequency of a vibrating body such as a vibrating tuning fork or a source of sound. They are classified as, **absolute measurement** and **comparative determination**. In absolute method, vibrations are recorded on paper or plate and number of vibrations from the record, occurring in unit of time with reference to a standard time recorder gives the value of the frequency. In the comparative method, the note whose frequency is to be determined is put in unison with the note whose frequency is known.

(i) Absolute Method

(1) Dropping plate method. A smoked glass plate is allowed to fall through a smooth vertical groove of a vertical stand. A light style is attached to the prong of a tuning fork which touches gently the smoked plate and traces out a wavy curve on the plate. The points A_1, A_2, A_3 are marked on the plate, such that the number of waves N between A_1 and A_2, A_2 and, A_3, are same. Let u be the velocity in cm/sec of the plate while crossing the point A_1. Let the distance $A_1 A_2$ be equal to l_1 and $A_2 A_3$ is l_2. Let the frequency of the fork be n.

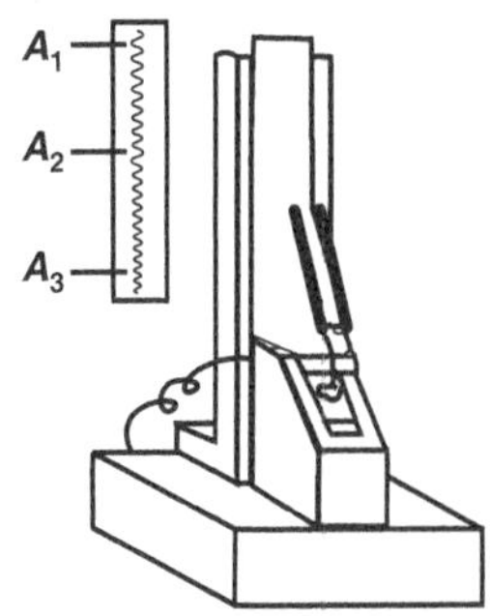

Fig. 1.

The time taken by the tuning fork N vibrations is $\frac{N}{n}$ seconds. So time taken by $2N$ vibrations is $\frac{2N}{n}$ sec. Since the plate is falling under gravity then

$$l_1 = u\frac{N}{n} + \frac{1}{2}g\left(\frac{N}{n}\right)^2 \quad \text{... (1)}$$

where g is acceleration due to gravity of the place in cm/sec^2.

Similarly
$$l_1 + l_2 = u\frac{2N}{n} + 2g\left(\frac{N}{n}\right)^2$$

$$= 2u\frac{N}{n} + 2g\left(\frac{N}{n}\right)^2 \quad \text{... (2)}$$

From (1)
$$2l_1 = 2u\frac{N}{n} + g\left(\frac{N}{n}\right)^2 \quad \text{... (3)}$$

Subtracting (2) from (3)

$$l_2 - l_1 = g\left(\frac{N}{n}\right)^2$$

or
$$n = N\sqrt{\frac{g}{l_2 - l_1}} \quad \text{... (4)}$$

So the frequency n is known.

The attachment of the pointer to the fork and the friction of the pointer on the plate lowers the frequency which requires correction. The method is not very accurate.

(2) Chronographic Method. A fine aluminium style is attached to a tuning fork *F* which will draw a weavy curve on a shooted paper *D* wrapped over a cylindrical drum of metal which can be made to rotate by hand with the help of the handle attached. The drum advances axially along the screw, while rotates, in order to avoid overlapping of the curve traced out over the smoked paper. An induction coil is introduced, the electrical circuit of which is completed through the style, smoked paper and the drum. This coil *S* serves as a secondary coil. The other coil *P* serves as a primary coil is connected to mercury in cup *C* through a battery of cells, and the other end to the seconds pendulum which comes in contact with mercury once in one second and makes the circuit complete. The moment the primary circuit is completed, induction in the secondary coil *S* takes place and spark passes through the style and smoked paper, producing a small spot by knocking out shoot at the point. The vibrating tuning fork produces weavy curve with small spot at the interval of one second, on rotating drum.

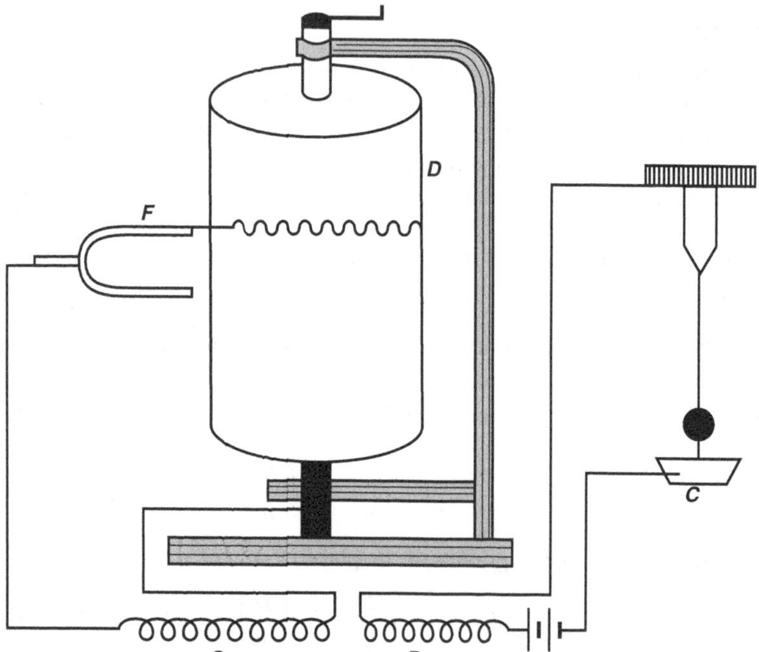

Fig. 2.

The number of waves between two consecutive spots is the number of vibrations per second which is the frequency of the fork. Similar defects as in the previous experiments are predominating, yet it gives a higher degree of accuracy.

(3) Stroboscopic Method. By this method, a moving thing appears stationary. A vibrating body is illuminated intermittently or glimpses of the motion are obtained intermittently. If the intermittence coincides with the period of motion, and the body will appear motionless.

This intermittent vision and illumination can be achieved by a rotating disc with radial slits, or bright dots, equally spaced round the circumference. The rotation is made by a suitable motor with arrangement for recording the speed of rotation.

Two light strips of tin foils are attached to ends of two prongs of the fork whose frequency is to be determined. One of the foils carries a narrow slit other foil just covers the slit such that the slit opens when the prongs move out. When the fork vibrates the slit opens once in each vibration. The fork F is placed vertically in front of the disc D. An arrangement for looking at the dots on the disc is made by the help of a telescope. The fork is set in vibration, the disc is made to rotate and the dots are seen through the slit on vibrating fork by the help of telescope T.

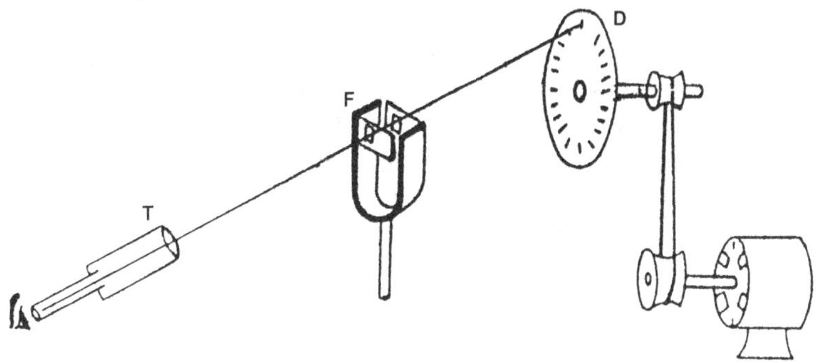

Fig. 3.

The speed of rotation is gradually increased while the fork is vibrating. If one dot takes up the position of the next in the period of vibration of the fork then the dot will appear stationary. The dot will appear to move slightly backward if the speed of rotation of the disc is lower, but when the speed is higher, dots appear to move slightly forward.

When the dots appear stationary time taken by one dot to move to the position of the previous dot will be $\frac{1}{n}$ sec, where n is the frequency of the fork.

Let r be the number of revolution of the disc per second, h be the number of dots in the ring of the disc. Then time taken by one dot to take the position of the previous dot is equal to $\frac{1}{rh}$.

Then $$\frac{1}{n} = \frac{1}{rh} \quad \text{or } n = rh.$$

The strips of two foils, attached to the prongs lower the frequency of the fork. The correction which will be necessary is obtained by attaching these foils to the prongs of a fork of approximately the same frequency and noting by which frequency is lowered.

Without attaching the tin foils to the prongs of the fork, the experiment can be carried out simply by marking a bright point on one of the prongs of the fork. The bright point which is marked on the vibrating fork is seen through the slit, radially cut, of the revolving disc by means of a telescope. The rest of the procedure is same as above. The bright spot appears stationary through the slit of the revolving disc at a given speed. Here no correction will be necessary. This renders the method more accurate.

(4) Phonic Wheel Method. The phonic wheel was devised by Rayleigh and De. la Tour independently.

It is similar to synchronous motor controlled by tuning fork in order to drive a stroboscopic disc D fixed on an axle S. The axel stands vertically on a metal base B in a ball bearing groove, in which it can rotate smoothly. A rotor R of soft iron disc which is cut in the form of stud or equidistant teeth. This disc R is fixed on the axel S while rotates, causes two diametrically opposite teeth move past a pair of electromagnets M_1M_2. These electromagnets are excited only at certain constant intervals by the fork, which are connected in series with a set of battery of cells as shown in Fig. 4. A speed counter Q is mechanically geared with the spindle S to a device which rings a bell after a particular number of revolution r.

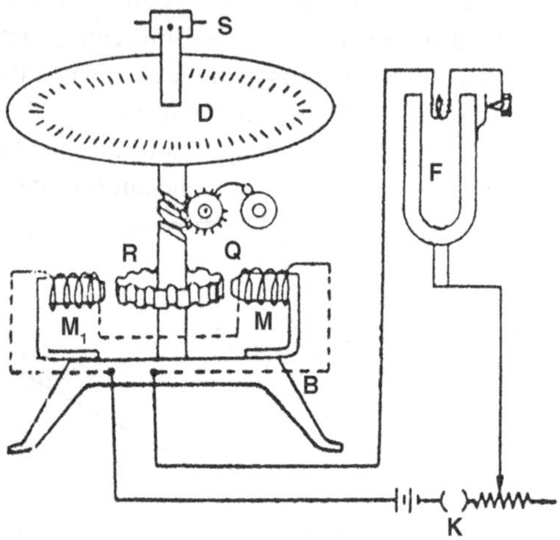

Fig. 4.

Such a phonic motor which consists of soft iron rotor having equidistant teeth and a stator consists of a pair of electromagnets magnetised by single coil of wire is not self-starting. So it must be run up by hand to synchronous speed. This is possible if the rotor rotates at such a speed that the teeth of the rotor are close to the magnetic poles when the current passing through the electromagnet is maximum. The electromagnetic force on each tooth of the rotor during the advance or retreat, synchronises with the time when current rises to maximum value and begins to fall. The motion of the rotor then will be maintained. If the rotor moves with a little lag behind the magnetic force of attraction at the time of maximum current in the electromagnet, it will increase the speed. On the other hand, if the motion of the rotor be quicker, the magnetic force in the electromagnet with increasing current, it retards the motion of the rotor. By self-adjustment of rotation of the teeth of rotor, it settles down to such a value that the middle of the time of excitement of the electromagnet synchronises with the time of passing of the teeth closed to the pole and the motor will be maintained.

Let the time interval between two ringing of the bell = t sec,

the number of rotation during this time = r

$$\text{the time for one rotation} = \frac{t}{r}$$

$$\text{number of teeth on the rotor} = m$$

$$\text{The interval between two successive excitation} = \frac{t}{m \cdot r}$$

$$\text{The periodic time of vibration of the fork of frequency } n = \frac{1}{n}$$

$$\frac{1}{n} = \frac{t}{rm}$$

or $$n = \frac{rm}{t}$$

so n is found out with the knowledge of r, m and t.

16.2. COMPARATIVE METHOD OF FREQUENCY DETERMINATION

The frequency of pure tone can be determined by comparison with a standard note, either by the method of resonance i.e., by adjusting the standard note in unison, or by counting the number of beats.

(i) Resonance Column Method

In previous chapter (Art. 15.3), we have seen the velocity of sound in air in the laboratory is determined with the help of a tuning fork of known frequency. When velocity is known the frequency can be determined or the frequency of a fork of unknown frequency can also be determined by comparing the resonant length of air column for a fork of known frequency with the length for a fork of unknown frequency.

(ii) Sonometer

The frequency of the fundamental note excited by a sonometer string under tension T of length l, linear density ρ is given by

$$n = \frac{1}{2l}\sqrt{\frac{T}{\rho}}$$ (vide Art. 10.1)

Any source of sound producing a note in unison with the note emitted by the string by adjusting the length tension or both has got the same frequency.

(iii) Siren

There are two types of siren.

(*a*) **Seebeck siren** is the simpler type.

(*b*) **De la Tour's siren** — where rotation is produced by air puff and speed is recorded by itself. The general principles of operation are identical.

(*a*) Seebeck's Siren. It is a circular metal disc D with a number of equidistant holes along concentric circles of different radii. The disc can be rotated in different speeds. A fine stream of air is blown through a jet of a narrow tube T held very close to the holes on a particular circle on rotating disc. A note is produced due to series of puffs coming out of holes at regular interval. If r be the number of revolution per second, and h be the number of holes in a particular row then the frequency n of the note is given by $n = rh$. By adjusting the speed of rotation of the disc or by changing the row of holes or by changing both, the frequency of the note emitted is adjusted for unison with the note under investigation.

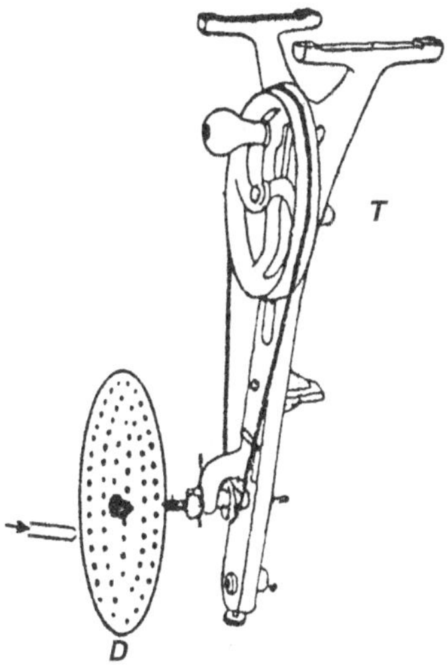

Fig. 5.

(*b*) Cagneard De la Tour's Siren is another improved form of siren. There is an air chamber A into which air is blown through the inlet T. The bed of the chamber is a disc D pierced with four series of the equidistant hole lying on four equidistant concentric circles. Immediately over the lid D there is a disc C with similar set of holes. (Shown separately below).

The upper disc C can rotate freely about the vertical spindle S. This is engaged to a speed counter at the upper end. The holes in two discs are inclined to each other at an angle. The air jets issuing through the holes in the fixed lower disc D strikes the holes on upper disc C obliquily. The horizontal component of impinging air jet will turn the movable disc C. The speed of rotation depends upon the pressure if air in A. As the upper disc rotates all the holes being in one circle are alternately opened and closed. This sends more powerful puffs of air into atmosphere than those produced by Seebeck's siren. As before, frequency n when in unison is given by $n = rh$ where h is the number of holes in one circle which is also equal to number of puffs in each revolution and r is the number of revolution per second.

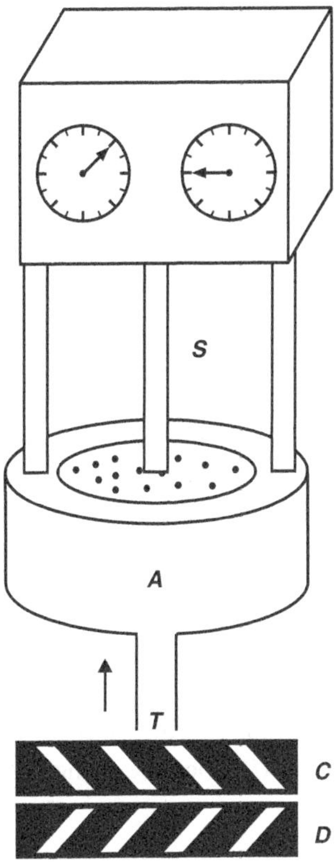

Fig. 6.

(iv) By the Method of Beats

Unlike the resonance of two notes beats produced by two notes may be used for determination of frequency of one when the other is known.

If two notes of frequencies n_1 and n_2 which are nearly equal but not exactly the same are sounded together then beats are formed. Let N be the number of beats per second then $n_1 \sim n_2 = N$ or $n_2 = n_1 \pm N$. Let the frequency n_1 of the note which is known is altered, slightly. If it is a tuning fork, loading by a bead of wax at the top of a prong lowers the frequency; if it is a siren, the speed of rotation when lowered, lowers the frequency. If now number of beats increases then unknown frequency $n_2 = n_1 + N$, and if decreases $n_2 = n_1 - N$.

16.3. HELMHOLTZ RESONATOR

If the sound given by a tuning fork produces resonance in Helmholtz resonator whose frequency is known, the frequency of the fork will be same as of the resonator. Helmholtz resonator is also used as a sharply tuned detector of sound vibration. There are two forms of resonators. In one, Fig. 7(a) they are hollow spheres of brass with a small opening or a neck P at one end and a small nozzle like opening N on diametrically opposite sides. Here the length of the nozzle is negligible compared to the diameter. A resonator of the type (b) which is cylindrical in shape.

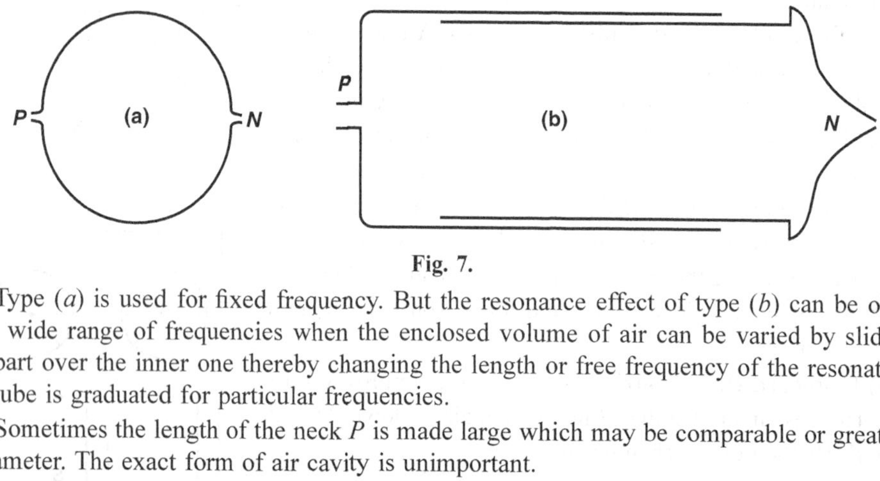

Fig. 7.

Type (a) is used for fixed frequency. But the resonance effect of type (b) can be obtained over a wide range of frequencies when the enclosed volume of air can be varied by sliding the outer part over the inner one thereby changing the length or free frequency of the resonator. The inner tube is graduated for particular frequencies.

Sometimes the length of the neck P is made large which may be comparable or greater than the diameter. The exact form of air cavity is unimportant.

The resonator speaks when a sound of relatively feeble intensity and of a appropriate frequency falls on the open end. The air inside vibrates with resonance. The smaller end N may be kept open and is used for the detection of feeble sound by putting it to the ear. Helmholtz showed that the vibration throughout the cavity is practically uniform and is negligible compared with the vibration of the neck. Rayleigh developed the theory. He regarded the air in the neck P is acting as a reciprocating piston compressing and rarefying the air contained in the cavity of the resonator.

Let V be the volume of the cavity of the resonator, l be the length and r be the radius of the neck. Let ξ be the displacement of the air piston of the sectional area $A = \pi r^2$. The displacement of the air plug causing the change of volume of air in the resonator being equal to

$$dV = A\xi = \pi r^2 \xi \qquad \text{... (1)}$$

for unit displacement the change of volume $dV = A$ ... (1.1)

(i) Since the change of pressure is adiabatic, volume elasticity

$$E = \gamma P$$

$$= -V\frac{dP}{dV}$$

or $$-\gamma P\frac{dV}{V} = dP \qquad \text{... (2)}$$

The restoring force F for unit displacement of air piston

$$F = AdP$$

$$= -A\gamma P\frac{dV}{V}$$

(From eq. 1.1 as $dV = A$)

$$F = -\frac{E}{V}A^2$$

Hence the eqn. of motion is

$$M\frac{d^2\xi}{dt^2} + \frac{E}{V}A^2\xi = 0 \qquad \text{... (3)}$$

where $M = \underline{lA}\rho$

Hence the angular frequency ω is given by

$$\omega^2 = \frac{EA}{Vl\rho}$$

$$= c^2\frac{A}{Vl}$$

$$c^2 = \frac{E}{\rho} \qquad \text{... (4)}$$

Hence the frequency n is given by

$$n = \frac{c}{2\pi}\sqrt{\frac{A}{Vl}} \qquad \text{... (5)}$$

where $\frac{A}{l}$ is know as conductivity.

As the correction for the open end of the tube is given by $l+\frac{16r}{3\pi}$, so the frequency is given by

$$n = \frac{c}{2\pi}\sqrt{\frac{\pi r^2}{\left(l+\frac{16r}{3\pi}\right)V}}$$

$$= \frac{cr}{2}\sqrt{\frac{3}{V(3\pi l+16r)}} \qquad \text{... (6)}$$

The smallness of damping and sharpness of resonance renders the resonators extremely sensitive detector of sound of particular frequency.

In order to detect the frequency of a note, a series of resonators of different natural frequencies are used. The nipple of the resonators are placed one by one in ear. The resonator will speak when the frequency of the note sounded near the neck of the resonator equals to its natural frequency.

The fundamental difference between the vibration of an air column in a pipe and in a Helmholtz resonator exists. The motion in a pipe which is due to stationary wave, set up in whole length of the pipe, and its length is comparable with the wavelength, whereas the dimension of a Helmholtz resonator is small compared with the wavelength, and the vibratory motion is confined in air in the neck which is vibrating as a reciprocating piston and the displacement amplitude is regarded small everywhere. The vibration in a resonator will persist owing to the smallness of dimension, and the length of the neck. The small damping cause the tuning very sharp which makes it very suitable analyser of musical sound. In the case of the tube, the damping is large so vibration does not persist.

16.4. ACOUSTIC FILTER

This is a device which filters out or alternates certain frequencies which are present in a complex sound. It is often done with the help of branch pipes open or closed, branch cavities and with tubes of constricted or expanded cross-section.

The simplest form is Quinck's acoustic filter from which the broad principle can be understood. A side tube B is placed in the path of the main tube A. A complex sound while passes through A, resounds the tube B to particular frequency, responds to the fundamental or overtones which the tube B can produce. So it will absorb energies associated with the wave of these frequencies.

The corresponding frequencies are $\frac{c}{4l}, \frac{3c}{4l}, \frac{5c}{4l}, \ldots$ etc.

where c is the velocity of sound wave, l is the length of the closed tube B.

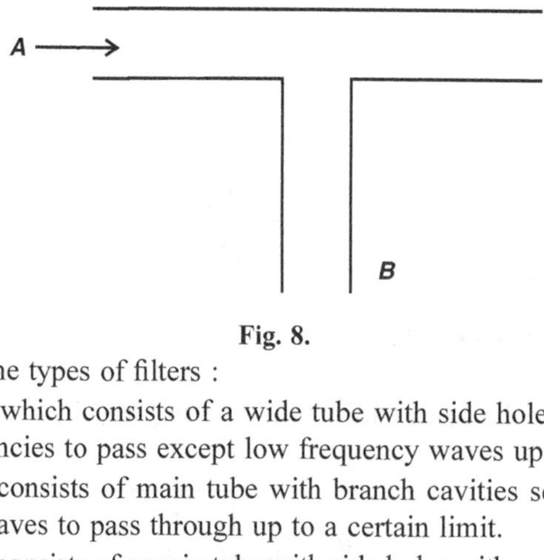

Fig. 8.

The following are the types of filters :

(*i*) **High pass filter** which consists of a wide tube with side holes having short necks. This allows all frequencies to pass except low frequency waves up to a particular limit.

(*ii*) **Low pass filter** consists of main tube with branch cavities set close together. It allows low frequency waves to pass through up to a certain limit.

(*iii*) **Band pass filter** consists of a main tube with side holes with neck but the necks are broken at the centre and communicated with closed vessels which surround them. Here waves of medium frequencies are allowed to pass through except low and high frequency waves. These are better explained with the help of electrical analogies.

SOLVED PROBLEMS

1. The volume of the vessel of a Helmholtz resonator is 1 litre. The cross-sectional area of its neck is $\pi^2 cm^2$ and the length of the neck is 1 mm. Determine the natural frequency of vibration of the resonator if the speed of sound in air is 320 m/sec.

The natural frequency of Helmholtz resonator is given by

$$n = \frac{c}{2\pi}\sqrt{\frac{A}{Vl}}$$

Given $c = 320$ m/sec,

$$A = \pi^2 \text{ cm}^2 = \pi^2 \times 10^{-4} \text{ m}^2$$

$$l = 1 \times 10^{-3} \text{ m and } v = l \text{ litre} = 10^{-3} \text{ m}^3$$

$$\therefore \quad n = n = \frac{320}{2\pi}\sqrt{\frac{\pi^2 \times 10^{-4}}{10^{-3} \times 10^{-3}}} \text{ Hz}$$

$$= 1600 \text{ Hz}$$

REVIEW QUESTIONS

1. Describe and discuss the principle and methods of measurements of the frequency of a tuning fork.
2. Explain the working of a phonic wheel for determining the frequency of a tuning fork.
3. What is stroboscopic method? How it is applied to find the frequency of a fork?
4. Derive an expression for the resonance frequency of a Helmholtz resonator in terms of its dimension and state its various applications with special reference to analysis of a musical note. [C.U. 1968]
5. What is a Helmholtz resonator? What measurement can be made with its help? [Burd. U. 1995]
6. Write a short note on
   (*a*) Helmholtz resonator [Pandit Ravisankar University 2010]
   (*b*) Seebeck siren
   (*c*) Acoustic filter.
7. Describe the rotating drum method for determination of frequency. [Barkatullah University 2011]

PROBLEMS:

1. Helmholtz resonator has a volume 500 cc. The radius of the neck is 1 cm and the length of the neck is 0.1 cm. Calculate the frequency of the resonator. Velocity of sound at room temperature is 330 m/sec. [C.U. 1986]

   [**Ans.** 1317 Hz]

17 CHAPTER INTENSITY AND ITS MEASUREMENT

17.1. ENERGY OF SOUND WAVES

Sound travels in waves. It is a form of energy and rate of flow of energy per unit area normal to the direction of propagation is called intensity of sound wave. Energy density is the energy per unit of volume of the wave.

Let us consider a plane wave moving along x direction is given by

$$\xi = A\sin\frac{2\pi}{\lambda}(ct - x) \quad \text{... (1)}$$

The total energy in the wave is the sum total of kinetic energy T and the potential energy V.

Kinetic energy of length dx $\quad \Delta T = \frac{1}{2}\rho dx\left(\frac{d\xi}{dt}\right)^2 \quad$ ... (2)

and unit cross section of the element dx.

Total K.E. $T = \frac{1}{2}\rho\int_0^\lambda\left(\frac{d\xi}{dt}\right)^2 dx$ of the tube of length λ the wavelength, where ρ is the density of the medium.

Put the value of $\left(\frac{d\xi}{dt}\right)$ from Eqn. (1)

$$T = \frac{1}{2}\rho\int_0^\lambda\left\{A\frac{2\pi c}{\lambda}\cos\frac{2\pi}{\lambda}(ct - x)\right\}^2 dx$$

$$= \frac{1}{2}\rho A^2\left(\frac{2\pi c}{\lambda}\right)^2\int_0^\lambda \cos^2\frac{2\pi}{\lambda}(ct - x)dx$$

$$= \frac{1}{2}\rho A^2\omega^2\frac{\lambda}{2} = \frac{1}{4}\rho A^2\omega^2\lambda \quad \text{... (3)}$$

$$= \lambda\frac{\rho}{4}v^2_{\max} \quad \text{... (4)}$$

where $\frac{c}{\lambda} = n$ the frequency of the wave and $2\pi n = \omega$ the angular frequency, and $A\omega = v_{\max}$.

Hence the K.E. per unit length of unit cross-section is given by

$$\frac{1}{4}\rho A^2\omega^2 = \frac{\rho}{4}v^2_{\max} \quad \text{... (5)}$$

Potential Energy

In art 8.10 eq. (4), it is found volume strain

$$= \frac{d\xi}{dx} = -S$$

where S stands for condensation, and $\frac{d\xi}{dx}$ is the fractional change of volume stress applied

$$dp = -E\frac{d\xi}{dx}$$

$$= -c^2\rho\frac{d\xi}{dx}$$

$$= c^2\rho S \quad \text{... (6)}$$

where dp is the change of pressure over normal pressure P_0

So work done to produce infinitesimal condensation dS is

$$dp\ dS = c^2\ \rho S dS \quad \text{... (7)}$$

Work done for condensation S is

$$= c^2\rho\int_0^S S dS = c^2\rho\frac{S^2}{2}$$

$$S = -\frac{d\xi}{dx} = +A\frac{2\pi}{\lambda}\cos\frac{2\pi}{\lambda}(ct - x)$$

Hence maximum condensations

$$S_m = +A\frac{2\pi}{\lambda} = \frac{A\omega}{c} = \frac{v_{\max}}{c} \quad \text{... (7.1)}$$

Potential energy of the tube of gas of unit cross-section and of length λ is

$$V = \int_\partial^\lambda + \frac{c^2\rho}{2}S^2 dx$$

$$= \frac{c^2\rho}{2}\int_\partial^\lambda\left(\frac{d\xi}{dx}\right)^2 dx \quad \text{... (8)}$$

$$= \frac{c^2\rho}{2}A^2\left(\frac{2\pi}{\lambda}\right)^2\int_\partial^\lambda \cos^2\frac{2\pi}{\lambda}(ct - x)dx$$

$$= \frac{\rho}{2}A^2\omega^2\frac{\lambda}{2}$$

$$= \lambda\frac{\rho}{4}A^2\omega^2 \quad \text{... (9)}$$

$$= \lambda\frac{\rho}{4}v^2_{\max} \quad \text{... (10)}$$

Thus total energy

$$T + V = \frac{\lambda}{4}\rho v^2_{\max} + \frac{\lambda}{4}\rho v^2_{\max}$$

$$= \lambda\cdot\frac{\rho}{2}v^2_{\max}$$

within the tube of length λ.

Energy density E = energy per unit vol.

$$= \frac{\rho}{2} v^2_{max}$$

Intensity $$I = Ec$$

$$= \frac{1}{2} \rho c v^2_{max}$$

Again $$(\sigma p)_{max} = -c^2 \rho S_{max}$$

$$= -c^2 \rho \frac{v_{max}}{c} = c \rho v_{max}$$

From eqns (6) and (7.1).

or $$v^2_{max} = \frac{(\delta p)^2_{max}}{c^2 \rho^2}$$

Hence $$I = \frac{1}{2} \rho c (v_{max})^2$$

$$= \frac{1}{2} \frac{(dp)^2_{max}}{\rho c}$$

$$= \frac{(\delta p)^2_{r.m.s}}{\rho c}$$

as for sinusoidal variation of p

$$\frac{p_{max}}{\sqrt{2}} = p_{r.m.s.}$$

Intensity is measured in watt per square centimetre. Definition of relative intensities is of much practical importance, than the absolute value. The ratio of the value of the intensity of sound to a standard value of intensity is known as intensity level. The standard value is taken as 10^{-12}watt/m^2 for a note of frequency 1000 c.p.s. which is due to feeblest audible sound. A sound having 10 times this energy is said to have an intensity level 1 Bel or 10 decibel. A sound of 100 times of this energy is said to possess an intensity level 2 Bel. The unit in common use is *decibel* which is 1/10 of a Bel or 0.1 Bel.

Intensity level $$N = \log_{10} \frac{I_1}{I_0} \text{ Bel}$$

$$= 10 \log_{10} \frac{I_1}{I_0} \text{ decibel}$$

I_0 is the standard value of intensity which is equal to 10^{-16} watt/cm^2 = 10^{-9} ergs/cm^2/sec corresponding nearly to root mean square sound pressure 0.000204 dynes/cm^2 at a frequency 1000 c.p.s. at 0°C and 76 cm. of mercury pressure.

If the intensity level increase by 1 decibel i.e. 0.1 bel, the intensity will increase as given below.

$$10 \log_{10} \frac{I_1}{I_0} = 1 \text{ decibel}$$

or $$\log_{10} \frac{I}{I_0} = \frac{1}{10} = 0.1 \text{ gives}$$

$$I_1 = (\text{anti log of } 0.1) \times I_0$$

$$= 1.26\, I_0$$

Thus the intensity increases 26% when intensity level increases by 1 decibel.

Sound Pressure Level

It is found that intensity level is given by

$$N = 10\log_{10}\frac{I_1}{I_0}$$

$$= 10\log_{10}\frac{(dp)^2_{rms}}{(dp_0)^2_{rms}}$$

when dp_0 is the r.m.s. pressure at the standard value I_0, dp is the *r.m.s.* value of pressure at the value of the intensity I_1

Then
$$N = 10\log_{10}\frac{(dp)^2_{r.m.s.}}{(dp_0)^2_{r.m.s.}}$$

$$= 20\log_{10}\frac{dp_{r.m.s.}}{dp_{0r.m.s.}} \text{ decibel}$$

$$= 20\log_{10}\frac{dp}{0.0000204 \text{ dynes/cm}^2}$$

17.2. LOUDNESS LEVEL

Loudness depends upon intensity and frequency. So loudness level is measured with reference to a standard frequency. The sound concerned of any frequency and another sound of pure tone of frequency 1000 c.p.s. are heard alternately. The pure tone is adjusted so that sound appears equally loud. The intensity level of pure tone of frequency 1000 c.p.s. is measured. If the quantity be N decibel, the other sound is said to possess an equivalent loudness of *N phons*.

But loudness is not always and at all frequencies proportional to intensity. The intensity level required for various loudness is shown in Fig. 5 Art. 18.4. The lowest curve of 0 phons is the threshold curve. The other curves show the relation between intensity and loudness.

17.3. SOUND INTENSITY MEASUREMENT

Absolute sound intensity measurement is difficult but comparative sound intensity measurements are possible with almost all sound receivers. There are several methods, which are:

(*i*) Rayleigh disc with or without resonator,

(*ii*) Sound radiometers,

(*iii*) Hot wire microphones.

(*i*) **Rayleigh disc.** When a light small disc, capable of rotating freely about a vertical axis suspended in a stream of air, will tend to set up in a stable position with its plane perpendicular to the flow. If the direction of flow is reversed, there will be no change in the stable position. The alternating stream as in sound wave will produce the same effect as a steady stream.

The study of the flow of the stream line across the disc will exhibit that these are completely stopped at points A and B Fig. 1 which become the position of maximum pressure. If the direction of flow is reversed there will be no alteration of lines of flow. Konig from hydrodynamical consideration proved that a couple acts on the circular plate of radius r and is placed in such a way that the normal to the circular disc making an angle θ with direction of undisturbed stream line of velocity v, the torque T tending to reduce the angle θ, the magnitude of which is given by

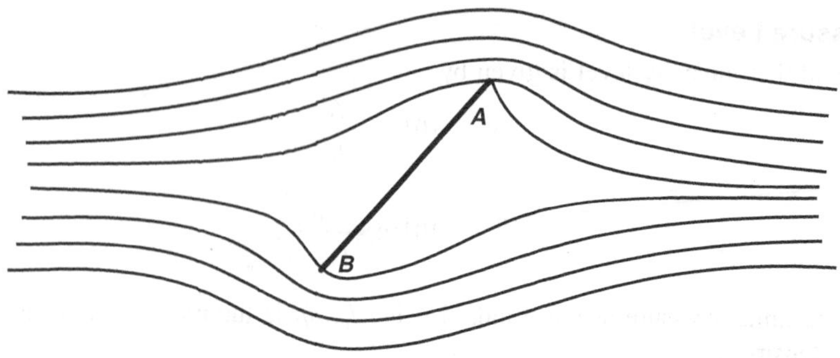

Fig. 1.

$$T = \frac{4}{3}\rho v^2 r^3 \sin 2\theta$$

whence
$$I = \frac{3}{4}\frac{Tc}{r^3 \sin 2\theta}$$

as I is given by $\rho c v^2$

The moment of the couple will be maximum if $\theta = 45°$. In case of alternate flow of air, the mean value of the square of the velocity of the stream which is the particle velocity should be taken into consideration.

The apparatus consists of the tube whose length is adjusted to $\frac{3}{4}\lambda$ of the sound concerned, of wavelength λ. The tube is closed at one end by glass plate W_1 and open at the other. The tube acts as a resonator whose length can be adjusted. The small circular disc with mirror is suspended by a torsion fibre at a distance $\frac{\lambda}{4}$ from the closed end which is the antinodal distance. The suspension is so adjusted that the normal to the disc is at an angle 45° with the axis of the tube. A ray of light from the source S falls on the disc and is reflected which passes through the side window and W_2 fans on a scale at a distance outside. The source of sound is placed at the mouth of the tube and length is adjusted for resonance. When resonance sets in, the alternate flow round the disc develops a torque. This causes to rotate the disc with the mirror through small angle. This small angle of deflection is proportional to the intensity of sound wave in the tube. The magnitude of the torque can be determined by torsion head, which brings back the disc to the initial position when absolute measurement of intensity is required. The relevant measurement of deflection will be sufficient when comparison of intensities are to be made.

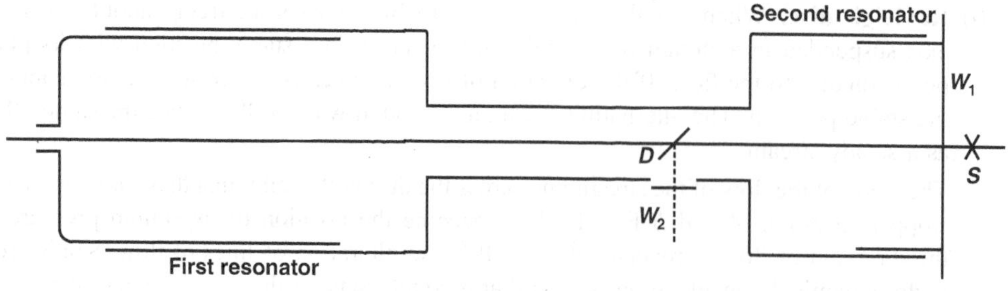

Fig. 2.

In order to make the arrangement very sensitive double resonators are used as shown in Fig. 2. Konig's theory ignores the diffraction of sound by the disc as its dimension is small compared to the wavelength of sound. The viscosity effect of the medium cannot be neglected in case of heavy medium like liquid. In this case

$$I = \frac{3Tc}{4r^3(1-\beta)^2 \sin 2\theta}$$

$$= \frac{3Tc}{4r^3(1-\beta)^2}$$

when $$\theta = 45°$$

$$\beta = \frac{u}{v}$$

u is the velocity amplitude of the disc and v of the medium.

17.4. SOUND RADIOMETER METHOD

The sound wave exerts radiation pressure. Let a plane wave is incident normally on a perfectly reflecting wall. Let the energy density in the incident wave train of velocity c be E. When stationary wave is considered the energy density is $2E$, considering both incident and reflected wave. Let the wall be pushed with velocity u to meet the advancing wave train. The length of the incident wave train that hits the wall in one second is $c + u$, the reflected wave train is compressed a length $(c-u)$. Consequently the energy density of the reflected wave will be greater by the amount dE.

As $$(c + u)E = (c-u)(E + dE)$$

or $$\frac{E + dE}{E} = \frac{c+u}{c-u}$$

or $$1+\frac{dE}{E} = \frac{1+\frac{u}{c}}{1-\frac{u}{c}}$$

$$= \left(1+2\frac{u}{c}\right) \text{ approx as } \frac{u}{c} << 1$$

or $$dE = 2\frac{u}{c}E$$

The total increase of energy in a column of length c is

$$cdE = 2uE$$

This is the work done by the wall in compressing the radiation. If P be the radiation pressure, the work done is

$$Pu = 2\,uE \qquad \therefore\ P = 2E$$

The radiation pressure is equal to the mean energy density in the medium in front of the reflector.

The pressure exerted by the sound wave on reflection from a plane solid wall was measured. A hole in the wall was closed, by a loosely fitting piston which is attached to one end of an arm of a torsion pendulum, this helps to measure the pressure on the piston by sound waves. Another balancing weight is attached to the opposite arm. The arm is suspended at its centre of gravity by a torsion wire. The deflection of the piston is noted with the help of a mirror and, lamp and scale. The torsion head adjustment brings it to zero position.

If A be the area of the piston r the effective distance from the point of suspension then the torque T is given by

$$T = Ar.P$$

This is balanced by torsion of the suspension

Then $$ArP = k\theta$$

where k is the torsion constant, θ is the angle of twist.

The intensity $$I = Ec = \frac{Pc}{2} = \frac{k\theta c}{2Ar}$$

so the sound intensity is measured.

Rayleigh rigorously calculated the relation between the energy density E and radiation pressure and found

$$P = (\gamma + 1)\, E \text{ (adiabatic) } \gamma, \text{ ratio of sp. heats of the gas.}$$
$$= 2E \text{ in case of isothermal process.}$$

Hence $$I = \frac{k\theta c}{(1+\gamma)Ar}$$

17.5. BY HOT-WIRE MICROPHONE METHOD

It is mainly a Helmholtz resonator R, with a grid G consists of platinum wire of 0.0006 cm diameter held in a glass enamel rod A mounted on an angular mica disc M. On two opposite faces of the mica disc M there are two silver discs S and S_1 fixed. One end of the grid wire is connected with one silver disc SS and the other end with another disc S_1S_1, on the back of the mica disc. The whole system is placed at the entry of the neck as a collar of the resonator. Very slight motion of air in the neck is sufficient to change the resistance by measurable current which is about 30 milli-amperes to produce just below the dull heat. Average resistance of the grid varies from 140 to 300 ohms owing to the heat developed. When the resonator responds to a tone of particular frequency, the platinum grid is cooled and resistance falls due to air draught in the neck.

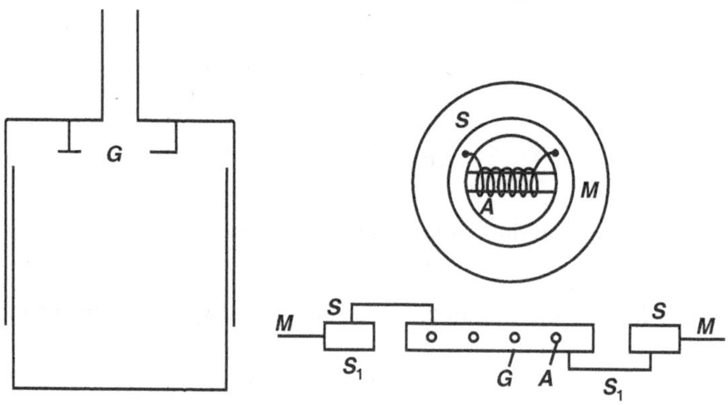

Fig. 3.

The average resistance of the hot wire during motion is measured by simple Wheatstone bridge. The resistance velocity curve of the grid in a steady draught can be used as a calibration curve to determine the amplitude of the motion at the neck from the measured resistance. The amplitude indicates the extent of response of the resonator.

It is found that the drop of resistance is proportional to the square of the velocity that is proportional to the intensity of the source therein.

The instrument is a sensitive detector of sound of frequency of the resonator.

17.6. ELECTRODYNAMIC METHOD

A very thin strip of aluminium alloy of 3μ thickness and a few millimetre wide is placed in a very strong magnetic field without appreciable tension. The plane of the strip is kept parallel to the magnetic field. Electric current is induced when the strip is set in vibration by incident sound wave. The useful electromagnetic damping is equal to the radiation clamping when the instrument becomes most sensitive. The forces on the strip due to incident sound waves are determined by balancing with a known an alternate current adjustable in magnitude and in phase, allowed to pass through the strip. The point of zero vibration is detected by a small sensitive microphone. The force on the strip due to sound wave is known in terms of dimension of balancing alternate current through it and the magnetic field.

The instrument is applicable at any frequency even when it is not resonance.

Thus it provides a method of absolute measurement of intensity of sound.

SOLVED PROBLEMS

1. Calculate the average energy density in a plane progressive harmonic wave in air if the intensity level of sound is 140 dB relative to the threshold intensity of 10^{-16} watt/cm^2. Given speed of sound in air = 340 m/sec [Burd. U. 1994]

Intensity level $$N = 10\log_{10}\frac{I}{I_0} \text{ decibel decibel}$$

Given I_0 = standard value of intensity

$$= 10^{-16} \text{ watt/cm}^2$$

$$N = 140\ dB,\quad c = 340 \text{ m/sec}$$

$$\therefore \quad 140 = 10\log_{10}\left(\frac{I}{10^{-16}}\right)$$

$$\frac{I}{10^{-16}} = 10^{14}$$

or $$I = 10^{-2} \text{ watt/cm}^2 = 10^2 \text{ watt/m}^2$$

$$\text{Average energy density} = \frac{10^2}{340} \text{ Joule/m}^3$$

$$= 0.2941 \text{ Joule/m}^3$$

2. A plane sound wave in air has an intensity 16 ergs/cm^2 per sec. Calculate the sound pressure assuming density of air 0.0012 gm/cm^2 and velocity of sound in air 330 metres/sec. [C.U.1954]

Given $$I = 16 \text{ ergs/cm}^2 \text{ per sec}$$

$$= 16 \times 10^{-3} \text{ watt/m}^2$$

$$\rho = 0.0012 \text{ gm/cm}^2$$

$$= 0.012 \text{ kg/m}^2$$

$$c = 330 \text{ metres/sec}$$

$$I = \frac{(\delta\rho)^2 r.m.s}{\rho c}$$

or $$(\delta P)^2_{max} = I\rho c = 16 \times 10^{-3} \times 0.012 \times 330$$

$$(\delta P)^2_{max} = \sqrt{2}(16 \times 10^{-3} \times 0.012 \times 330)^{1/2} \text{ N/m}^2$$

$$= 3.56 \text{ N/m}^2$$

REVIEW QUESTIONS

1. Derive an expression for the intensity of a plane progressive harmonic sound wave. Explain clearly what you mean by Bel and Phon. [C.U. 1993]
2. Define the intensity and energy density at a point in a plane wave of sound. How are they related? Obtain an expression for the intensity of a plane wave in terms of acoustic pressure.
3. Discuss in detail the method by which the intensity of sound wave is measured by Rayleigh disc mentioning the advantage and disadvantage of this method.
4. Write a note on the following:

   Intensity measurement by (*a*) Sound radiometer (*b*) Hot wire microphone (*c*) Electrodynamic method.

PROBLEMS

1. In ordinary conversion the intensity level is 70 *dB* above the threshold level. Calculate the absolute intensity and acoustic pressure. Take datum level for intensity as 10^{-12} watt/m^2 and for the pressure as 0.00002 N/m^2.

   [**Ans.** $I = 10^{-5}$ watt/m^2, $\rho = 2 \times 10^{-15}$ N/m^2]
2. A microphone emits a 1 KHz pure tone having an intensity level of 65 *dB*. Calculate the actual intensity (reference = 2×10^{-5} N/m^2) and loudness level of sound. When another source of the same intensity is switched on, calculate the increase in the intensity level in *dB*. [C.U. 1998]

   [**Ans.** 3.16μ N/m^2 ; 3*dB*]

18 CHAPTER

SPEECH AND HEARING

18.1. HUMAN VOICE

The following organs help to produce sound (i) lungs, (ii) vocal cord in the larynx and (iii) cavities formed by the larynx, mouth and nose which act as a resonator. By the expansion and contraction of lungs, air gushes out through the narrow slit of the vocal cord Fig. 1, which is made

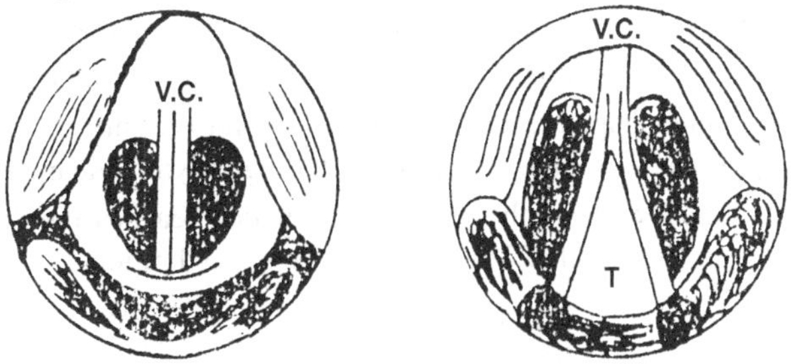

(a) (b)

Fig. 1.

open by a membranous reed under tension of vocal cord and vibrates with the desired frequency. This vibration is communicated to the cavities of mouth and nose Fig. 2 which function as resonators. The volume of the resonator, mouth and the larynx is capable of adjustment for resonance. Sound so produced in larynx is modified as articulated pitch in mouth and pharynx by the movement of the tongue and lips. One can whisper without forcing air through vocal cord simply by the motion of the lips and tongue. But without the movement of lips and tongue one can sing by changing the frequency of the vocal cord where it is not converted into speech, like musical instrument which can produce tones but not speech.

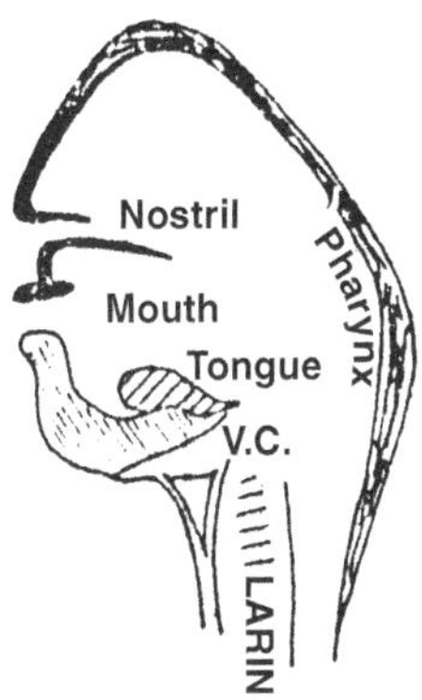

Fig. 2.

Vowel Sound. The resonant cavities of mouth, nose and throat and their modulations of size and shape, play a dominant part in the production of vowel sound. The shape of the mouth is changed to produce different vowels.

Harmonic Theory of Helmholtz. The vowel sound is due to combination of harmonic component tones produced by larynx tone reinforced by resonance in the cavity of mouth, nose and throat. Helmholtz suggested that amplification of specific group of harmonics of larynx tone by specific set of resonance called 'format' is different for different vowels. Any vowel sound is characterised by the presence of two or three definite groups of partial tone. Any one of these group of partials is known as 'format' of which lower one identifies the vowel and is enriched in quality by addition of third.

D.C Miller made optical recording of vowel sound. The experimental results supported Helmholtz theory. He concluded that vowel can be grouped in two classes–

(*i*) Some of those which have single characteristic frequency of resonance as in 'a' in 'father'.

(*ii*) The other have two characteristic frequencies of resonance as in 'e' in 'met'.

Consonants are recognised by the changes of resonance which are set up. These may be differently classed –

(*i*) Complete temporary stoppage of sound such as '*B*' 'hard *C*' '*D*';

(*ii*) Incomplete cessation, the remaining sound anticipating the following vowels by a weak note as *F*, *V*, *L*, *M*, *N*;

(*iii*) By a small drop of intensity with anticipation of following vowels as *R*, *S*.

18.2. HEARING–EAR AND ITS STRUCTURE, MECHANISM OF HEARING

Ear is the most sensitive receiver of sound. Two such receivers are located one on either side of the head connected with brain with nervous system. Ear, qualitatively can analyse a complex note. Any simple harmonic motion in air is perceived by ear as a simple tone, all others are resolved by ear into a series of simple tones of different periods. This is known as Ohm's law of acoustics.

Human ear consists of three parts–external ear, middle ear and internal ear. Fig.(3).

External ear is a funnel-shaped receiver *P* with a hole for the entrance of sound called pinna. This is connected to a tapering tube called the canal *C* ending in tympanic membrane or eardrum *D*.

The external ear Fig. 3 collects the sound wave which reaches the drum after successive reflection on the inner wall of the tube. This ear drum sets in vibration.

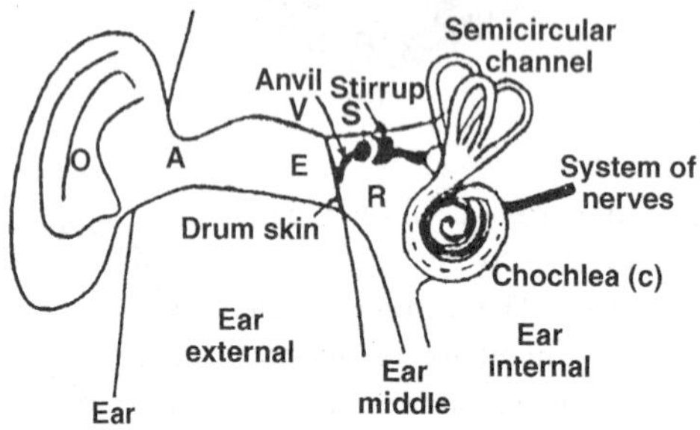

Fig. 3

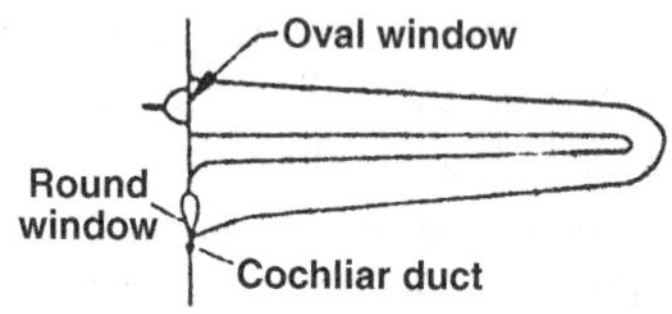

Fig. 4

The *middle ear* Fig. 3 is a cavity containing three small bones called *Hammer, Anvil* and *Stirrup*.

The chamber is connected to the inner side of the eardrums asymmetrically. The vibration at the eardrum *D* sets the hammer *H* to vibrate which in turn set the anvil *A* into vibration which causes stirrup *S* to vibrate. These bones act as levers to transmit vibration to the middle ear to the inner ear. The last part of the stirrup is connected to the oval window *O* or *Fenestra Ovalis* which forms the upper part of the wall of the inner ear. On the lower part there is a *circular window R*.

There is a tube *E* called *Eustachean* tube which connects the cavity of the middle ear to the throat. This keeps the pressure of the middle ear equal to that of the air in the mouth cavity. The ear drum is not ruptured therefore due to loud explosive sound as air in the cavity gets into mouth cavity, thus inside pressure is not increased. The mode of vibration of the foot plate of the stirrup during loud sound changes and vibrates parallel to the oval window due to special arrangement of suspension of these three bones by muscles.

Inner ear Fig. 3 is a cavity in bone and is divided into two parts (*i*) semicircular canal Se and cochlea *Co*. The semicircular canal *S.C.* is fitted with a fluid and placed mutually at right angles. These coils do not take part in hearing. It is concerned with the 'balance'.

The *cochlea* is a bony spiral tube of two and half turns about 20 to 30 mm long. The Fig. 4 represents a longitudinal section of uncoiled cochlea. This is divided along the length by *Lamina spiralis* and *Basilar membrane* into two compartments, the '*scala tympani*' and '*scala vestibule*'. The two compartments enter, communicate at the tip of the spiral at the end of the *Lamina spiralis.*

A membranous labyrinth, called *cochlear* duct nearly triangular in section is passing all through its length, formed by *membrane Reissner* and *membrane basilaris* which is kept under tension. The cochlea is filled with a liquid called *Perilymph,* and the labyrinth is filled with *Endolymph.* The organ of corte is mounted over the portion of basilar membrane. These convert the sound energy into nervous energy. Hair cells about 25,000 in numbers project from *Corti-organs*. Another membrane–*membrane tectoria, TM* overhang over these nerve endings. The pressure vibration in perilymph will cause minute vibration, press the hair cells against the *Tectarial membrane* and stimulate the nerve ending which causes the nerve impulses to transmit by auditory nerves to the brain which interpret the sound signal in a way not yet known.

18.3. HELMHOLTZ THEORY OF AUDITION

Helmholtz assumed that the nerve fibres are behaving like a heavily damped resonators. These fibrous structures are stretched like strings of a harp across the canal, which are identified with cortis rod. Later on Helmholtz expanded his views and assumed that the various parts of basilar membrane behave like resonators. A sound of particular frequency advancing along cochlea, excites into resonance the particular radial strip of basilar membrane which is in tune with it. The vibration thus sets up is communicated to corresponding hair cells and spiral nerve fibres lying between rods of cortis and basilar membrane.

Waver and Bray (1930) arrived at a conclusion from their experiments on the auditory nerves that the vibration on one part of basilar membrane produces stress on hair cells connecting the portion of tractoreal membrane. This develops potential difference which may play a basic part in producing the sensation of hearing in the brain. This throws some light in the mechanism of inner ear.

18.4. THRESHOLD OF HEARING

The sensation of sound depends both on frequency and intensity within certain approximate range, which varies from one person to another, also upon the age of a particular person.

Frequency threshold. The frequency range of audibility extends from 20 c.p.s. to 20,000 c.p.s. Some persons can hear sound of vibration of frequency lower than 20 c.p.s. Young persons can hear frequency up 20,000 c.p.s. but with age audibility reduces to a frequency 12,000 c.p.s. or less.

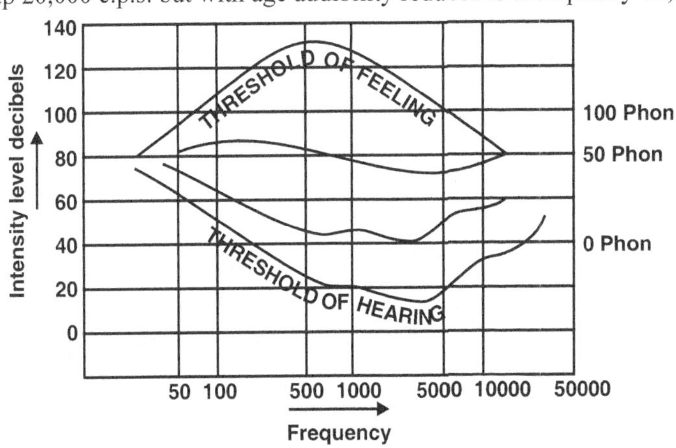

Fig. 5.

Intensity threshold. Besides frequency range, there are intensity range in the detection of sound. It becomes painful to the sensation in ear when very intense sound is heard. Above certain level of intensity, ear falls to recognise it as a sound.

The least audible intensity i.e. *sensibility of sound by ear* depends upon and varies with frequency. For a particular frequency within the range there is a minimum intensity for audibility. This is called *Threshold of audibility*.

Wegel (1922) obtained an interesting diagram, showing the range of response of average human ear with respect to intensity and frequency.

The lower curve Fig. 5, represents the audibility, that is lower limit of amplitude at which audibility of sound can be detected. The upper one is the diagram representing more or less arbitrary upper limit to the range of audibility sensation. It is regarded as *the threshold of feeling* at which the loudness is so great so that the effect is somewhat painful and any further increase of amplitude cannot be perceived as increase of loudness.

The area enclosed between these two curves has been described as auditory sensation area. Outside this area sound is not perceived whatever be the intensity but ear will indicate like a pressure indicator.

The sensitiveness of ear is found to be as a function of frequency. Minimum sensitivity of a normal ear occurs at a frequency near about 2,000 c.p.s. to 25,000 c.p.s.

18.5. WEBER FETCHNER LAW

It has been observed that loudness depends upon intensity only for sounds of frequencies lying on either side of the middle of audible range of frequencies. But near the limit of this range sound of large intensities and of large amplitudes are very feebly audible. Hence the perception of loudness is not proportional to the intensity of sound.

Weber has established the law that the increase of stimulus to produce the minimum perceptible increase of sensation is proportional to stimulus. It is applied to the case of change of intensity of sound to produce a change perceived in loudness.

If dL is the change perceived in loudness when dI is the change produced in original intensity I then we have

$$dL \propto \frac{dI}{I} \quad \text{or} \quad dL = k\frac{dI}{I}, \; k \text{ is a constant}$$

or

$$L = k \log I + A$$

where A is a constant.

Let L and L_0 be the loudness corresponding to sound of intensities I and I_0 then the difference of loudness say P is given by

$$P = L - L_0 = k \log \frac{I}{I_0}$$

I_0 is defined as the standard intensity the value of which is taken 10^{-12} watt/m^2 at a frequency 1,000 c.p.s. This is same as discussed previously.

REVIEW QUESTIONS

1. Describe the structure of Human ear and give the Helmholtz theory of audition.
2. What is threshold of hearing and threshold of feeling? Discuss the different factors on which they depend.

   What is Weber Fetchner law?
3. Give the harmonic theory of Helmholtz for vowel sound.

19

CHAPTER

Musical Sound

19.1. MUSICAL SOUND

All sounds can be classified in two groups (1) Musical sound and (2) Noise.

Musical sound produces pleasant sensation to ear. It consists of regularly sustained notes produced by periodic type of motion. This distinguished by three subjective characteristics: (*a*) pitch, (*b*) loudness (*c*) quality. The source of sound is always in a stage of vibration. The objective character of vibration may have variation of the frequency of vibration, the amplitude and complexity of vibration.

Noise is an abrupt sound and cannot be maintained for a long time. It produces unpleasant sensation in our ear due to abruptness and sharpness. It has no definite regularities.

In practice, however, musical sound are seldom free from irregularities, the vibration of the source not being maintained with perfect periodicities. On the other hand, even in the noise the phases of the vibrating motions are repeated at regular intervals, thus producing regular wave motion in the irregular disturbances. Thus two classes of sounds therefore shade into one another, and it is not possible to draw a sharp line of demareation between them.

It is generally said that several notes owing harmonic relations to each other when sounded together produce musical sound, but when they do not possess any harmonic relation between them are said to produce noise.

Principal Features of Musical Sound

Pitch

It is physiological process by which one can distinguish a grave note from a sharp note. It is not a physical quantity but depends upon frequency so ordinarily, it is measured by frequency. Pitch also depends upon intensity and composition, when the sound is a complex one. The unit of frequency is represented by cycles per second (c.p.s.) but unit of pitch is defined as 'mel'. The exact knowledge of pitch still remains as a complex problem, though some experimental observations of subjective character have been made.

Quality or Timber

It is also subjective in character, which makes it possible for us to distinguish one sound from other sounds having the same loudness and pitch. The physical cause of loudness is amplitude of the vibration of the source, and that of pitch is frequency. The frequency and amplitude being the same, the only other aspect in which the sounding bodies can differ is the mode of vibrations at each instant which, therefore, is a cause of physical character.

A musical *note* is generally a mixture of simple tones comprising of simple harmonic vibrations. Of these mixture of tones, the one which has the lowest frequency is called fundamental, those which have the higher frequencies are called *overtones* or partials. The frequencies of these overtones when they are the multiples of the fundamental are called Harmonics. The note having the frequency double the fundamental frequency is known as *octave* or Second Harmonics.

The quality of musical note depends upon the number, the phase, and intensities of these overtones, which will modify the form of the time-displacement curves.

The time displacement curve for a tuning fork (a), violin string (b) and pianoforte string (c) are shown in Fig. 1. The frequencies and amplitudes are remaining the same, the curves show how they differ in quality.

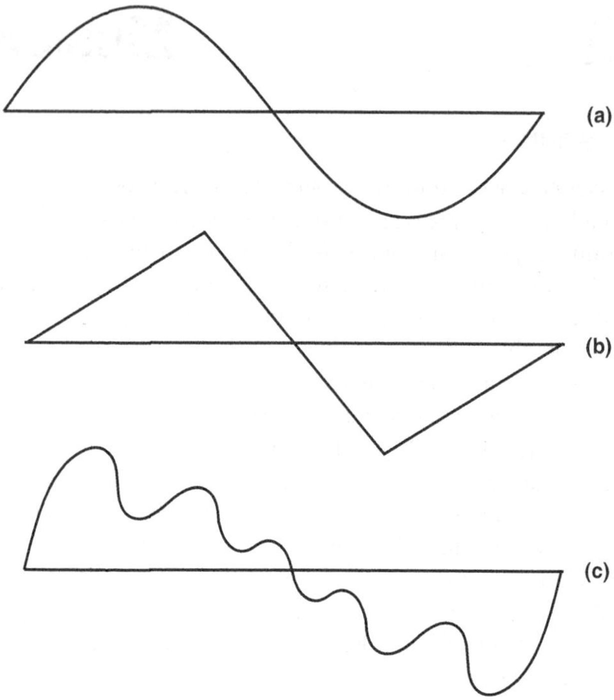

Fig. 1.

Loudness depends on intensity of sound but is not proportional to it. The intensity of sound depends upon (*i*) density (*ii*) wave velocity (*iii*) square of the amplitude of the wave. Again the amplitude of the wave depends on the amplitude of the vibration of the source. The intensity is also proportional to the distance of the source from the observer.

19.2. MUSICAL SCALE

It consists of a series of notes having certain relation to one another as regards the frequency of vibration. Between a note and its *octave*, the human ear can distinguish a number of notes of definite frequencies. The note of lowest frequency at such a series is called *key note* or *tonic*. The physicist prefer 256 Hz as a key note while musicians concert, the pitch is based on frequency 264 Hz.

Musical intervals. The ratio of frequencies between two notes in the musical scale is known as *musical intervals*.

Chord. A chord is said to be produced when two or more notes are sounded together.

Concord and consonance. When two or more notes are sounded together and combination produces a pleasing effect to the ear it is called concord and notes are said to be in consonance. But if they produce unpleasant effect to the ear it is said to a *Discord* and notes are said to produce *Dissonance.*

Harmonic triad. Any three notes within the musical scale whose frequencies are in ratio 4 : 5 : 6 are said to be harmonic triad.

Major chord. A harmonic triad together with an octave to the lower, constitute a major chord.

Minor chord. Any three notes whose frequencies are in ratio of 10 : 12 : 15 when sounded together are slightly dissonance but not altogether disagreeable. Three such notes with the octave to the lower, constitute minor chord.

Harmony and melody. When two or more notes of the scale are sounded together the pleasing effect due to concord is known as *harmony* while the succession of notes producing a pleasing sensation is known as *melody*. The musical scale should be of such notes so that they are able to produce both *Harmony* and *Melody*. Indian music is generally based upon melody while the continental music is based upon harmony.

19.3. HELMHOLTZ THEORY OF CONSONANCE AND DISSONANCE

All dissonances or discords are due to unpleasant beats formed by compound notes. Consonances or concords are formed by two notes which fail to produce unpleasant beats.

The beats are unpleasant, as the effect of beats to ear is analogous to that of the flickering light to the eye. Periodic variation of intensity of sound is due to the formation of beats formed out of the resultant of two notes. The ear is fatigued during the loudest phase of the beat and ear finds rest during the feeblest phase of the beat. This state of the ear is distressed due to the recurrence of the loudest phase of beats. So there is a disagreeable sensation or dissonance. A few beats per second say 2 or 3 do not produce any unpleasantness. They simply produce tremulous sound which is common to musicians. The disonance will start as numbers of beats per second increases. It gradually attains maximum and when beats become rapid it begins to diminish. When beats become too rapid, ear fails to respond to the rise and fall of intensity of sound and the dissonance disappears. Maximum irritation to the ear is produced when beat frequency is 33 per second and diminishes to nil when beat frequency is 80.

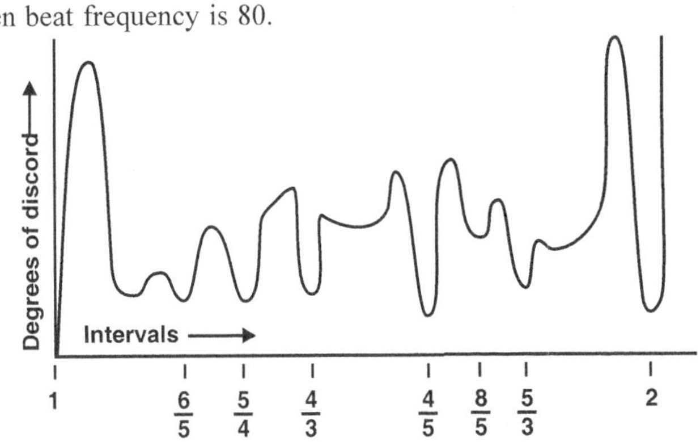

Fig. 2.

The adjoining curve is the reproduction of Helmholtz estimation of the degree of discord plotted on ordinate against the interval within the octave plotted as abscissae. The two notes are initially in unison, then pitch of one is kept constant and other is gradually raised to one octave higher than the condition is represented in the diagram. It is found that all well known intervals are represented by minima of the curve and are more or less closely bounded by strong dissonance. But the beat frequency alone is not the only determinant factor in producing discord between two notes but it also depends upon the following factors:

(*i*) The beat frequency.

(*ii*) The quality of the sound.

(*iii*) Absolute pitch of the sound.

(*iv*) Presence of combinational tone.

(*v*) Interval or the frequency ratio.

It has been observed that two tones having frequency ratio which may be called "simple ratio of small numbers" be sounded together then the chord produced will be consonance otherwise dissonance.

Consonant Interval. The degree of minimum discord was estimated by Helmholtz. The order of consonance is given below with their intervals and names.

Name	Interval
Unison	1 : 1
Octave	2 : 1
Fifth	3 : 2
Fourth	4 : 3
Major third	5 : 4
Minor third	6 : 5
Major sixth	5 : 3
Minor sixth	6 : 3

19.4. MUSICAL SCALE

A series of notes will constitute a musical scale which have consonant intervals. There are ordinarily two types of musical scales (*i*) Diatonic scale and (*ii*) Tempered scale.

Diatonic scale. It is the scale of music in vogue in almost all countries. It consists of eight notes. The interval between the eighth and the first is 2 : 1. They are divided into suitable smaller intervals as to produce melody when tones are successively sounded. Melody is produced between any two tones while going from lowest to the highest of the frequencies or from highest to the lowest. The frequencies of this scale are as follows.

Symbol	Western name	Indian name	Proportional frequency	Frequency in the basis of 256	Interval
C	Do	Sa	24	256	
					$> \frac{9}{8}$ Major
D	RE	Re	27	288	
					$> \frac{10}{9}$ Minor
E	MI	Ga	30	320	
					$> \frac{16}{15}$ Limma
F	FA	Ma	32	341.3	
					$> \frac{9}{8}$
G	SOL	Pa	36	384	
					$> \frac{10}{9}$
A	LA	Dha	40	326.7	
					$> \frac{9}{8}$
B	SI	Ni	45	480	
					$> \frac{16}{15}$ Limma or Semitone
Ci	do	Sa	48	512	

The lowest of the tones is called '*Tonic*' or '*Key note*' and the highest is called *Octave*. In music seven such octave extending on both sides of the 'Key note' are required in lowest is taken from C to C_1 then C_1 to C_2 and highest from C_6 to C_7. Any other note is represented with the subscript to the octave to which it belongs. In Europe nowadays, the frequency of A_4 note is taken 440 c.p.s. which makes the frequency of C_4 261.60 c.p.s.

$$C = 32.703 \text{ c.p.s.}$$

$$C_7 = 4186.0 \text{ c.p.s.}$$

The interval between two successive frequencies in a diatonic scale are not equal. The outstanding defect of this scale is that, it is not possible to change the tonic, that means that, it lacks in the power of modulations, as it is the characteristic of modern music to change frequency of the tonic. Any change of tonic is not possible in diatonic scale, as new notes, not given in diatonic scale, will arise. This is the case with all musical instruments except the stringed instruments where frequency can be changed adjusting the position of the finger, pressing the string.

Thus shortcomings of diatonic scale led to develop a new scale known as *Tempered scale.*

Tempered Scale. In this scale, the following considerations are taken into account, (*i*) Possibility of changing tonic, where a large number of notes slowly rising in pitch within the octave, will be necessary, (*ii*) but too many notes should not be included for sake of practical convenience (*iii*) dissonance should be avoided. Temperament is the compromise between these conflicting demands. The adjusted scale is called tempered scale. The one generally used is the tempered scale of equal temperament. Here the temperament divides the interval between the tonic and the octave into 12 equal intervals. If x be the interval between the two consecutive note then

$$x^{12} = 2$$

or

$$x = 2^{\frac{1}{12}} = 1.05946$$

The notes with their relative frequencies are given as follows.

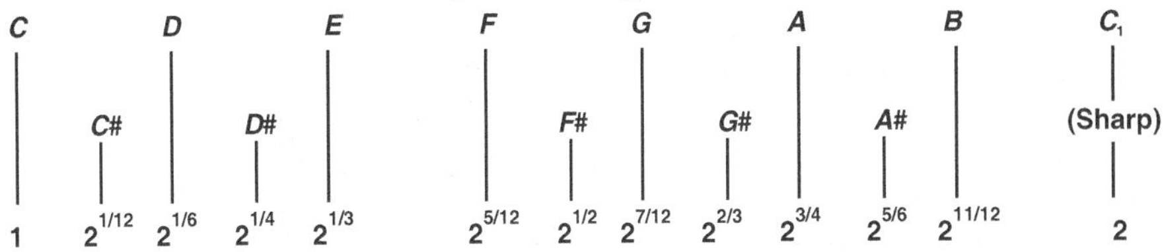

The white keys in piano or harmonium correspond to original $CDEFGABC_1$ in diatonic scale. The black keys give $C\#$ $D\#$ $F\#$ $G\#$ called C sharp D sharp etc.

REVIEW QUESTIONS

1. Distinguish between Musical sound and Noise. What are the principal features of Musical sound?

2. What is concord and discord? Describe the physical cause of consonance and dissonance

3. Distinguish between

(*a*) Diatonic scale and Tempered scale (*b*) Consonance and Dissonance

4. Define the following terms

(*i*) Pitch (*ii*) Quality or Timber

(*iii*) Loudness (*iv*) Musical intervals.

5. Explain loudness and intensity of sound. [Barkatullah University 2011]

CHAPTER 20

MICROPHONES AND LOUD SPEAKERS

20.1. INTRODUCTION.

Microphones convert acoustical energy into electrical energy. The wave forms in these two systems are substantially similar. It is an electro-acoustic transuder. The transuders used in air are known as microphones, those operating in water are hydrophones.

The microphones which are very widely used, are classified into two groups (1) Pressure microphones (2) Pressure gradient or velocity microphones. The following are the pressure microphones (*a*) Carbon microphones (*b*) Hot wire microphones (*c*) Condenser microphones (*d*) Moving coil electro-dynamic microphones (*e*) Crystal (Piazo-electric) microphones. Pressure gradient microphones where difference of pressure acting on two sides of the moving element i.e., diaphragm, is responsible for driving force, as Ribbon microphones.

The essential character of a good microphone:

(*i*) The sensibility should be high *i.e.,* open circuit voltage developed per unit sound pressure, at a given frequency, should be high.

(*ii*) The response should be uniform throughout the whole range of audio-frequency,

(*iii*) The output must be free from any self-noise, generally known as background noise.

20.2. (I) CARBON MICROPHONE

The operation of this microphone depends on the variation of resistance of carbon contact. The simplest form is used in telephones.

Carbon grains are packed between two thin polished parallel carbon plates *PP*. The first plate is attached to a boss *S* which is connected to the centre of a thin metal diaphragm *D*. The pressure variation in air produced by sound wave causes the metal diaphragm to vibrate, forces the front carbon plate to vibrate. The current flows, from a battery of cells, through the carbon granules, from one plate to other. There are two terminals for battery connections. The pressure change on front carbon plate will produce the change of resistance of carbon grains, when pressure increases, resistance decreases and *vice-versa*. Hence current in the circuit fluctuates according as there is fluctuating pressure on the diaphragm *D*, and hence the front plate *P*. It may be further amplified by step up transformer *F*.

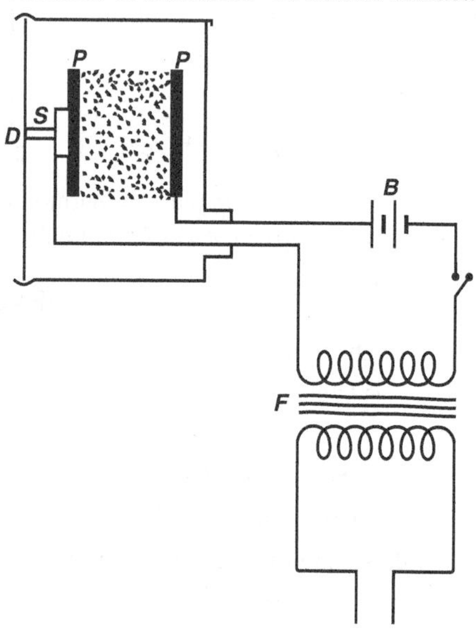

Fig. 1.

Let x be the displacement of the centre of the diaphragm.

R_x and R_0 are the resistances of the carbon unit when the centre of the diaphragm is displaced by x and 0 respectively.

Then $$R_x = R_0 + hx$$
where h is the resistance for unit displacement.

The amplitude of oscillation of the diaphragm is very small at the frequencies other than the fundamental resonant frequency. The motion of the diaphragm is stiffness controlled, so the force $= \mu x = AP_0 \sin \omega t$, where P_0, A and ω are the pressure amplitude, area of the diaphragm and angular frequency of the incident wave respectively, or we have

$$x = \frac{AP_0}{\mu} \sin \omega t \quad \text{... (2)}$$

$$= x_0 \sin \omega t$$

$$x_0 = \frac{AP_0}{\mu} \quad \text{... (3)}$$

So $$R_x = R_0 + \frac{hAP_0}{\mu} \sin \omega t$$

and the alternating current is given by

$$I_x = \frac{E_0}{R_0 + \frac{hAP_0}{\mu} \sin \omega t} = \frac{E_0}{R_0 + hx_0 \sin \omega t}$$

$$= \frac{E_0}{R_0}\left(1 + \frac{hx_0}{R_0} \sin \omega t\right)^{-1} \quad \text{... (4)}$$

$$= \frac{E_0}{R_0}\left\{1 - \frac{hx_0}{R_0} \sin \omega t + \left(\frac{hx_0}{R_0}\right)^2 \sin^2 \omega t ...\right\} \quad \text{... (5)}$$

$$= \frac{E_0}{R_0}\left\{\left(1 + \frac{h^2 x_0^{\ 2}}{2R_0^{\ 2}}\right) - \frac{hx_0}{R_0} \sin \omega t - \frac{h^2 x_0^{\ 2}}{2R_0^{\ 2}} \cos 2\omega t ...\right\}$$

The first harmonic of the alternating current is

$$= -\frac{E_0 hx_0}{R_0^{\ 2}} \sin \omega t$$

$$\frac{E_0}{R_0}\left(1 + \frac{h^2 x_0^{\ 2}}{2R_0^{\ 2}}\right) = \frac{E_0}{R_0} \text{ as } hx_0 << E \text{ which represents steady current}$$

E, M, F of alternate current $E_x = I_x R_0 = -\frac{E_0 hx_0}{R_0} \sin \omega t$ ... (6)

The amplitude of the voltage developed E_a

$$= \frac{E_0 hAP_0}{R_0 \mu}$$

Hence $$\frac{E_a}{P_0} = \frac{E_0 hA}{R_0 \mu} \quad \text{... (7)}$$

$\frac{E_a}{P_0}$ is a measure of sensitivity of microphone and is called open circuit voltage response.

The frequency response curve is given in Fig. 2. The straight portion represents the region of uniform response. The response increases with the polarizing battery voltage E_0, area of the diaphragm A and decreases with the increase of circuit resistance R_0 and stiffness μ.

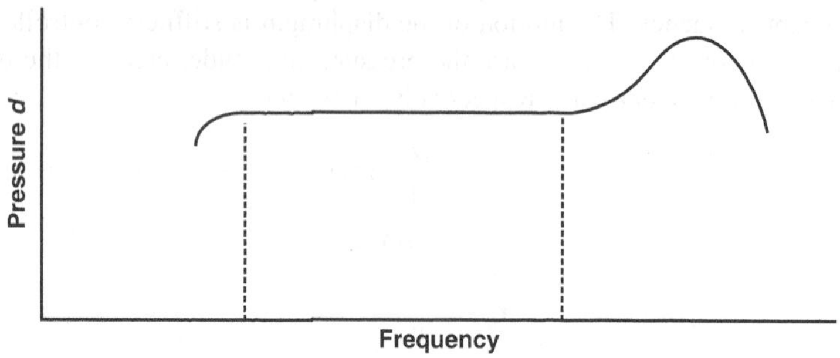

Fig. 2.

But the increase of E_0 gives rise to heating effect producing internal noise again sensitivity cannot be increased by increasing the ratio $\frac{h}{R_0}$ in which case the second harmonics (given by the second term Eqn. (5)) will be more pronounced to produce distortion.

The sensitivity may be increased by using two carbon packs one on each side of the diaphragm. The diaphragm while moves the resistance of one increases and the other decreases. In such case, the even harmonics generated in microphone cancel out. This type is known as double button push-pull type microphone. The main difficulty in this type is, the 'self noise'.

(*ii*) **Hot wire microphone vide Art. 17.5.**

20.3. CONDENSER MICROPHONE

The change of capacity of a condenser, due to variation of distance between two parallel plate condenser is the fundamental principle of a condenser microphone. The diaphragm B is a steel plate forming one plate of the condenser, has a thickness of 0.002 inch (1 inch = 2.54 cm), and of diameter of about 2 inches, is placed under tension being stretched to elastic limit, at a distance 0.004 inch from the other plate of the condenser, the other plate known as back plate is embedded in ebonite insulation. The back plate which is also metal is grooved radially, to give the desired damping. The air gap between the diaphragm and the back plate forming the condenser is very small of the order of 10^{-3} cm. On speaking to the mouth piece, the diaphragm vibrates, the thickness of air layer changes and produces change in the capacity of the condenser.

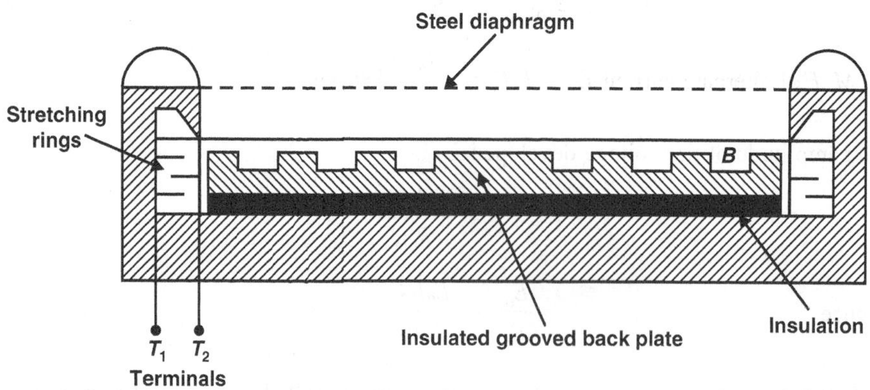

Fig. 3.

The microphone diaphragm and back plate are connected through two leads $T_1 T_2$ respectively, to high resistance of several megohms, and with a battery of cells of about 300 volts. The motion of the diaphragm alters the capacitance which causes the corresponding fluctuation in potential difference across the high resistance giving rise to a small momentary current in the circuit. This current, owing to smallness, needs amplification for any practical use. This microphone does not produce any hissing sound or noise.

Output Voltage of the Microphone. The value of the capacity C_0 at equilibrium stage in c.g.s. system is given by

$$C_0 = \frac{A}{4\pi d}$$

where A is the area of the plate and d is the thickness of air dielectric.

Hence $$\frac{1}{C_0} = \frac{4\pi}{A} d \qquad \text{... (1)}$$

If the thickness of dielectric changes due to vibration by the amount $\pm W$, the value of the capacity C_1 will change as given by

$$\frac{1}{C_1} = \frac{4\pi}{A}(d \mp W)$$

$$= \frac{1}{C_0}\left(1 \pm \frac{W}{d}\right) \qquad \text{... (2)}$$

Due to vibration of the diaphragm, the potential difference E_1 between the plates at and the capacitance C_1 will change with time. E_0 is the average value of potential difference and C_0 the capacitance at rest but the charge remains constant

Hence $$E_1 C_1 = E_0 C_0$$

or $$E_1 = E_0 \frac{C_0}{C_1}$$

From Eq. (2)

$$= E_0\left(1 \mp \frac{W}{d}\right) \qquad \text{... (3)}$$

Hence $$E_1 - E_0 = \left|\frac{E_0 W}{d}\cdot\right| \text{(magnitude)} \qquad \text{... (4)}$$

The maximum displacement at any time is given by

$$W = \frac{pA}{8\pi T} \qquad \text{... (5)}$$

where p is the excess of pressure amplitude acting on the area A of the diaphragm. T is the tension per unit length of the boundary of the plate. The maximum value of the output E_ω is given by

$$E_\omega = (E_1 - E_0)_{max}$$

$$= \frac{E_0 pA}{8\pi T d} \qquad \text{... (6)}$$

From Eq. (4) and (5)

$$= \frac{E_0 pa^2}{8Td}$$

when the plate is of circular in cross-section $A = \pi a^2$, a being the radius of the plate.

The sensitivity of the open circuit is given by

$$\frac{E_\omega}{P} = \frac{E_0 a^2}{8Td}$$

The instrument has uniform response for wide range of frequency which is about 0 → 10,000 Hz and response beyond which falls to a very low value. It has directional response and is free from back ground noise. Owing to stray capacitance due to connection cables, they should be as short as possible, otherwise sensitivity will be reduced. A special moisture resisting coating is used, as humidity will cause leakage across the insulating surfaces. The instrument has high internal impedance, which is reduced by using a pre-amplifier before voltage amplification.

20.4. MOVING COIL ELECTRODYNAMIC MICROPHONE

This microphone is based upon the principle of electromagnetic induction in a coil, moving in strong magnetic field.

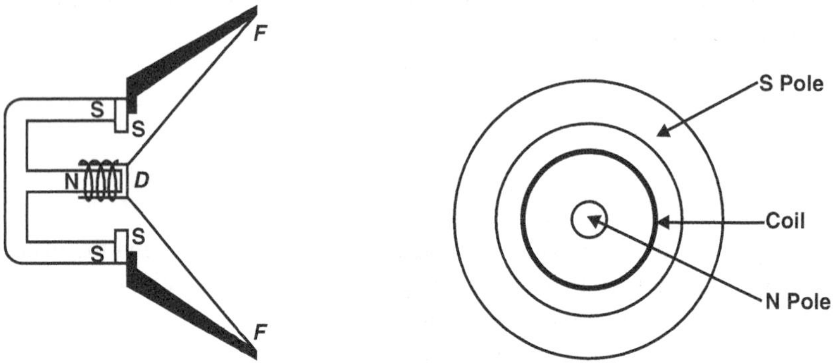

Vertical Section
(a)

Horizontal Section
(b)

Fig. 4.

The instrument consists of a permanent strong pot magnet section of which is shown in Fig. 4. A coil of insulated copper wire is wound on light cylindrical frame which is rigidly attached to diaphragm D of waxed paper or very thin aluminium sheet. The coil can move backward and forward freely into the annular space between north pole and south pole, but it is not capable of movement in any other direction and the movement of the coil generates an induced E.M.F., which is amplified by valve amplifier. The open circuit response

$$\frac{E}{P} = \frac{BlA}{Z_m}$$

where B is the strength of the magnetic field, l is the total length of the coil, A is the area of the diaphragm and Z_m is the mechanical impedance of the system. Z_m is obtained by considering the mechanical system as equivalent to a simple oscillator of mass m stiffness μ, and resistance R and p is the angular frequency of the varying force acting (vide Eq. 34 Art. 3.6)

$$Z_m = \sqrt{\left\{R^2 + \left(\frac{\mu}{p} - mp\right)^2\right\}}$$

The response is more or less flat within a wide range of frequency 50 Hz to 7,000 Hz. Between 0° and 90°, the response is constant to frequency of about 2,000 Hz.

20.5. CRYSTAL MICROPHONE

The mechanical stress produced by sound is transformed into electrical oscillation by piezo-electric properties of piezo-electric crystals like Quartz, Rocksalt, Barium titanate, Lithium sulphate, Ammonium dihydrogen sulphate, etc. Rochel salt is preferred over other crystals as it produces larger E.M.F. for same stress. This property falls rapidly with temperature exceeding 100°C so it is not successful in hot countries like India.

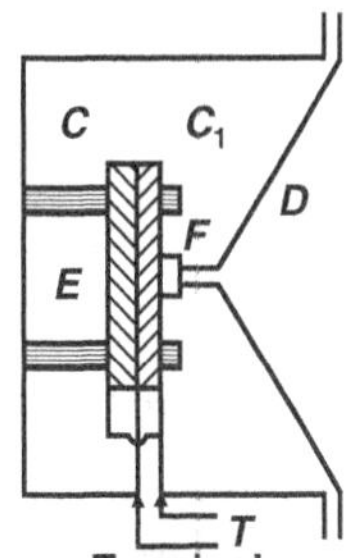

Fig. 5.

There are two types of this microphone :

(*i*) Diaphragm type;

(*ii*) Sound cell type.

(*i*) Two specially cut crystals slabs of Rochel salt crystal–C_1 and C are arranged differentially and cemented together with foil electrodes EF. The motion of the diaphragm D causes the crystal to vibrate. A mechanical bending of such a composite slab causes the other crystal to expand and inner crystal to contract. This produces piezo-electric E.M.F. which is amplified connecting through the terminals T, and finally passed through loud speaker.

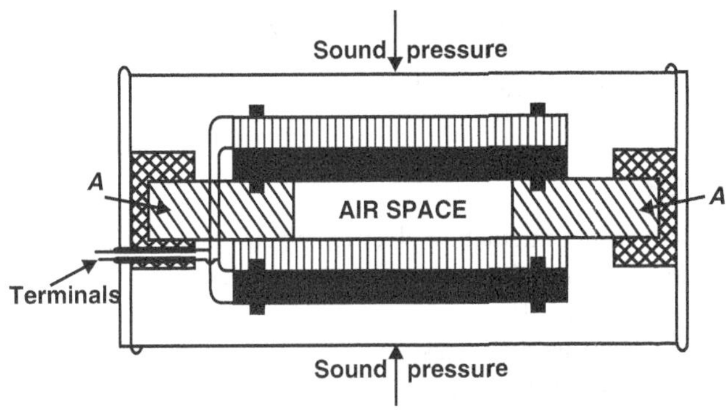

Fig. 6.

(*ii*) The crystal plates used here are 0.5 inch square and 0.2 inch thick (1 inch = 2.545 cm). Two plates as above are cemented together to form one unit. Two such units are required for this type of microphone. The resonant frequency is over 12000 c.p.s. Two such units are mounted on a rectangular frame of insulating material A, an enclosed air space is being left between the crystal units in order to provide freedom of movement. Finally the space containing the crystal unit is made air tight and moisture proof. The variation of sound pressure acts on suitable diaphragms which transmits the variation to the crystal units. The voltage developed across the crystal faces of two units are arranged to be in phase. Here in this arrangement the output is increased and response is made more perfect. In practice a number of such sound cells are used together. The output may be connected in series or in parallel or a combination of the two. The whole sound cell assembly is usually surrounded by a grille.

20.6. RIBBON OR VELOCITY MICROPHONE

Ribbon microphone is a special variety of moving coil microphone. A very light ribbon R usually of aluminium or its alloy 2 inches long 0.25 inch wide, 0.0001 inch thick is suspended between the poles of a strong magnet, the electrical connections are made at its ends. It is usually foiled like a concertina or corrugated in a direction perpendicular to its length in order to make the ribbon sufficiently flexible to allow free movement in backward and forward direction and to prevent any movement in sideways in the direction of pole pieces. The action of the ribbon may be compared with that of the single straight wire moving at right angles to the direction of the magnetic flux. The motion of the ribbon, by sound pressure will induce *E.M.F.* which is much smaller in amount than in the case of the moving coil. It is considerably amplified for loudspeaker. It has better response curve.

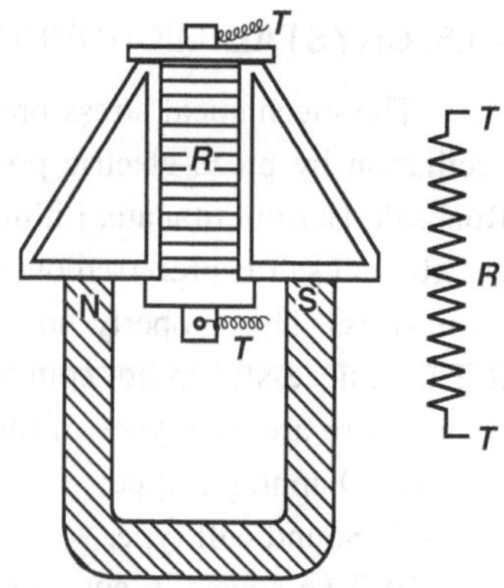

(*a*) (*b*)
Fig. 7.

20.7. HYDROPHONE

It is similar to microphone with important changes in design. It is used to find the direction of sound under water. It is a massive lens-shaped metal disc to which a metal diaphragm D is rigidly fixed making it watertight. There is a small carbon microphone M. It has two parallel carbon disc C, C, supported in a small brass capsule on mica disc other soldered to the brass body. A soft ring of felt surrounds the carbon and presses tightly the mica disc. Two-third of the cavity is filled with carbon granules. The varying pressure of sound wave in water causes the diaphragm to vibrate which produces the change of resistance between carbon plates, which produces the change of polarising current. The hydrophone kept under water from a ship is connected to battery of cells, transformer and telephone by cables which are kept at the test centre.

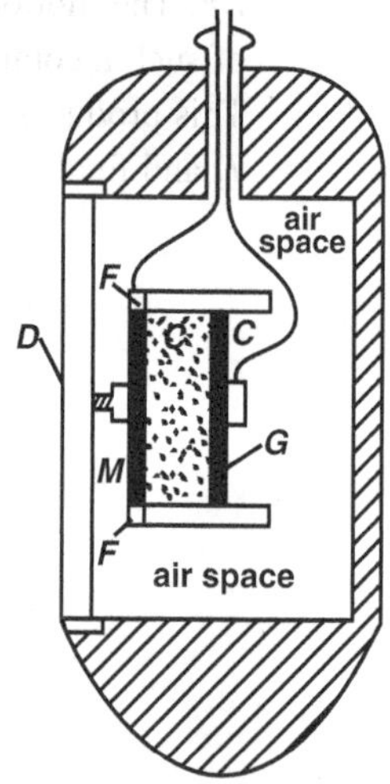

Fig. 8.

20.8. LOUD SPEAKER

It is a device to convert the fluctuating electric current produced in a microphone by varying pressure of sound wave, into sound energy for radiation in a room or open space. Thus a loud speaker is a microphone worked back.

Loud speakers are mainly of two types

(*i*) Fixed coil or moving iron type

(*ii*) Moving coil type.

(*i*) **Fixed coil or moving iron type.** It is similar to telephone receivers. It consists of a permanent horseshoe magnet with two pole pieces. A steel diaphragm is fixed in the throat of a horn in front to the pole pieces. The fluctuating current corresponding to the original sound is made to pass through the coil, wound round the pole pieces. It produces corresponding movement of the diaphragm and sound is thus produced.

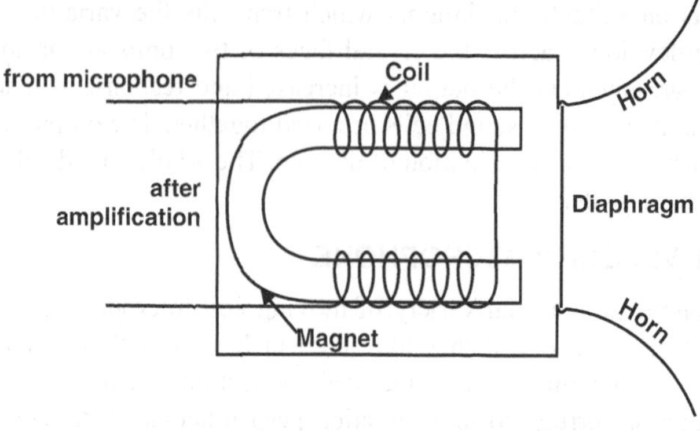

Fig. 9.

(*ii*) **Moving coil type loud speaker**. The construction is similar to moving coil microphone. In such a microphone voltage induced in the voice coil is proportional to the velocity of the coil which must, in turn, proportional to the sound pressure, whereas in the loud speaker it is proportional to the amplitude and hence pressure of sound is being reproduced.

Description of loud speaker is same as moving coil microphone Art, 20.4. Fig. 4. When fluctuating current flows through the coil, it with its diaphragm, vibrates hence sound is produced.

REVIEW QUESTIONS

1. Describe the different parts and explain the action of

(*i*) Carbon microphone

(*ii*) Condenser microphone

(*iii*) Moving coil electrodynamic microphone

(*iv*) Crystal microphone

(*v*) Ribbon microphone

(*vi*) Hydrophone

2. Describe a loud speaker

(*a*) Moving iron type

(*b*) Moving coil type

3. What is a crystal microphone? Explain its principle of operation.

21 CHAPTER RECORDING OF SOUND AND REPRODUCTION

21.1 RECORDING OF SOUND : MILLERS PHONODEIK

H is a horn, neck of which is fitted with thin glass or mica diaphragm of thickness 0.00762 cm which is held tightly between soft rubber rings and closes the neck of the horn. One end of a thin platinum wire *W* of 0.00127 cm thick is attached perpendicularly to the centre of the diaphragm. The wire wraps on the small pulley *P* fixed with a small steel spindle, mounted on jewelled

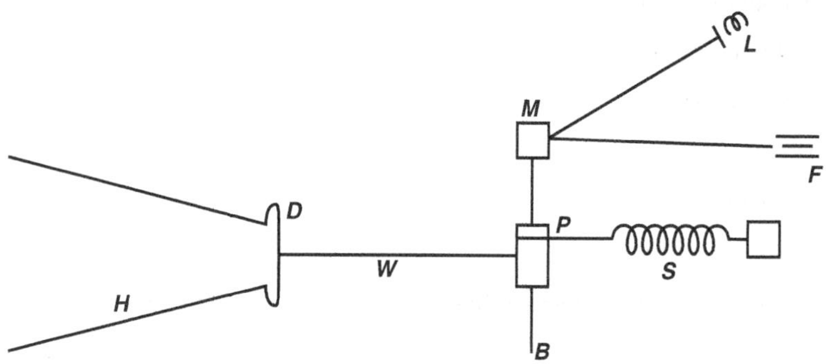

Fig. 1.

bearing, and is connected to a spring *S*. The spindle carries a very tiny mirror *M* and the plane of the mirror *M* is kept vertical. Light from a pin hole *L* is focussed by lens fall upon the mirror and is reflected to a moving film *F* or a rotating mirror. As sound waves produced before the horn, which falls on diaphragm, it vibrates. This will cause the platinum wire resting upon the pulley to move in the forward and backward direction. This will give rise to rotation of the spindle and mirror. The spot of light reflected from the mirror, falling on a distant screen performs a horizontal oscillation when the diaphragm is disturbed. The film running in the vertical direction will record the wave form of sound. The natural frequency of the diaphragm should be high so as to eliminate resonance effect but difficulty arises to eliminate the resonance effect of the horn.

The curves, so obtained may be analysed by making use of Fourier's theorem to find the intensities of different harmonic components present.

21.2. CATHODE RAY OSCILLOSCOPE (CRO)

This is one of the modern way of recording the sound wave of any musical note for analysis of frequency and quality of the note. It is discussed in details in Art. 2.12(*b*).

21.3. PHONOGRAPH—RECORDING AND REPRODUCTION

A cylinder *C* covered with a thick layer of wax is mounted, so that it can be turned mechanically by a handle *H* fitted to its axis, part of which is a screw, working in a nut so that as the cylinder rotates, it also moves endwise. The sound to be recorded is directed into the funnel *P* which converges the waves upon a thin glass diaphragm *D*, to the centre of which is attached a small style *S* with a sharp cutting edge. The vibration of diaphragm causes the style to vibrate longitudinally at right angles to the surface of the cylinder, and there, cut a groove of variable depth on the wax of the rotating cylinder. This groove thus formed is a record of the vibration of diaphragm.

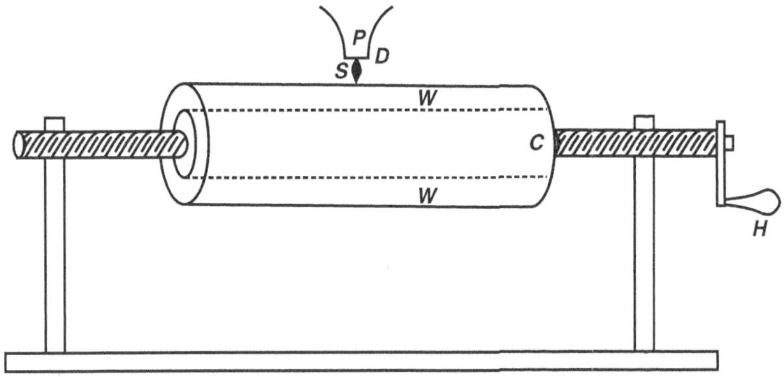

Fig. 2.

To produce the sound, the cylinder is rotated back to the original position and cutting style is replaced by a blunt point. On rotating the cylinder in the same direction and with the same speed as in the previous case, the blunt point is caused to pass along the grooves and vibrates exactly in the same way as the recording style and its motion being communicated to the diaphragm, the latter repeats the movements it executed under the influence of the incident sound waves. The vibrating diaphragm thus sets up waves, which produces the original sound.

21.4. DISC–RECORDING AND REPRODUCTION-GRAMOPHONE

For sake of convenience, recording is made on highly polished hard wax disc *L*-shaped style *S*, moves gradually in a spiral path on the rotating disc. The forward and backward motion of the style due to the diaphragm owing to varying pressure of sound is converted into lateral displacements and cuts the record on the revolving disc in weavy spiral groove.

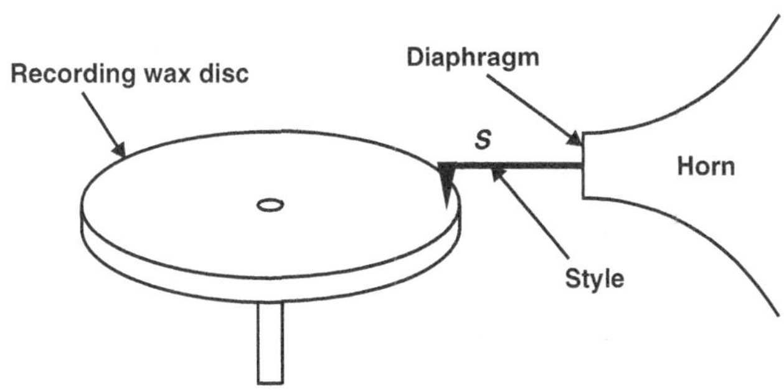

Fig. 3.

As wax is soft, the surface is dusted with graphite powder and copper is deposited electrolytically. The copper plate with projections corresponding to depression in the wax disc is, called 'negative' record or master shell. Hot disc of plastic material when pressed upon the master shell will form a 'positive' record with depression similar to wax disc. This becomes hard on cooling. Any number of such records on plastic disc can be obtained.

(*ii*) **Electrical method of recording.** The fluctuating amplified current from a microphone is passed into a coil whose plane is perpendicular to the plane of revolving wax disc. The transverse vibration of varying amplitude is produced in the soft iron armature by amplified current through microphone which causes the stylus to cut a weavy groove in a spiral path on the revolving disc. The rest of the processes are same as described before.

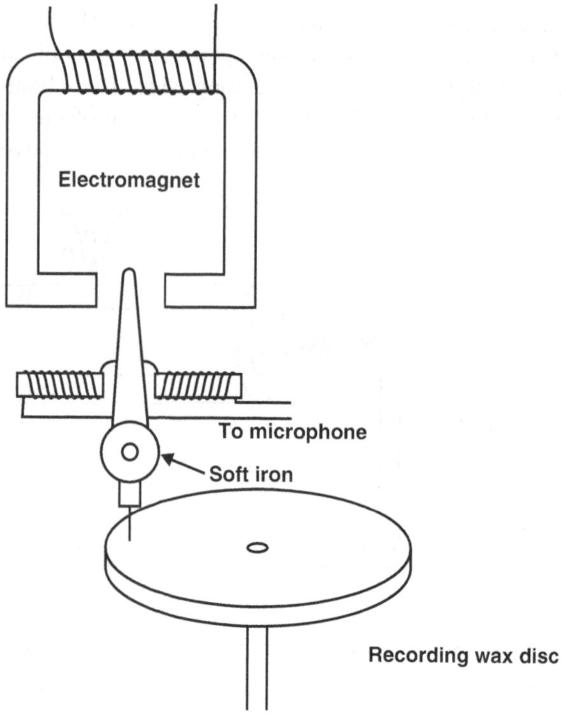

Fig. 4.

21.5. GRAMOPHONE

Reproduction of sound from record. This consists of a *sound box* and a horn Fig. 5. The sound box contains a diaphragm of aluminium alloy which sets in vibration by means of a lever *L* attached to a needle *N*. The needle moves to and fro in the groove of the uniformly revolving plastic record. The lateral motion of the needle is magnified by the lever *L* and is communicated to the air chamber of the diaphragm, in order to avoid different amplitude of vibration in the different parts of the diaphragm. Rubber packing *R* is used at the edge to fix the diaphragm in its position in order to avoid any sound produced by friction between diaphragm and its casing. The air inside the box also the air inside the horn is set in vibration due to the vibration of the diaphragm. Hence the sound recorded on the disc is reproduced.

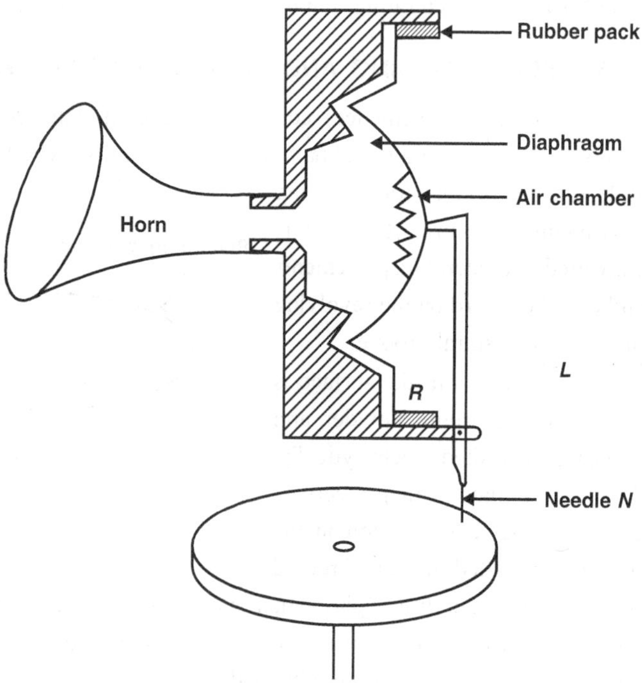

Fig. 5.

21.6. RADIOGRAM OR RECORD PLAYER

This is an electrical reproduction of sound. Unlike the above mechanical method 'Pick up' replaces the sound box and loud speaker replaces the horn. These two together with an amplifier constitute a radiogram. There are different types of pick up. (1) The magnetic pick up is commonly used Fig. 6. It consists of a small permanent horseshoe magnet M with four pole pieces N_1N_2 S_1S_2. A soft iron armature in the form of magnetic needle is pivoted at the centre P, is placed inside the pole pieces of the magnet M. This is filled with a needle N. A fixed coil of insulated copper wire C is placed inside the gap of the magnet pole. As the needle N moves side to side in the groove of the record plate revolving with constant speed, the soft iron armature moves and changes the magnetic flux lined with coil. The e.m.f. so developed is amplified and fed on loud speaker. The armature is suitably damped by rubber packing in the gap of the magnet. This type is known as Electromagnetic pick up.

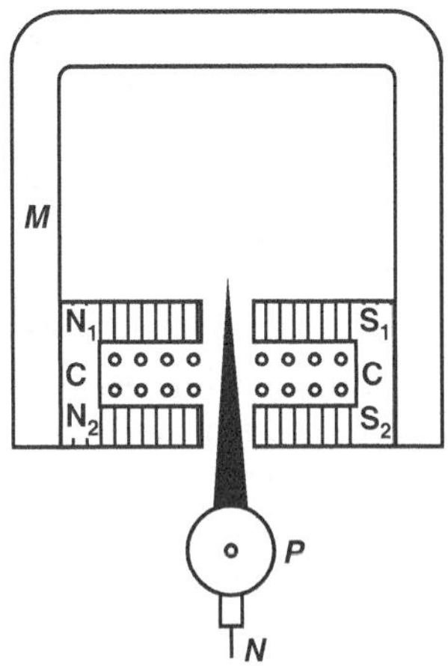

Fig. 6.

The electrodynamic pick is similar to electrodynamic microphone. The motion of the coil (vide art. 20.4) is brought about by the motion of the needle groove of the rotating record. The induced e.m.f. in the coil is amplified and fed in the loud speaker.

In the crystal pick up, the peazo-electric properties of a Rochelle salt is utilised. The movement of the needle produces strain particularly by torsion as in crystal microphone. The potential difference developed on two faces of the crystal is amplified and fed into loud speaker.

21.7. FILM RECORDING AND REPRODUCTION

In talking cine-film, there are two different methods of recording and reproduction of sound. The sound track occupies a space of width 2.5 mm, in a film of 35 mm width.

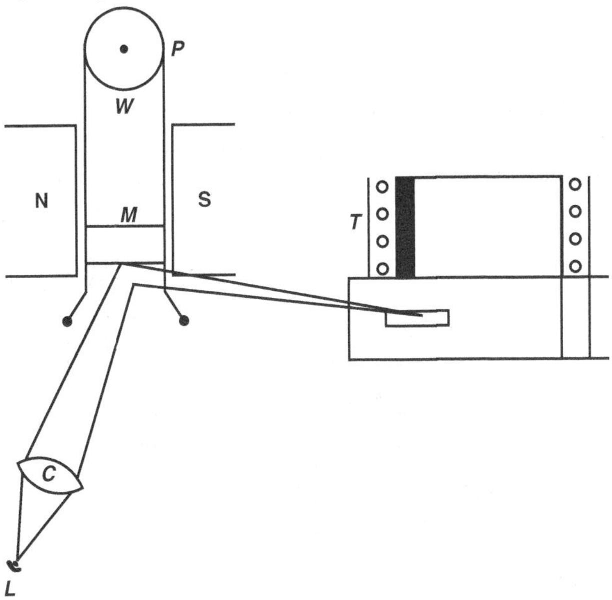

Fig. 7.

The methods are (*i*) *Variable width method* and (*ii*) *Variable density method.*

(*i*) **Variable Width Method.** The amplified current from the microphone is passed through a loop of wire W passing through a small pulley P placed in a strong magnetic field transversely. The current enters through one arm of the loop and goes out through the other arm. A small mirror M is attached to the loop. The beam of light from a powerful recording lamp L falls on the mirror after concentration by a lens C. The light reflected from the mirror M falls on the slit T on the track of the film moving with uniform speed of 0.4572 metre per second. The mirror M vibrates as the loop vibrates due to the passage of fluctuating current from the microphone. The vibrating mirror illuminates a length of the slit which varies with the tilt of the mirror. The tilt varies with strength of the current from the microphone in the loop placed in the magnetic field. Thus a variable width on the film represents the amplitude of the sound vibration along its length.

(*ii*) **Variable Density Method.** The sound record in this type consists of series of parallel lines of different transparency. These are the images of a narrow slit illuminated by a source of variable intensities by amplified microphone current. The intensities are varied either (*i*) by modifying the intensity of the source or (*ii*) by modifying the width of the slit through which passes the light beam.

(*a*) By the modification of intensity. Fig. 8. The amplified microphone current is applied to a gas discharge lamp G. The charging current produces the change in brightness of the lamp light

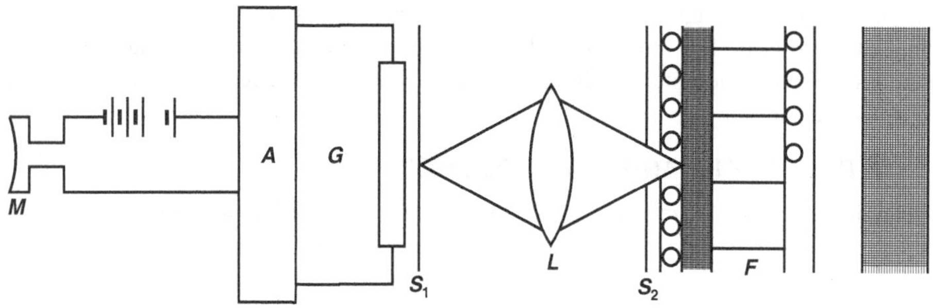

Fig. 8.

through S_1 and lens L illuminates the longitudinal slit S_2 in contact with the film moving in the vertical direction with constant speed. The film will be most opaque in places of maximum brightness of the lamp. The situation will be reversed when positive will be made for reproduction.

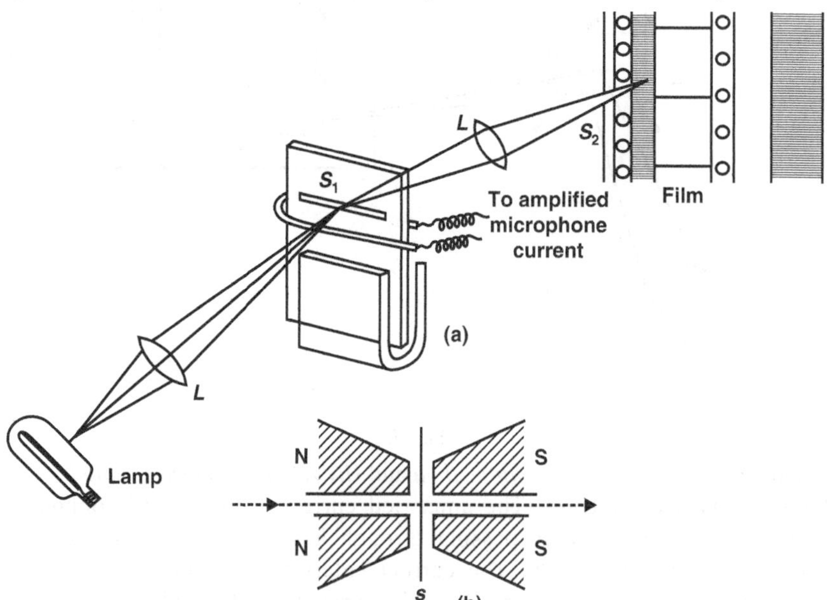

Fig. 9.

(b) By modifying the width of the slit. Fig. 9(a) A beam of light from a strong source of light passes through a shutter consists of a loop of duralluminium alloy strip or ribbon (0.5 mm thick 6 mm wide and 25 mm long) through which the amplified microphone current is passed.

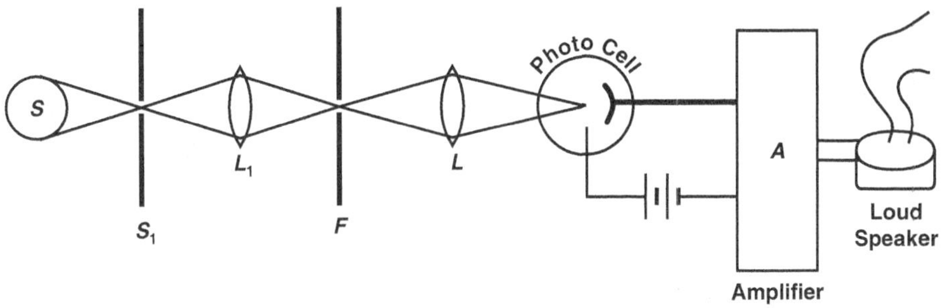

Fig. 10

The two arms of the loop are placed on two sides of a slit S_1. They are mounted so that *one is lightly behind the other* Fig. 9(b), to leave a gap normally 1 mm between the edges. They are supported in the field of a powerful magnet. The force is exerted by the magnetic field on the current from microphone passing through the ribbons which set in vibration, that is, they either come closer or diverge depending on the direction and magnitude of the current flowing. The movement of the two parts of the strips causes the change of the width of the slit. Thus it acts as a light valve. Light through various widths of the slit S_1 passes through another slit S_2 of constant width. The image of the slit falls on a film running at a constant speed. A continuous series of parallel lines of various capacity and number per unit length appear on the sound track of the film. The number of lines per unit length depends on frequency, and capacity depends on intensity.

21.8. REPRODUCTION OF SOUND

A slit S_1 is illuminated by a strong source of light S Fig. 10. The real image of the slit is formed on the sound track of the film F by a lens L_1. The light passes through the sound track of the film and converges to the sensitive surface of a photoelectric cell by a second lens L. The transmitted light through the film varies in intensity in both types of records (variable density or area), produces feeble current due to emission of electrons by the sensitive plate impinging the second plate of photoelectric cell. This current is amplified, it then actuates a loud speaker and corresponding speech or music is produced.

21.9. MAGNETIC TAPE RECORDING

The representation of the wave form by longitudinal variation of magnetisation in a magnetic tape is the fundamental principle of this type of recording. Rapid improvement is made in the preparation of the tape. Initially thin steel tape or wire was used. But it is being replaced by plastic tape 0.05 mm, impregnated with magnetic oxide of iron.

The tape is drawn from one reel to another at a constant speed of about 90 metres per minute and has a total length of 4000 metres. The tape, while unwound from one spool to another, moves over three heads of electromagnets, successively each having chisel-shaped poles. The first one is known as washing or erasing head. Here the tape is subjected to a strong high frequency (ultrasonic frequency of 100 kilocycle per second) magnetic field which wipes out any residual field in the tape and thereafter becomes suitable for fresh record.

The second electromagnet is known as recording head. The amplified current from microphone superimposed upon high frequency current is allowed to pass through the coil of the electromagnet. The tape while passing over the chisel-ended poles of this magnet with uniform speed is magnetised transversely varying along the length which fluctuates like the microphone current and is used in impressing the magnetic record.

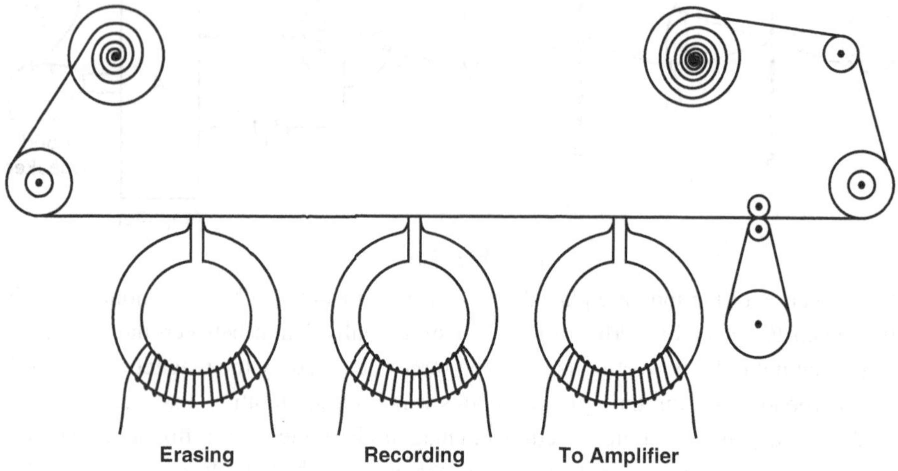

Fig. 11.

Such a magnetised tape is passed between two chisel-shaped poles of the third electromagnet. The flux variation induces corresponding current variation in the coil of the electromagnet. This current is amplified and is fed to loud speaker which reproduces the original sound.

REVIEW QUESTIONS

1. Describe Millers Phonodeik and how the same is used to photograph sound wave. What improved method is used in order to avoid mechanical inertia and response?
2. How Gramophone plates are prepared and how the sound wave recorded in the plate is reproduced?
3. Describe sound recording and reproduction in a film.
4. Describe magnetic recording of sound and reproduction stating the principle on which it is based.

CHAPTER 22

SHOCK WAVES – EXPLOSIVE SOUND

22.1. SHOCK WAVES-EXPLOSIVE SOUND

So far we have considered sound waves of small amplitudes. The wave form is not changed with propagation. The condensation produced is found to be very small. The region of high and low pressures are moving with the same velocity as that of the velocity of sound. The wave propagates without the change of forms. But sound generated suddenly, by explosion, pressure increases to a very high value. The high pressure region moves faster than the region of low pressure, consequently the wave changes its form. The pressure, velocity and density change on the shock front, and become discontinuous. Such a wave is called *shock wave.* Owing to discontinuity, differential equation for wave propagation is not applicable in this case. A set of difference equation is to be used.

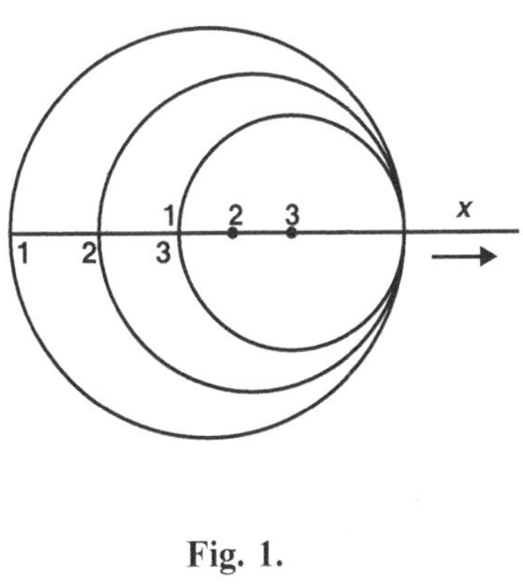

Fig. 1.

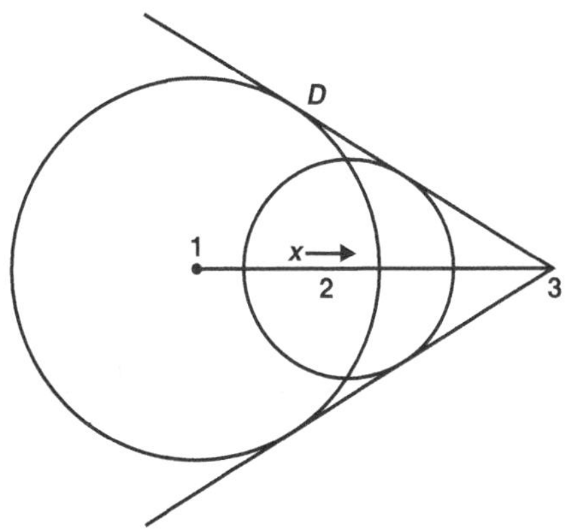

Fig. 2.

A peculiar phenomenon occurs if the source moves with the speed of sound or when moves with the speed greater than the speed of sound. In the first case, the speed of spreading out of the wave being equal to the speed of the source, they will move together. The pressure waves are generated by the body while in motion, at x_1, x_2, x_3, x_4 which spread out as shown in Fig. 1. In this case, the body and the wave move together, and relative motion of the compression wave to the body is zero. The waves pile up and the medium in front, will receive them as a shock. This is also called *shock wave* which produces sudden rise of pressure and density of air.

Let the source of wave be moving faster than wave velocity or phase velocity. At a given moment, sound wave is generated from the source at x_1 extends by radius r_1 smaller than the distance that source moves. Another wave starts from x_2. When the sound source moves still further to x_3 and a wave is starting there. The wave from x_2 has expanded to r_2 and the one from x_1 has expanded to r_3. the thing is done continuously and not in steps so we have a series of wave circles with a common tangent line which goes through the centre of the source. Here the source generates a wave front which forms a cone in three dimensions instead of spherical waves by a source standing still. All the pressure waves will reach the tangent x_3D when the body starting from x_1 reaches x_3. x_1 x_3 and x_2 D are proportional to the speed of the body v and the speed of the wave c. The *shock wave front* lies on a cone with apex at the source.

Let us assume a tube of unit cross-section through which shock wave front or transition region is moving to right.

Let p_1, ρ_1 and u_1 are the pressure, density and particle velocity on the high pressure side of the shock front and p_0, ρ_0 and u_0 are the corresponding values on the low pressure side. Fig. 3.

The rate of mass flow m into the front is equal to, that out of it

$$m = u_1\rho_1 = u_0\rho_0 \quad \text{... (1)}$$

where $p_1 > p_0$

$\rho_1 > \rho_0$

and $u_1 < u_0$

The rate of change of momentum is

$$m(u_1 - u_0) = \text{discontinuous pressure difference.}$$

From (1),

$$p_0 - p_1 = mu_0\left(\frac{u_1}{u_0} - 1\right)$$

$$= \rho_0 u_0^2\left(\frac{u_1}{u_0} - 1\right) \text{ as } m = \rho_0 u_0$$

$$= \rho_0 u_0^2\left(\frac{\rho_0}{\rho_1} - 1\right)$$

$$= \frac{u_0^2}{v_0^2}(v_1 - v_0) \quad \text{... (2)}$$

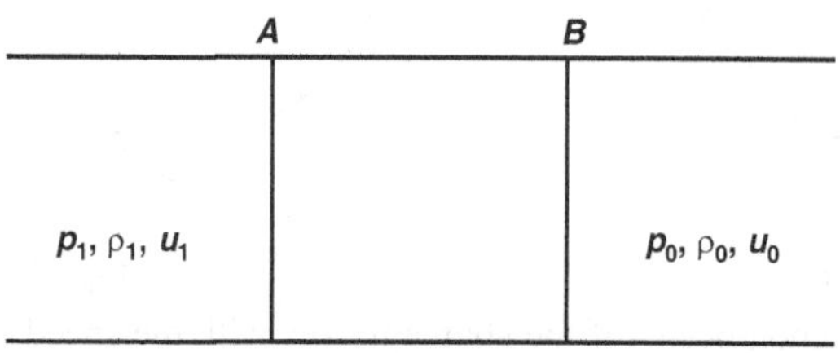

Fig. 3.

where v_1 stands for $\frac{1}{\rho_1}$ v_0 for $\frac{1}{\rho_0}$

or we have

$$\frac{p_0 - p_1}{v_1 - v_0} = \frac{u_0^2}{v_0^2}$$

or

$$u_0 = v_0\sqrt{\frac{p_0 - p_1}{v_1 - v_0}} \quad \text{... (3.1)}$$

Similarly

$$u_1 = v_1\sqrt{\frac{p_0 - p_1}{v_1 - v_0}} \quad \text{... (3.2)}$$

These are the first equation of *Rankin-Hugoniot*

or

$$u_0 - u_1 = \sqrt{(p_0 - p_1)(v_1 - v_0)} \quad \text{... (4)}$$

This is the second equation as given by Rankin-Hugoniot.

As the shock wave front progresses the work done per second is $p_1u_1–p_0u_0$.

Hence $$p_1u_1 - p_0u_0 = \frac{1}{2}m(u_1^2 - u_0^2) + m\Delta E \quad \text{... (5)}$$

where ΔE is the change of internal energy per unit mass per second.

From Eq. (5),

$$m\Delta E = p_1u_1 - p_0u_0 - \frac{m}{2}(u_1{}^2 - u_0{}^2)$$

On simplification remembering

$$u_0 = \frac{m}{\rho_0} = mv_0,$$

$$u_1 = \frac{m}{\rho_1} = mv_1$$

$$\Delta E = p_1v_1 - p_0v_0 - \frac{1}{2}(u_1{}^2 - u_2{}^2)$$

$$= \frac{1}{2}(p_0 + p_1)(v_0 - v_1) \quad \text{... (6)}$$

with the help of Eq. (3)

This is the third equation of Rankin-Hugoniot.

From thermodynamic consideration for ideal gas in adiabatic state

$$PV^\gamma = \text{Const. } C$$

$$dQ = 0 = dE + pdv$$

or $$E = -\int pdv = -C\int\frac{dv}{v^\gamma} = \frac{pv}{1-\gamma}$$

Now $$\Delta E = \frac{p_1v_1 - p_0v_0}{1-\gamma}$$

$$= \frac{1}{2}(p_0 + p_1)(v_0 - v_1) \text{ from Eq. (6)}$$

From Eq. (6) on simplification

$$\frac{p_1}{p_0} = \frac{(\gamma+1)v_0 + (\gamma-1)v_1}{(\gamma+1)v_1 + (\gamma-1)v_0} \quad \text{... (7)}$$

$$\frac{v_0}{v_1} = \frac{(\gamma+1)p_1 + (\gamma-1)p_0}{(\gamma+1)p_0 + (\gamma-1)p_1} \quad \text{... (8)}$$

The ratio of absolute temperatures on two sides are given by

$$\frac{T_1}{T_0} = \frac{p_1v_1}{p_0v_0} = \frac{c_1^2}{c_0^2}$$

where c_0 and c_1 are the wave velocities as

$$c = \sqrt{\frac{vp}{\rho}}$$

or $$\frac{c_1}{c_0} = \sqrt{\frac{p_1\rho_0}{\rho_1p_0}} = \sqrt{\frac{p_1v_1}{p_0v_0}}$$

22.2. MACK NUMBER OF FLOW

Mack number M is defined as the ratio of the mass velocity $(u_1 - u_0)$ and the small amplitude of sonic velocity $c_1 = \sqrt{\frac{\gamma p_1}{\rho_1}}$ at shock pressure.

So

$$M = \frac{u_1 - u_0}{c_1} = \frac{1}{c_1}\{(p_0 - p_1)(v_1 - v_0)\}^{\frac{1}{2}}$$

$$= \left\{\frac{1}{\gamma}\left(\frac{P_0}{P_1} - 1\right)\left(1 - \frac{v_0}{v_1}\right)\right\}^{\frac{1}{2}}$$

By the help of Eq. (8), we have

$$M = \left[\frac{2(p_r - 1)^2}{\gamma p_r\{(\gamma - 1) + p_r(\gamma + 1)\}}\right]^{\frac{1}{2}}$$

where p_r stands for the ratio $\frac{p_1}{p_0}$.

The flow behind the shock front is supersonic or subsonic accordingly as $M >$ or < 1.

In an ideal gas, the passage of shock front increases the gas temperature, pressure and density but reduces the Mack number.

For very strong shock, the larger change of temperature through shock will change the value of γ and will cause dissociation and ionisation. At 3000° K dissociation of air occur at 5000° K ionisation occur. The following table shows the variation of shock velocity, mass velocity, sonic velocity and temperature with the change of shock pressure in air.

Pressure in atm	Shock velo. in m/sec	Mass velo. m/sec	Sonic velo. m/sec	Temp. in ° K
20	1430	1140	680	300
100	3130	2730	1110	3440
200	3320	3920	1470	5090
500	6850	6270	2160	7940

22.3. ANOMALOUS PROPAGATION OF EXPLOSIVE SOUND

The following anomalies are noticed in the propagation of sounds from large explosions:

(*i*) Abnormally large velocities are developed in the neighbourhood of explosion. This produces considerable mechanical movement of air.

(*ii*) A sound zone of normal velocity is developed.

(*iii*) A zone of complete silence is developed after the zone of normal velocity.

(*iv*) Thereafter a zone is developed where sound reappear with unusual intensity which occur after a long interval.

The occurrence of sound at a greater distance after the zone silence is curious. Apart from the normal limitation of intensity due to damping, the waves experience a powerful source of damping in crossing the broken surface of the earth, the tops of forests etc. so that sound becomes inaudible after some distance. Audibility of sound with renewed intensity after the zone, is due to, wave travelling up into the higher part of the atmosphere and being bent down gradually to the surface of the earth. This explains the abnormal lapse of time between the occurrence of explosion and arrival of sound in the fourth zone. Such bending takes place where there is sudden discontinuity in the atmosphere, from the layer of atmosphere where sudden change of density prevails due to sudden change of temperature, or the sudden or the gradual change of composition of the atmospheric gas. Various assumptions are made to explain the recurrence of the sound in the fourth zone but no definite conclusion could be arrived at.

22.4. SOUND RANGING

The problem of sound ranging means the location of distant explosive sound from the instant of reception at three or more listening stations, with microphone. If the distance of the source is large compared to the distance between the microphones in different stations, the sound rays are taken to be parallel.

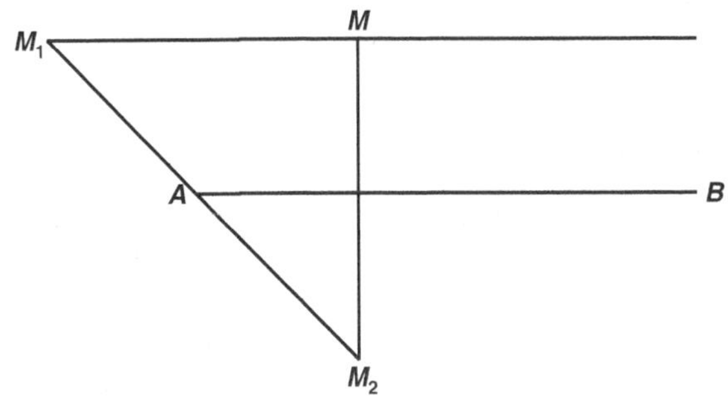

Fig. 4.

Let M_1 and M_2 be two microphones. Join M_1 and M_2 and bisect at A, AB being the direction of the source. M_1M is drawn parallel to AB and M_2M perpendicular to AB. Sound is heard at M_1 at t_1 and M_2 at t_2. The interval of time between the instants, sound is heard in two stations is

$$t_1 \sim t_2 = t \text{ say}$$

Then
$$M_1M = ct$$
$$= M_1M_2 \cos\angle M_2M_1M$$

$$\therefore \qquad \cos\angle M_2M_1M = \frac{ct}{M_1M_2}$$

or
$$\cos\alpha = \frac{ct}{d},$$

d is the distance between M_1M_2, c is the velocity of sound and $\angle M_2M_1M = \alpha$

For a second pair of microphone

$$\cos\alpha_1 = \frac{ct'}{d_1}$$

The angle α gives the direction of the source.

The intersection of two such lines of two pairs of microphone gives the position of the source.

REVIEW QUESTIONS

1. Discuss the anomalous propagation of Explosive sound.
2. What are shock waves? Deduce Rankin-Hugoniot formulae. What is Mach number and find the conduction of subsonic barrier of shock waves?
3. Discuss the principle of sound ranging in locating the position of the source.
4. What is a shock wave? Mention anomalous properties of Explosive sound.

23

CHAPTER

ULTRASONICS

23.1. ULTRASONICS.

The upper and lower limit of frequency of which sound wave ceases to produce the sensation of sound in human ear depends on the individuals. But the average value is taken 20 Hz to 20,000 Hz. which a normal ear can respond. This range within which the sound is audible is called sonics. Below the audible range *i.e.* below 20 Hz it is called Infrasonic, above 20,000 Hz is called Ultrasonics up to the range of 10^9 Hz above which they are termed Hypersonics.

Similar to sonic range, ultrasonic range requires a source of energy. A device which transfers energy from one system to another generally with the change of the form of energy mainly from acoustical to mechanical or *vice-versa* is known as Transuder. A medium is also necessary for transmission of ultrasonic waves.

Ordinarily the following types of *transuders* are used. They are

(*i*) Magnetostriction method. Joul (1847) observed that a rod of ferromagnetic substance such as nickel undergoes a change in length when magnetic field is applied along the length. The increase in length is proportional to the field applied. This change of length is greater in the case of nickel, than to any other ferromagnetic substances. The increase in length is independent of the direction of the field acting. So an alternating magnetic field applied parallel to the length of the nickel rod or tube, as field grows the length increases and the vibration sets up in the rod with a frequency double that of the frequency of the field. But if a permanent steady polarising field of suitable strength is applied from Direct Current supply, and a changing alternate current supply giving rise to alternating field is superimposed on it, the longitudinal vibration of the nickel rod will set up having frequency of vibration same as the frequency of the alternating field. But resonance will occur when the frequency of alternation of the field be equal to the natural frequency of the bar depends on the elastic constant and the dimension of the bar. The amplitude of vibration will be large. The rod vibrates with ultrasonic frequency and will emit ultrasonic waves.

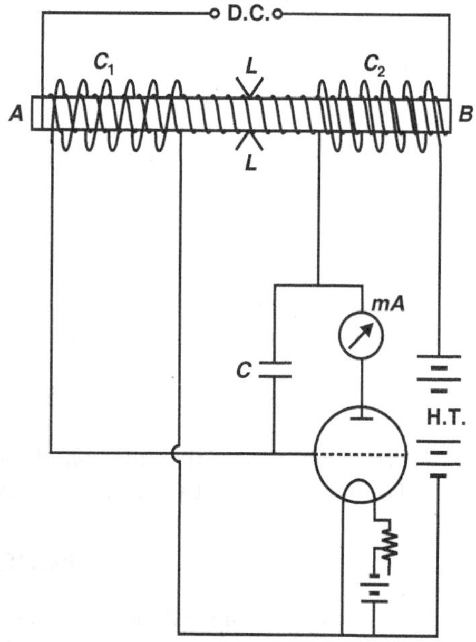

Fig. 1.

AB is a nickel rod clamped at the middle Fig. 1. A coil *L* wound round the rod carries a steady current, which polarises the rod. One coil C_1 in the grid circuit of a valve oscillator is wound on one side of the rod and another coil C_2 in the anode circuit is wound on the other side of the rod. The frequency of the valve oscillator is adjusted with the help of the variable condenser so that, it is equal to the natural fundamental frequency of the longitudinal vibration of the nickel rod. The application of anode potential causes a current to flow through the coil C_2. This will cause the rod to increase its length. This change in length causes a change of magnetism (Inverse Joule effect or Vellary effect) which produces change of flux in C_1 which is connected so as to raise the potential of the grid. This further increases the anode current which helps the maintenance of vibration of the rod. The natural frequency of the fundamental is

$$n = \frac{1}{2l}\sqrt{\frac{Y}{\rho}}$$

where l is the length Y is the Young's modulus and ρ the density of the material of the rod. A nickel rod of 5 cm length can produce an ultrasonic of frequency 50,000 Hz.

A common type of powerful magnetostriction transuder is shown in the Fig. 2. A number of nickel tubings act as a driving element, each having a length equal to one quarter of the wavelength of sound to be radiated. One end of these tubings is free and the other end is embedded in a circular sheet plate P whose resonant frequency is very nearly equal to that of the nickel tube. Each tube is surrounded by its own driving coil C and all are driven in phase by the current from a powerful oscillator. Optimum polarising field are supplied by permanent magnets M mounted inside the water tight housing containing the tubes. The alternating forces exerted on the plate by the reaction to the stresses in the nickel tubes are transmitted by the plate.

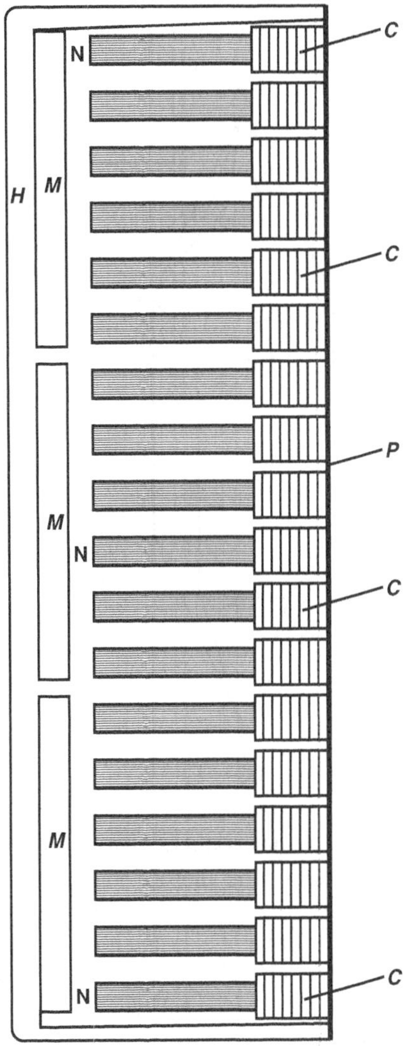

Fig. 2.

This type of transuder has higher power handling capacity. It has a sharper resonance curve which helps in reducing the disturbing effect of the background noise. But it is difficult to produce optimum polarising field in this type of instrument.

The same instrument can be used as a receiver, as magnetostrictive effect is reversible. Sound waves impinging on the plate P set into vibration, causes the nickel tubes to vibrate in magnetic field. The resulting alternating currents, induced in the surrounding coils which may be amplified for detection. The frequency range of magnetostriction is 5-60 KHz beyond which rods of very small lengths are needed but handling becomes difficult. Generally 25 KHz are used for good results.

23.2. PIEZO-ELECTRIC PROPERTIES OF CRYSTALS AND ITS APPLICATION TO PIEZO-ELECTRIC OSCILLATOR TO GENERATE ULTRASONICS

Two brothers J and P. Curie discovered (1880) that electric charges are developed on two opposite faces of quartz crystal cut in a perpendicular way when two opposite faces are subjected to pressure or compression perpendicularly. The sign of the charge developed on two opposite faces will change with the change of tension on compression. The electric charges developed is proportional to the amount of pressure or tension. Conversely, if a properly cut piezo-electric crystal is placed in an electric field, compression or extension, depending on the nature electric field occurs the amount of compression or extension being proportional to the potential difference between the faces.

Hence when alternating electric field is applied, it contracts and expands periodically and sets up mechanical vibrations in any acoustical medium in which it is placed. The frequency of the vibration is within the ultrasonic range. The amplitude ordinarily is very small. But when the frequency of the alternating current is equal to the natural frequency of vibration of the plate, resonance occur and amplitude becomes considerably large. This is adopted in the construction of piezo-electric transuder.

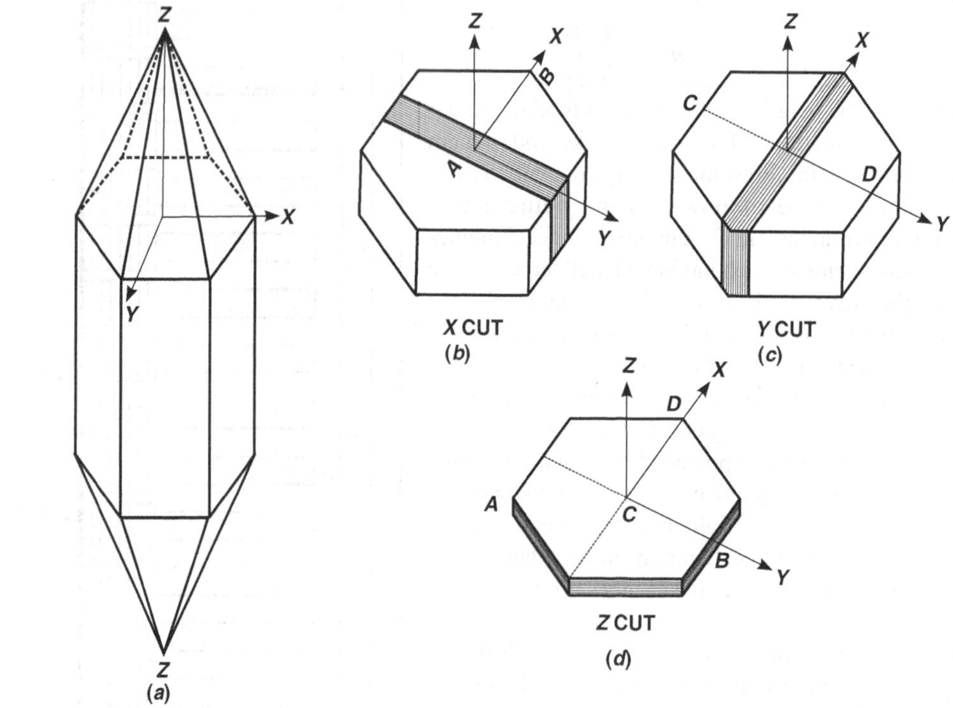

Fig. 3.

A single crystal of quartz is an uniaxial hexagonal prism bounded by two six sided pyramids at two ends. The line *ZZ* joining the two opposite vertices of the pyramids is its *optical axis*, Fig. 3(*a*). The Fig. 3 (*b,c,d*) represents the sections of the crystal perpendicular to *Z* axis. Line perpendicular to *Z* axis and joining two *opposite corners* of hexagonal slice of the crystal as shown in AB Fig. (3*b*) *the electric axis known as x-axis*. A line represented by *CD* Fig. (3*c*) joining the *opposite faces* of the hexagonal slice and perpendicular both to *x* axis and *z* axis represents *y*-axis or *mechanical axis*.

Y-cut crystal is obtained by cutting the hexagonal slice along *x-axis i.e.*, along the corners, such that the largest faces are perpendicular to *Y*-axis and *Z*-cut crystal is obtained by cutting the crystal perpendicular to *Z*-axis.

X-cut crystal is obtained by cutting the hexagonal slice along *y*-axis such that the largest face is perpendicular to *x*-axis. Differently crystals are shown by shaded portions in figures. They are generally used for piezo-electric oscillators. Its faces are silvered to ensure proper electrical contact. The alternate electric field applied by the metal plates pressing the faces, is along *x*-axis. The crystal vibrates in the direction of *y*-axis.

Pressure along the directions of *x*-axis produces the charges on two surfaces normal to this axis but tension will produce opposite charges, opposite to what is produced when pressure is applied.

Tension along the direction of *y*-axis charges two surfaces normal to *x*-axis as is produced when pressure is applied along *x*-axis and pressure along the direction of *y*-axis charges oppositely two opposite surfaces normal to *x*-axis, as the charge is produced when tension is applied.

The pressure or tension along *Z* axis produces no effect at all.

Electric field along x-axis only causes the crystal to expand along X-axis and to contract along Y-axis, reversal of the field produces the reversal of the effect. The deformation along Y-axis will produce longitudinal vibration and mechanical deformation along X-axis will produce surface vibration which is generally used as a source.

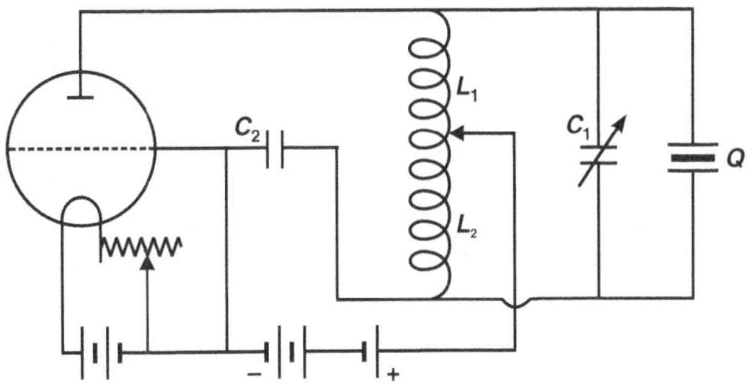

Fig. 4.

A piezo-electric crystal plate of quartz is placed between two plates of a condenser, which is connected to a source of A.C supply. It will set the crystal into elastic vibration of the frequency of A.C. supply. When the resonance occur between this frequency and the natural frequency of the plate, which depends upon dimension, the amplitude of vibration will be large. Such vibratory system is used as standard of frequency in electrical circuit.

The simplest form of the circuit is Hartly circuit as shown in Fig. 4 to maintain the oscillation by triode valve. The variable condenser C_1 help to tune the coil of inductance L_1 to the natural frequency of the quartz. Resonance is obtained. The electrodes for the quartz placed parallel to the plane perpendicular to X-axis.

Sound wave is emitted from these faces which is called thickness vibration and also from the face perpendicular to Y-axis called length vibration. Here very high frequencies can be obtained by exciting higher harmonics of the fundamental.

23.3. PIEZO-ELECTRIC DETECTORS

The principle of action of piezo-electric detectors is very simple. The waves fall on the crystal produces alternating stresses, hence alternating charge which develops alternating voltage at the faces perpendicular to those along which waves are received. This voltage is amplified and detected.

23.4. MEASUREMENT OF VELOCITY OF ULTRASONIC WAVES

The velocity of ultrasonic wave is determined by ultrasonic interferometer adopting a method of stationary wave. The surface of piezo-electric crystal C is made perfectly parallel to the Reflector R which is mounted and can move parallel to itself by micrometer screw. The wide vessel is filled up with the medium through which the plane wave will propagate when the crystal is in a state of vibration by the electric circuit. A micro-ammeter A (or a valve tube voltmeter) is placed in the plate circuit Fig. 5. The reflector R moved up and down. The reflected wave from R falls on C will help or oppose the vibration of C, according as the reflected wave fall on C in phase or in opposite phase. The reading of the microammeter will fluctuate with the motion of the reflector. The reading of microammeter is recorded and plotted against the position of R. The reading of the ammeter will pass through maxima and minima as shown in the Fig. 6. The distance between the two maxima or two minima is $\frac{\lambda}{2}$. Average value of λ being known, the velocity may be calculated knowing the frequency of the ultrasonic wave.

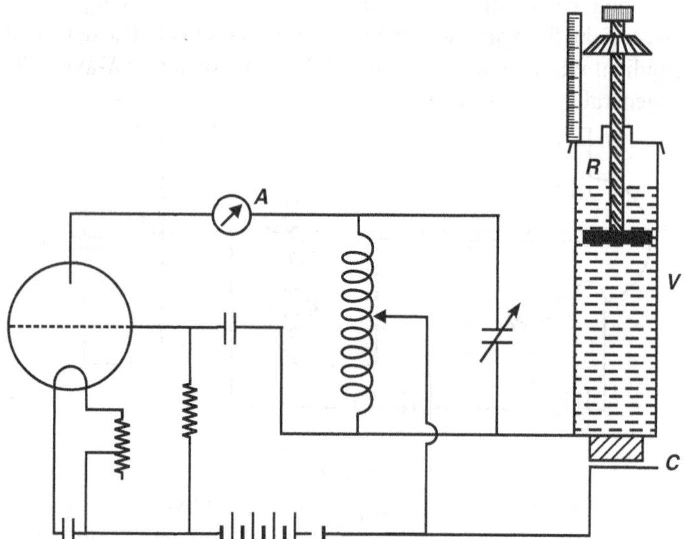

Fig. 5.

By this method of *absorption coefficient* may be measured by measuring the ratio of maximum or minimum amplitudes (vide Art. 3.2 eq. (1)).

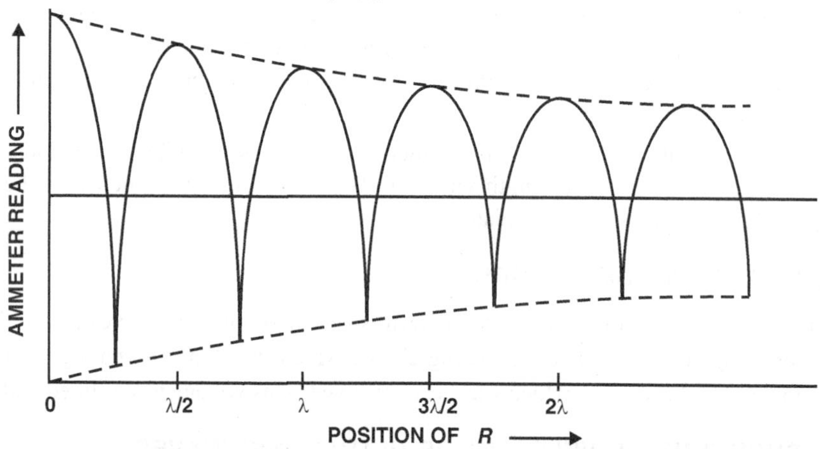

Fig. 6.

In using the ultrasonic interferometer the precautions to be taken are as follows:

1. The reflector *R* and the crystal surface of *C* should be perfectly parallel.
2. In order to obtain the consistent value of the observations, the following errors are to be avoided (*i*) the reflection of waves from the walls of the vessel, (*ii*) diffraction due to finite size of the source (*iii*) excitations of other modes of vibration than what is used.
3. Care should be taken so that there be no change of frequency of vibration on during observations.
4. The area of the crystal vibrating should be large enough in order to produce plane wave.

23.5. DIFFRACTION EFFECT AND ITS APPLICATION TO DETERMINE THE VELOCITY OF THE WAVE

When ultrasonic wave of three mega cycle per second passes through a liquid it will effectively produce a diffraction grating for light wave. The compressions and rarefactions produced thereof, will have high and low densities. Here a beam of light travelling normal to the direction of propagation of the plane ultrasonic wave in a liquid will find these compressions and rarefactions as parallel rulings of diffraction grating and is affected as a grating.

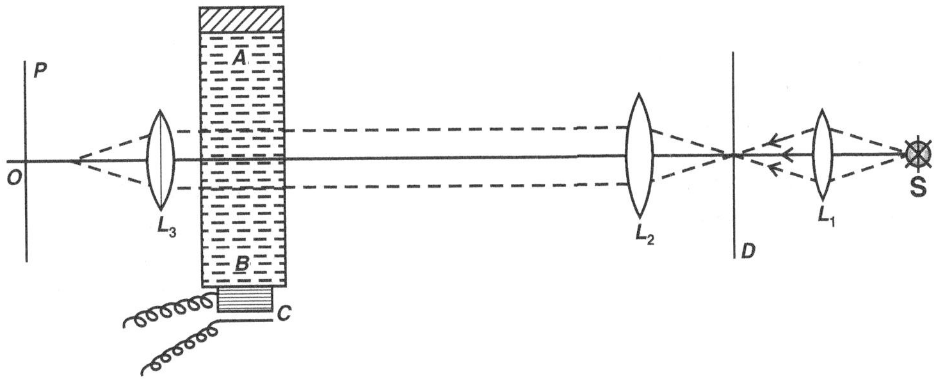

Fig. 7.

In order to find the velocity of wave a x-cut crystal of quartz C is placed at the bottom of a vessel B. There is absorber A preventing the formation of stationary waves by reflection. D the slit is illuminated from the source S of monochromatic light. The rays of light is collimated and is focused by L_1 and L_2. The image of the slit S is focused on the photographic plate P by lens L_3 and forms the diffracted image of different orders on either side of the central image.

For Nth order spectrum of the beam of light, of wavelength λ, having a grating constant $(a + b)$ suffers a deviation θ which is given by

$$(a + b)\sin\theta = N\lambda \qquad \text{... (1)}$$

But $(a + b)$ in this case is the width of one compression and one rarefaction which comprises the full wavelength λs of ultrasonic wave.

So Eq(1) becomes

$$\lambda_s \sin\theta = N\lambda$$

or

$$\lambda_s = \frac{N\lambda}{\sin\theta} \qquad \text{... (2)}$$

Knowing the value of the wavelength of light λ and deviation θ of the order of spectrum, the wavelength of ultrasonic wave λ_s is determined. The *velocity* is obtained, from the knowledge of λ_s and the frequency of ultrasonic wave.

23.6. THE ULTRASONIC WAVES AND ITS USE

(*i*) Determination of the depth of the sea. Owing to its extremely short wavelength it can be obtained in the form of narrow directional beam. The depth of the sea is found by recording the time, that elapses between the emission of ultrasonic wave and their reception from the bottom of the sea. The transmitter and receiver are placed both in water. Indicator consists of a drum rotating at constant speed, with a paper soaked in solution of potassium iodide and starch solution, with a metal style touching the paper. The stylus is connected with the arm of the ultrasonic generator. A pulse of ultrasonic is sent by the transmitter as soon as this stylus presses the zero of the paper scale. The echo is received by the stylus and the sensitised paper after a short time. Iodine is liberated at the point of contact of the stylus and the paper, and makes a bluish brown mark on the paper, indicating the depth of the sea.

The above principle is adopted for the detection of aircraft, submarine shoals of fish etc.

***(ii)* Detection of flaws in metal.** The metal is deprived of its usual strength by the presence of flaws. The presence of flaws changes the character of the medium and becomes responsible for the reflection of ultrasonic wave. The reflection occurs from the flaws as well as from the back surface of the metal. A piezo-electric crystal receiver is placed closed to the transmitter to receive the reflected pulse, which after amplification and rectification is allowed to pass one of the pairs of the plates of a cathode ray, and other pair of the plates are connected to linear time base. The transmitted signal, and the reflected signal from the flaw, and the back surface of the metal produce vertical trace. The position of the second vertical trace on the time base of the oscillograph will give the distance of the flaw and height will give the size of the flaw.

***(iii)* Mechanical, chemical and biological effect of ultrasonic waves.** When ultrasonic waves pass through a substance, intense heat is produced. A rod even when kept immersed in a liquid through which ultrasonic wave is passing becomes so hot that it becomes difficult to hold by hand. Water at 0°C with a piece of floating ice will boil without melting ice, when ultrasonic wave passes through it.

A glass rod oscillating with ultrasonic frequency will bore a hole through steel, glass even through diamond. It produces a hole by its to and fro motion about a thousandth of an inch with each stroke.

It is also used for removing grease, dust metal filling etc. in the manufacture of delicate instrument. It is also used in cleaning clothes, by shaking of the dust particles to detergent. It is also used to clean watches and hypodermic needles etc.

Ultrasonic waves are used to cure neuralgic and rheumatic pain as it produces soothing message action in the affected area. These waves penetrate the tissues and relieves the pain of arthritis.

When ultrasonic wave is directed in fine beam into the diseased tissue inside the body, they are destroyed without loss of blood.

Weak ultrasonic waves are able to produce good emulsion of two immiscible liquids, so it is used for the manufacture of good paints.

Ultrasonic wave is used along with electric current in soldering of metals like aluminium. It removes the oxide film formed, hence helps soldering.

Ultrasonic wave is used in telephony in sea water through which electromagnetic waves do not propagate. The speech wave is superimposed over ultrasonic wave which behaves as a carrier wave and transmitted through an electrostriction oscillator. The transmitter is used as a receiver by changing the connections.

It breaks up the molecular chain of high polymers hence hastens the reactions.

It kills the fish in water when transmitted from submarine.

Strong beam of ultrasonics kill bacteria, destroy blood corpuscles and protozoa. It is applied in sterilization and in preparation of vaccines.

SOLVED PROBLEMS

1. Bats emit ultrasonic wave. The shortest wavelength in air by a bat is about 0.33 cm. What is the highest frequency a bat can emit? (Velocity of sound in air is 330 m/sec)

[*Osmania University* 2010]

Shortest wavelength in air emitted by a bat

$$\lambda = .33 \text{ cm} = .33 \times 10^{-2} \text{ m}$$

We know $V = n\lambda$

$\therefore$ $330 = n \times .33 \times 10^{-2}$

or $n = 10^5 \text{Hz}.$

$\therefore$ Highest frequency a bat can emit $= 10^5$Hz.

REIVEW QUESTIONS

1. What are ultrasonic vibrations? How are they produced?
2. Describe and discuss the method of measurement of ultrasonic velocity in a liquid.
3. Discuss the relative advantages and disadvantages of magnetostrictive and piezo-electric ultrasonic oscillators.
4. How ultrasonic waves can be produced by a quartz plate? How ultrasonic wave is detected by piezo-electric detectors?
5. Give the principle of the diffraction effect for determination of the velocity of the ultrasonic waves.
6. What are Transducers? [Barkatullah University 2011]
7. What is Piezo-electric effect? Explain the production and working of this effect. [Pandit Ravisankar University 2010]

24 CHAPTER

ARCHITECTURAL ACOUSTICS

24.1. ARCHIETECTURAL ACOUSTICS AND SABINES'S EMPIRIAL FORMULA

In a good auditorium particularly theatre, concert halls, school rooms etc. sound produced by the speaker or by a source should be heard with sufficient loudness and clarity. The science that deals with the design of a good auditorium and buildings so as to reduce the external and internal noise and maintains the conditions of sufficient loudness and clarity of the source of sound inside, is known as Architectural Acoustics.

When a source is continuously sounding in a closed room intensity at a point will vary inversely as the square of the distance. But some reflection from the surrounding wall will reach the point. At first, the energy density will increase and ultimately it will reach a steady state when the rate of the production, is equal to the rate of loss due to dissipation.

The sound emitted by the source will reach the point directly along with successive reflections from the walls by travelling a greater length. These waves may interfere at the point. The effect of interference may be neglected as it is equivalent to sum of the separate intensities. The velocity of sound is low, so sufficient time will elapse between the arrival of the first wave and the last, which travel a large distance due to successive reflections. This will contribute much to the increase of loudness but at the same time clarity is impaired. Further at the resonant frequency of the hall the interference will produce distortion.

When the source is cut off sound reception by the receiver will not stop instantaneously, but will continue to pick up the reflected sound wave till it becomes too weak to be received due to absorption.

When the source is cut off sound reception by the receiver will not stop instantaneously, but will continue to pick up the reflected sound waves till it becomes too weak to be received due to absorption.

The general dying out of sound, after the source is cut off is known as *Reverberation.* The *time of reverberation* is the time for which sound remained audible after the source of sound had been cut off. It is measured as the time in seconds required to reduce the intensity from a level 60 db., above the threshold of audibility to the threshold. This is equivalent to the time required in second for a sound to diminish from its initial intensity to one millionth of that intensity. This time depends upon the size of the room and absorption coefficient of the material of the surrounding wall.

For a given room, if the absorption coefficient is small (*i.e.*, less than 0.4) the reverberation time will be large which is the case of a *Live room*. The reflected waves will not be reduced sufficiently in intensity and they will interfere with the succeeding syllables, coming directly from the speaker, so clarity will be destroyed. But on the other hand, if the absorption of the room is too great (greater than 0.4) the reverberation time will be short, intensity of the sound will be lowered–such a room is said to be a '*Dead room*'.

The following are the requirements of a good auditorium.

(*i*) The sound heard should be loud;

(*ii*) Successive syllables uttered should be distinctly heard;

(*iii*) There must not be any distortion of the sound heard.

Prof. W.C. Sebine of the Harvard University while trying to improve the acoustical condition of the university auditorium where a speaker could hardly be understood by the audience made a fruitful research (1903–1911) and gave an empirical formula for of the time T of reverberation.

$$T = \frac{.05V}{\Sigma aS}$$

where V is the volume of the room in cubic feet, ΣaS absorption of various surfaces of the room, a is the absorption coefficients and S is the area of the surface in square feet. [1 foot = 0.3048 m]

The above empirical formula of Sebine was deduced by Franklin (1903) and Jager (1911) and others.

24.2. REVERBERATION TIME THEORETICAL TREATMENT

The following assumptions are made in deducing the formula.

(*i*) The sound energy is distributed within the rooms sufficiently uniformly, to make valid consideration in terms of an average energy density.

(*ii*) At steady state, the rate at which the energy is produced at the source is equal the rate at which energy is increased in the medium and the total rate at which the energy is absorbed by the walls and the surroundings.

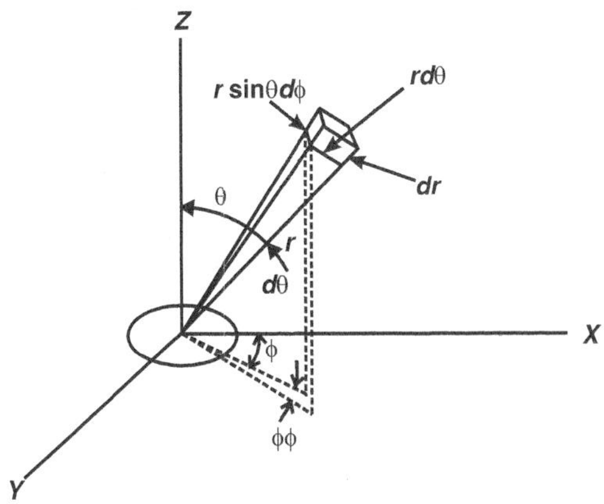

Fig. 1.

We consider a hemispherical shell of radius r and thickness dr, the centre of which is taken on an area dS.

The element of volume dv is given by

$$dv = r^2 \sin\theta \, d\theta \, d\phi \, dr \quad \text{... (1)}$$

Let E be the energy density so that the amount of energy on the elementary volume dv propagating in all directions, is

$$Edv = Er^2 \sin\theta \, d\theta \, d\phi \, dr \quad \text{... (2)}$$

The amount of energy propagated through unit solid angle is $\frac{Er^2}{4\pi}\sin\theta \, d\theta \, d\phi \, dr$. The solid angle subtended by ds at the elementary volume is $\frac{ds\cos\theta}{r^2}$ ... (3)

The fraction of energy in the elementary volume propagating towards ds is

$$\frac{ds\cos\theta}{r^2 4\pi} Er^2 \sin\theta \, d\theta \, d\phi \, dr \quad \text{... (4)}$$

The total energy incident on the upper face of ds from a distance between r and $r + dr$ is obtained by integrating θ and ϕ between the limits $0 \to \pi/2$ and $0 \to 2\pi$ respectively,

$$= \frac{Edsdr}{4\pi} \int_0^{\pi/2} \cos\theta \sin\theta \, d\theta \int_0^{2\pi} d\phi$$

$$= \frac{Eds}{4} dr \quad \text{... (5)}$$

Energy received in unit of time

$$= \frac{Eds}{4}\int_0^c dr$$

$$= \frac{Ec}{4}ds \qquad \text{... (6)}$$

where c is the velocity of propagation *i.e.* distance travelled in one second.

If 'a' be the absorption coefficient of the elementary surface ds, so energy absorbed in unit time is

$$\frac{Eca}{4}ds \qquad \text{... (7)}$$

The total rate of absorption of energy at any time

$$= \frac{Ec}{4}\Sigma\, ads$$

$$= \frac{EcA}{4} \qquad \text{... (8)}$$

where $A = \Sigma ads$ = total absorption of all the surfaces exposed to sound wave.

The total rate of energy increase in the medium within the whole volume V of the room is

$$V\frac{dE}{dt} \qquad \text{... (9)}$$

If P be the rate at which energy is produced by the source, then from Eqns. (8) and (9)

$$V\frac{dE}{dt} + \frac{Ac}{4}E = P \qquad \text{... (10)}$$

or $$\frac{dE}{dt} + \alpha E = \frac{P}{V}$$

where $$\alpha = \frac{Ac}{4V} \qquad \text{... (10.1)}$$

or $$\frac{dE}{dt} + \alpha\left(E - \frac{P}{V\alpha}\right) = 0 \qquad \text{... (10.2)}$$

put $$E - \frac{P}{V\alpha} = Z$$ Eqn. (10.2) becomes

$$\frac{dZ}{dt} + \alpha Z = 0$$

or $$\int\frac{dZ}{Z} = -\int \alpha\, dt$$

or $$\log Z = -\alpha t + C \qquad C = \text{constant}$$

or $$Z = E - \frac{P}{V\alpha} = C_1 e^{-\alpha t}$$

or $$E = \frac{4P}{Ac} + C_1 e^{-\frac{Ac}{4V}t}$$

At $t = 0$ $E = 0$ gives $$C_1 = -\frac{4P}{Ac}$$

or $$E = \frac{4P}{Ac}\left(1 - e^{-\frac{Ac}{4V}t}\right) \qquad \text{... (11)}$$

This expression shows the *growth* of E with time t after the start of the sound.

24.3. REVERBERATION TIME OF A LIVE ROOM

The source is cut off when E has reached maximum, so that

$$P = 0 \text{ at } t = 0$$

and at $t = \infty$ $$E = E_{\text{max}} = \frac{4P}{Ac} \qquad \text{... (11.1)}$$

Eqn. (10) becomes

$$V\frac{dE}{dt} + \frac{Ac}{4}E = 0 \qquad \text{... (12)}$$

gives $$E = E_{\text{max}}e^{-\frac{Ac}{4V}t} \qquad \text{... (13)}$$

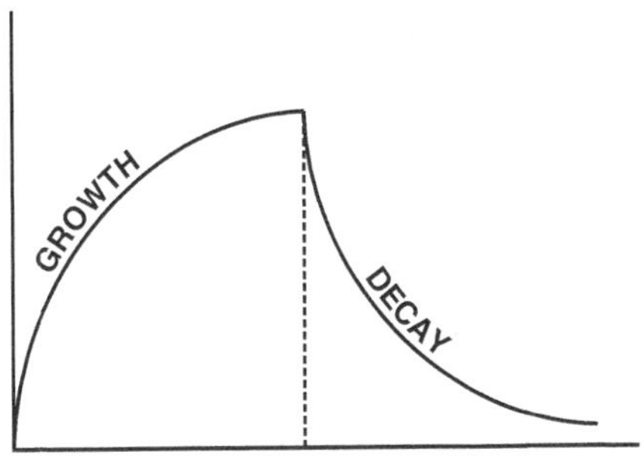

SOURCE CUT OFF

Fig. 2.

This expression represents the *decay of energy*. The Fig. 2 represents the growth and decay of sound energy. (This may be compared with the variation of potential difference across a condenser during charge and discharge through a resistance).

Let T be the time taken to reduce the energy density and therefore intensity to decay by 60 db. i.e. one millionth part of the value just before cut off, is given by

$$\frac{E}{E_{\text{max}}} = e^{-\frac{Ac}{4V}T} = 10^{-6} \qquad \text{... (13)}$$

$$\frac{Ac}{4V}T = \log_e 10^{+6} = 6 \times 2.303$$

$$T = 6 \times 2.303 \times \frac{4V}{Ac}$$

$$= 0.16\frac{V}{A} \qquad \text{... (14)}$$

when dimensions are in metre and sec.

$$c = 343 \text{ metre/sec}$$

Again $$= 0.05\frac{V}{A} \qquad \text{... (15)}$$

when dimensions are in feet and second c = 1120 feet/sec. = 341.376 metre/sec.

This is in good agreement with experiment which suggests that the assumptions made are correct, that the energy distribution is uniform, the interference effect can be neglected. The absorption coefficient is independent of intensity, the power emitted by the source is independent of energy density, the loss of energy in air is negligible.

24.4. REVERBERATION TIME OF A DEAD ROOM

When the absorption coefficient of the wall of the room exceeds 0.4 sabine, the reverberation time will be different from what is obtained before. Such a room is known as *dead room.*

In such a case, multiple, instead of single reflection, is considered.

At steady state let I_0 be the intensity of sound in such a room. After successive reflections, that is after first, second, third.... nth reflections the intensities become

$$I_0(1-\alpha),\ I_0(1-\alpha)^2,\ I_0(1-\alpha)^3 \text{ .... } I_0(1-\alpha)^n.$$

when θ is the absorption coefficient of the walls.

At steady state, let the source of sound be suddenly cut off, the intensity I of sound after nth reflection becomes

$$I = I_0(1-\alpha)^n$$

or

$$\frac{I_0}{I} = \frac{1}{(1-\alpha)^n} \qquad \text{... (1)}$$

It is assumed that the reflection of a particular wave of sound is confined within the four walls of the room having total area S of a rectangular room of volume V. The average path λ traversed by sound wave between two successive reflection is given by

$$\lambda = \frac{V}{S/4} = \frac{4V}{S} \qquad \text{... (2)}$$

as successive reflection of a particular ray of sound is confined to four walls only for a rectangular room.

From Eqn. (1)

$$\log\frac{I_0}{I} = -n\log(1-\alpha)$$

or

$$n = \frac{\log I_0/I}{-\log(1-\alpha)} \qquad \text{... (3)}$$

Total path traversed after n reflections

$$n\lambda = \frac{\log I_0/I}{-\log(I-a)}\cdot\frac{4V}{S} \qquad \text{... (4)}$$

Let the time taken be T in order that the intensity decreases by 60 *db* i.e.

$$\frac{I}{I_0} = 10^{-6}$$

or

$$\log\frac{I_0}{I} = \log 10^6$$

$$= 6 \times 2.303 \qquad \text{... (5)}$$

So eq (4) becomes

$$n\lambda = \frac{6\times 2.303}{-\log(I-\alpha)}\frac{4V}{S} \qquad \text{... (6)}$$

But after n reflection the intensity decreases by 60 db for which time required is T. The total path traversed in this time T is $c.T$ which is equal to $n\lambda$.

Hence
$$cT = \frac{4 \times 6 \times 2.303}{-\log(1-a)} \cdot \frac{V}{S} \quad \text{... (7)}$$

Taking c the velocity of sound = 343 m/sec we have from eq. (7)

$$T = \frac{0.161V}{S\{-\log(1-\alpha)\}} \quad \text{... (8)}$$

This is the Eyring's formula for the reverberation time of a 'dead room'. When α is small, denominator becomes $S\alpha$ (= A the total absorption power of the walls) neglecting higher powers of α eq. (7) becomes same as {cq. (14) art 24.3} given by Sabine.

24.5. OPTIMUM REVERBERATION TIME

In a hall where a natural voice is to reach the listener's ear, some reverberation is necessary to produce requisite loudness so that clarity is ensured. Nowadays voice is amplified by loud speakers, the reverberation in this case should be shorter for clarity. It has been experimentally found that the reverberation time of 0.8 sec in an auditorium of 141.6 cu. m (50000 cu. ft) in volume gives best intelligibility. For a large hall of 1132.6 cu. m (40000 cu. ft) value should be between 1.5 to 2 seconds in order to give good results.

If the absorption in the room is large the reverberation time will be small. The members present in the auditorium will cause further absorption. Each member will absorb about 4.7 sabine while a woman will absorb somewhat greater than that of a man due to differences in the nature and extent of the clothings used.

Elaborate experimental work has been done in connection with the studies for broadcasting and gramophone recording. On account of different directional response of a microphone in comparison with two ears, the acoustic defect due to wrong reverberation time are accentuated in these cases. A careful speaker who speaks slowly can produce, good intelligibility even with the reverberation time depends upon the type of speech, music and the frequency.

There are other factors which influence the acoustic properties of buildings. Other than reverberation time, the correct position of reflectors and absorbing substances play an important role in the matter. A good absorber at the back of the hall will minimise the distinct echo being heard. A good reflector behind the performers will direct the energy where it is wanted. The room with extensive reflecting surface will set stationary waves, and will produce bad intensity distribution. The concave surfaces which cause focusing effect, are undesirable from acoustic point of view. There will be structural response if the structure is not rigid, naturally, there will be uneven frequency response.

Thus the *requirements of a good auditorium are:* The reverberation time should be suitable for the purpose of music or speech for which the hall is designed. Resonance of different parts of the hall such as the wall, the space in the hall, sound boards etc. should be avoided. A small room gives better result. In order to avoid focusing effect curved walls, domed ceiling etc. are to be avoided. The interference effect should be minimised, else it will give maximum and minimum at different places. Proper design of the room and proper use of sound absorbing materials on the walls will greatly minimise the most of the defects.

24.6. MEASUREMENT OF TIME OF REVERBERATION

Sabine measured the time of reverberation by chronograph. The sound was produced by an organ pipe in the room. The time during which the sound intensity decayed from a level 60 db above the threshold of audibility to threshold was recorded on chronograph drum.

Improvement has been made in recent days. A cathode ray oscillograph is used which has slow sweep rate and long persistent fluorescent screen. The record is made photographically from which reverberation time is computed.

An audiofrequency oscillator connected to a loud speaker is used as a steady source of sound of constant frequency. To measure the reverberation time after the source is suddenly shut off, a microphone connected to an amplifier and a valve voltmeter with a pen recording system over a calibrated chart moving with constant high speed are used. The reverberation time is computed from the record which indicates the variation of intensity level with time.

24.7. MEASUREMENT OF ABSORPTION COEFFICIENT

There are different methods for the determination of absorption coefficient of materials.

(*i*) The reverberation time T_1 in a particular hall is first determined, taking a source of sound of particular frequency say 512 c.p.s. by an audiofrequency oscillator, then the material whose absorption coefficient is to be determined is taken within the hall. Let the area be S and absorption coefficient be α and T_2 is the reverberation time of the hall with the material inside, then

$$T_1 = \frac{0.05V}{\Sigma aS} \quad \text{... (1)}$$

S is the area of the walls, a is the absorption coefficient of the walls.

When the substance is taken in the hall having absorption coefficient α the reverberation time,

$$T_2 = \frac{0.05V}{\Sigma aS + \alpha S} \quad \text{... (2)}$$

From eqs (1) and (2)

$$\alpha = \frac{0.05V}{S}\left(\frac{1}{T_2} - \frac{1}{T_1}\right) \quad \text{... (3)}$$

(*ii*) The reverberation times T_1 and T_2 for two sources emitting power P_1 and P_2 are found out. Two powers are not required to be known but the knowledge of their ratios will be sufficient. The loud speaker with measured audio frequency current are used as the sources.

The steady energy densities maintained by two sources are (vide eq. (11.1).

$$\frac{4P_1}{Ac}, \frac{4P_2}{Ac}$$

During decay in time T_1 and T_2 respectively they will be

$$\frac{4P_1}{Ac}e^{-\alpha T_1} = \frac{4P_2}{Ac}e^{-\alpha T_2}, \text{ where } \alpha = \frac{Ac}{4V}$$

Hence
$$\frac{P_1}{P_2}e^{-\alpha(T_2-T_1)} = 1$$

or
$$e\alpha(T_2 - T_1) = \frac{P_2}{P_1}$$

$$\alpha = \frac{2.3}{T_2 - T_1}\log_{10}\frac{P_2}{P_1}$$

as
$$\alpha = \frac{Ac}{4V}$$

So
$$A = \frac{4V}{c}.\frac{2.3}{(T_2 - T_1)}\log_{10}\frac{P_2}{P_1}$$

where A is the total absorption by the areas of the room is given by aS, a being the mean absorption coefficient, S the area.

The experiment is carried out with empty chamber and then with various materials of different areas under test.

SOLVED PROBLEMS

1. A broadcasting studio measuring 25 × 12 × 7 metres has a reverberation time of 0.90 sec when empty. What will be the reverberation time when an audience of 250 persons is present?

Given absorption coefficient of a person = 0.4,

Average absorbing area of a person = 0.6 sq.m.

Volume of the studio = 25 × 12 × 7 c.m.
= 2100 c.m.

Initially if A be total absorption then

$$0.9 = \frac{0.16 \times 2100}{A}$$

or $A = 373.33$

Due to the audience A increases by 250 × 0.4 × 0.6 sabins
= 60 sabins

$$\text{The new reverberation time} = \frac{0.16 \times 2100}{(373.33 + 60)} \text{ sec } = 0.78 \text{ sec}$$

2. The interior walls of an auditorium 80 × 20 × 10 m have an average absorption coefficient 0.2. The carpeted floor has an absorption coefficient 0.25 and the ceiling has a coefficient 0.4. What is the reverberation time? Find the power output of a source required to produce an intensity level of 70 dB (Given intensity level = 10^{-12} W/m^2)

The volume of the auditorium = 80 × 20 × 10 m^3
= 16000 m^3

Area of the ceiling (or floor) = 80 × 20 sq.m
= 1600 sq.m

Area of the four walls = 2 (80 × 10 + 20 × 10) m^2
= 2000 m^2

Total absorption by the ceiling, floor and boundaries

= 0.4 × 1600 + 0.25 × 1600 + 0.2 × 2000
= 1440

The reverberation time
$$T = \frac{0.16 \times 16000}{1440} \text{ sec } = 1.78 \text{ sec}$$

If I be the desired intensity then

$$10 \log_{10}\left(\frac{I}{I_0}\right) = 70$$

or
$$I = I_0 \times 10^7$$
$$= 10^{-12} \times 10^7 \text{ W/m}^2$$
$$= 10^{-5} \text{ W/m}^2$$

But we know $I = \dfrac{P}{A}$ where P is the power output of the source

$\therefore$
$$P = I.\ A$$
$$= 10^{-5} \times 1440 \text{ watts} = 0.0144 \text{ watts}$$

REVIEW QUESTIONS

1. Give the theory of growth and decay of sound in a 'live room'. Find the reverberation time. Why the theory fails in a 'dead room' ?
2. What are the acoustic requirements of a good auditorium? Discuss the theory of reverberation. [C.U. 1952]
3. How is absorption coefficient of a material measured?
4. What are reverberation and reverberation time ? How does the latter affect audibility in a room ? How can reverberation time can be measured ? How can it be altered from one value to another?
5. Derive Sabine formula for reverberation time. What is the limitation of the formula? Discuss other formulae for reverberation time, giving the physical basis of each.
6. Define reverberation period and derived Sabine's reverberation formula.

   [N.H. Bahuguna Garhwal University 2010]

PROBLEMS

1. A cinema hall has dimensions 25 × 15 × 10 m. Its reverberation time is 1.3 sec when empty. Find the reverberation time when an audience of 300 persons is present. Assume that the sound absorption by each person is 4.5 sabine. [**Ans.** 0.332 sec]
2. A room 10 × 10 × 4 m has an average sound absorption coefficient 0.1. Calculate its reverberation time.

25 CHAPTER ELECTRICAL ANALOGY OF ACOUSTIC VIBRATION

25.1. ELECTRICAL ANALOGY OF ACOUSTIC VIBRATION.

There is great similarity between a mechanical system with the sinosoidal voltage applied to a series resistancse, inductance and capacitance circuit.

In equation (1) of Art 3.6, the equation of motion of a particle of mass m is given by

$$m\frac{d^2y}{dt^2} = F\sin pt - \mu y - R\frac{dy}{dt} \qquad \text{... (1)}$$

or

$$m\frac{d^2y}{dt^2} + R\frac{dy}{dt} + \mu y = F\sin pt \qquad \text{... (2)}$$

We know in an electrical RLC *a.c.* circuit when a sinusoidal voltage is applied the *emf* equation is

$$L\frac{di}{dt} + Ri + \frac{q}{c} = E_0 \sin pt \qquad \text{... (3)}$$

Putting $i = \frac{dq}{dt}$ the equation (3) transforms to

$$L\frac{d^2q}{dt^2} + R\frac{dq}{dt} + \frac{q}{c} = E_0 \sin pt \qquad \text{... (4)}$$

If we write $\mu = \frac{1}{c_m}$, c_m which is called compliance the equation (2) becomes

$$m\frac{d^2y}{dt^2} + R\frac{dy}{dt} + \frac{y}{c_m} = F\sin pt \qquad \text{... (5)}$$

The equations (4) and (5) are identical in form but the constants have different meanings in two cases. The correspondence between the equations (4) and (5) is that mass m corresponds to inductance L, mechanical resistance R is equivalent to electrical resistance, compliance c_m corresponds to capacitance c, velocity v corresponds to current i. E_0 is equivalent to F.

We know the electrical impedance

$$|Z| = \left[R^2 + \left(pL - \frac{1}{pc}\right)^2\right]^{1/2} = [R^2 + X^2]^{1/2}$$

where $X = pL - \frac{1}{pc}$ is called electrical resistance.

Mechanical impedance from (11.1) in Art. 3.6

$$Z = \left[\left(\frac{\mu}{p} - mp\right)^2 + R^2\right]^{1/2} \qquad \text{... (6)}$$

$$= [X^2 + R^2]^{1/2} \qquad \text{... (7)}$$

where $X = \dfrac{\mu}{p} - mp$ is called mechanical reactance.

25.2. SPECIFIC ACOUSTIC IMPEDANCE

The ratio of the acoustic pressure to the particle velocity at a point is called the specific acoustic impedance (Z_{sp}) and is given by $Z_{sp} = \dfrac{P}{u}$, where P is the acoustic pressure and u the particle velocity.

It is analogous to electrical impedance $Z_e = \dfrac{e}{i}$ where e is the emf and i is the electric current. But the main difference between specific acoustic impedance and electrical impedance is that specific acoustic impedance is a point property and electric impedance exists between two points in current carrying circuit.

25.3. POWER RELATION

The instantaneous power in watts supplied to the mechanical system by the driving force is equal to the product of the instantaneous driving force and the resulting velocity. From equations (1) and (14) vide Art 3.6 we have

$$Wi = F\sin pt.\frac{F}{Z}\sin(pt - \in) \qquad \text{... (8)}$$

$$= \frac{F^2}{Z}\sin pt\sin(pt - \in) \qquad \text{... (9)}$$

The average power supplied to the system

$$W = \frac{\int_0^T W_i dt}{\int_0^T dt} \qquad \text{... (10)}$$

where T is the time period. With the help of equation (9) the equation (10) becomes

$$W = \frac{1}{ZT}\int_0^T F^2 \sin pt\sin(pt - \in)dt$$

$$= \frac{F^2}{2Z}\cos \in \qquad \text{... (11)}$$

$\cos\in$ is the power factor.

The average power supplied to the system by the driving force is not permanently stored in the system but is, instead, dissipated in the form of work expended in the moving the system against the frictional force $\left(R\dfrac{dy}{dt}\right)$. It is to be noted that the expression for the average power is analogous in form to the power in a series electrical circuit containing resistance, inductance and capacitance.

25.4. TRANSIENT RESPONSE OF AN OSCILLATOR

We will consider the effect of superimposing the transient response upon the steady state condition. The equation (*i*) of Art 3.6 is

$$m\frac{d^2y}{dt^2} + R\frac{dy}{dt} + \mu y = F \sin pt \qquad \text{... (12)}$$

The general solution of equation (12) is

$$y = Ae^{-\frac{K}{2}t}\sin(\omega_1 t + \varphi_p) + \frac{F}{pZ}\sin(pt - \varepsilon) \qquad \text{... (13)}$$

[Vide equation (9) of Art 3.5 and equation (14) of Art 3.6]

The transient response has been so damped that the final steady state is nearly reached. The waveform is somewhat irregular immediately after the application of the driving force but it soon come down to the steady state.

25.5. FILTER

One of the most important use of acoustics is acoustic filter and it is analogous to electric filter. It is discussed in detail in Art 16.4.

Analogous Electrical and Mechanical Quantities

Mechanical quantities	Electrical equivalent
Force	Voltage
Displacement	Charge
Velocity	Current
Mass	Inductance
Mechanical resistance	Resistance
Mechanical impedance	Impedance
Mechanical reactance	Electrical reactance

REVIEW QUESTIONS

1. Write a note on analogy between electric vibration and mechanical vibration.

26 CHAPTER SHORT BIOGRAPHY OF NOTED ACCOUSTICS PHYSICISTS

26.0. INTRODUCTION

In this chapter the author wishes to present the short biographies on the life and contribution of noted pioneer in the field of Vibration and Acoustics, Sir C.V. Raman Indian Scientist and Lord Rayleigh British Scientist. Both the scientists were the receipient of Nobel Prize in Physics.

26.1. PROFESSOR SIR CHANDRASEKHARA VENKATA RAMAN

Professor C.V. Raman was born on the 7th November 1888 in Thirvaikkaval near Trichinapally in old Madras in India and died in Bangalore in Karnatak at the age of 82 on November 21, 1970. Raman matriculated at the age of 11 and won scholarship at the age of 13 and at 15 he passed his B.A. examination with gold medal for physics and for English from Presidency college in Madras. He passed M.A. examination in 1907 at the young age of 19. He took the civil services competitive examination for Finance Department and he topped the score in that examination and joined the Finance Department as Assistant Accountant General in Calcutta.

In 1906, at the age of 16, Raman as a student of M.A. class first published two papers on "diffraction band" in Philosiphical Magazine and in Nature in 1908 [Phil. Mag. 12, 475, 1906 and Nature 78, 55, 1908].

In 1907 on his way to office at the Finance Department in Calcutta Prof. Raman noticed the sign board of "The Indian Association for Cultivation of Science (I.A.C.S.) and felt interested in doing research. He took permission for doing research in the "Indian Association for Cultivation of Science" as a part-time worker. His research work got interrupted due to his transfer to Rangoon and Nagpur from 1909 to 1910. In 1991 he subsequently transferred back to Calcutta and could continue his research work again in I.A.C.S.

In 1917 by the request of Prof. Asutosh Mukherjee Vice Chancellor of Calcutta University C.V. Raman left his job at the Account Department and joined at Calcutta University as prestigious Taraknath Palit Professor of Physics. In 1919, Prof. Raman became the Honorary Secretary of the Indian Association of Cultivation of Science. In 1921, Prof. Raman went to Oxford as a delegate at the Universities Congress. In 1933 Prof. Raman was in Bangalore as the Director of Tata Institue, now known as Indian Institute of Science. Later Prof. Raman founded the Raman Institue in Bangalore and stayed there till his death.

Prof. Raman was an excellent research worker. He worked in different branches of Physics viz., vibration and sound and musical instruments, Interference, diffraction and other optical phenomena and emission spectra, elastic properties of solids, on colloids, Molecular scattering of light, electrical, Magnetic and Mechanical Berefringeneces, Raman effect and Raman spectra and X-rays studies and X-ray diffraction of solids. Raman worked an acoustics from 1909 to 1921 on vibration and sound and the theory of musical instruments Raman worked on different aspects of vibration in strings analytically and experimentally. Raman discovered the mode of operation of violin bowed string. The acoustical properties of the Tanpura and Vina was studied by Raman. Raman has twenty five research publications on acoustics and musical instruments. Raman starte a Bulletin of the Indian Association of Cultivation of Science. This Bulletin Grew in 1917 into a full science journal "Indian Journal of Physics".

In 1924 Professor Raman was elected as a Fellow of the Royal society. Prof. Raman received Nobel Prize in physics in 1930 for his work on the scattering of light.

26.2. JOHN WILLIAM STRUTT, 3rd BARON LORD RAYLEIGH

Lord Ryleigh was born on 12th November, 1842 and died in Withem on 30th June, 1944 at the age of 76. He started eductaion in Horrow School. He studied mathematics at Trinity College and obtained a Bachelor of Arts degree.

He was the senior Wranglor and First Smitth Prize holder in 1865 and a Master of Arts degree in 1868. He was awarded a Fellowship from Trinity College. He was the second Cavendish Professor of Physics at the University of Cambridge from 1879 to 1884. From 1887 to 1907 he was Professor of Natural Philosophy at Cambridge University.

Around the year 1900, Lord Rayleigh developed the duplex theory of human sound localization using two binatural cues, interaural phase difference and internal level difference. The theory posits that we use two primary caues for sound lateralization using the difference in the phases of sinusoidal components of sound and the difference in amplitudes between two ears. The 'rayl' unit of acoustic impedance in named after him.

Lord Rayleigh was elected Fellow of the Royal society on 12th June, 1873 and served as President of Royal Society from 1905 to 1908. Surface was known as Rayleigh wave was named in his name.

He was honoured with Smith Prize in 1864, Royal Medal in 1882, Matheucci Medal in 1894, Member of the Royal Swedish Academy of Science in 1897, Copley medal in 1899, Order of merit in 1902, Nobel Prize for Physics in 1904 for his discovery in Argon, Elliot Cresson Medal in 1913 and Rumford medeal in 1914. Rayleigh published two books on "Theory of sound Vol. I and Vol. II in 1877 and 1878 from Macmillon publication, London. Rayleigh had large number of scientific papers published from Cambride Universty Press Vol. I to Vol. VI.

INDEX

USEFUL PHYSICAL CONSTANTS

Speed of light in vacuum	c	$= 2.99792 \times 10^8$ metre per second
Charge of electron	e	$= 1.6 \times 10^{-19}$ coulomb
Planck's constant	h	$= 6.6 \times 10^{-34}$ joule sec
Electron's rest mass	m_e	$= 9.1 \times 10^{-31}$ kg
Avogadro number	N	$= 6.023 \times 10^{23}$ per mole
Boltzmaun's constant	K	$= 1.35 \times 10^{-23}$ joule per kelvin
One hertz (Hz)		= 1 cycle per second (cps)
One electron volt		$= 1.6 \times 10^{-19}$ joule
One Watt		= 1 joule per sec
Universal gas constant		$= 8.314\ JK^{-1}\ mole^{-1}$
		$= 1.986$ cal $K^{-1} mole^{-1}$

UNIT PREFIXES

Multiple and Submultiple	Prefixes	Symbols
10^9	Giga	G
10^6	Mega	M
10^3	Kilo	K
10	Deca	da
10^{-1}	Deci	d
10^{-2}	Centi	cm
10^{-3}	Milli	mm
10^{-6}	Micro	μ
10^{-9}	Nano	n
10^{-12}	Pico	p

ENERGY CONVERSION

1 J	= 1 Ws = 10^7 erg
1 eV	$= 1.6 \times 10^{-19}$ J
1 cal	= 4.187 J
1 J	= 0.2388 cal
1 inch	= 2.54 cm
1 foot	= 30.48 cm
1 Torr (T)	= Pressure of 1 mm of Hg
1 Pa (Pascal)	= 1 newton/metre^2
1 Fathom	= 1.8288 m